Research on Polyoxometalate Materials

Research on Polyoxometalate Materials

Editor

Xiaobing Cui

Basel • Beijing • Wuhan • Barcelona • Belgrade • Novi Sad • Cluj • Manchester

Editor
Xiaobing Cui
Jilin University
Changchun, China

Editorial Office
MDPI
St. Alban-Anlage 66
4052 Basel, Switzerland

This is a reprint of articles from the Special Issue published online in the open access journal *Molecules* (ISSN 1420-3049) (available at: https://www.mdpi.com/journal/molecules/special_issues/polyoxometalate_materials).

For citation purposes, cite each article independently as indicated on the article page online and as indicated below:

Lastname, A.A.; Lastname, B.B. Article Title. *Journal Name* **Year**, *Volume Number*, Page Range.

ISBN 978-3-0365-9313-5 (Hbk)
ISBN 978-3-0365-9312-8 (PDF)
doi.org/10.3390/books978-3-0365-9312-8

© 2023 by the authors. Articles in this book are Open Access and distributed under the Creative Commons Attribution (CC BY) license. The book as a whole is distributed by MDPI under the terms and conditions of the Creative Commons Attribution-NonCommercial-NoDerivs (CC BY-NC-ND) license.

Contents

Xiao-Bing Cui
Special Issue: Research on Polyoxometalate Materials
Reprinted from: *Molecules* 2023, 28, 4662, doi:10.3390/molecules28124662 1

Victoria V. Volchek, Nikolay B. Kompankov, Maxim N. Sokolov and Pavel A. Abramov
Proton Affinity in the Chemistry of Beta-Octamolybdate: HPLC-ICP-AES, NMR and Structural Studies
Reprinted from: *Molecules* 2022, 27, 8368, doi:10.3390/molecules27238368 5

Bo Qi, Luran Jiang, Sai An, Wei Chen and Yu-Fei Song
Detecting the Subtle Photo-Responsive Conformational Bistability of Monomeric Azobenzene Functionalized Keggin Polyoxometalates by Using Ion-Mobility Mass Spectrometry
Reprinted from: *Molecules* 2022, 27, 3927, doi:10.3390/molecules27123927 21

Manal Diab, Ana Mateo, Joumada El Cheikh, Zeinab El Hajj, Mohamed Haouas, Alireza Ranjbari, et al.
Grafting of Anionic Decahydro-*Closo*-Decaborate Clusters on Keggin and Dawson-Type Polyoxometalates: Syntheses, Studies in Solution, DFT Calculations and Electrochemical Properties
Reprinted from: *Molecules* 2022, 27, 7663, doi:10.3390/molecules27227663 31

Chong-An Chen, Yan Liu and Guo-Yu Yang
Designed Syntheses of Three {Ni$_6$PW$_9$}-Based Polyoxometalates, from Isolated Cluster to Cluster-Organic Helical Chain
Reprinted from: *Molecules* 2022, 27, 4295, doi:10.3390/molecules27134295 53

Yiran Wang, Fengxue Duan, Xiaoting Liu and Bao Li
Cations Modulated Assembly of Triol-Ligand Modified Cu-Centered Anderson-Evans Polyanions
Reprinted from: *Molecules* 2022, 27, 2933, doi:10.3390/molecules27092933 65

Tian Chang, Di Qu, Bao Li and Lixin Wu
Organic/Inorganic Species Synergistically Supported Unprecedented Vanadomolybdates
Reprinted from: *Molecules* 2022, 27, 7447, doi:10.3390/molecules27217447 79

Yu Fu, Yanyan Yang, Dongxue Chu, Zefeng Liu, Lili Zhou, Xiaoyang Yu and Xiaoshu Qu
Vanadium-Substituted Dawson-Type Polyoxometalate–TiO$_2$ Nanowire Composite Film as Advanced Cathode Material for Bifunctional Electrochromic Energy-Storage Devices
Reprinted from: *Molecules* 2022, 27, 4291, doi:10.3390/molecules27134291 93

Xiaoxia Li, Ni Zhen, Chengpeng Liu, Di Zhang, Jing Dong, Yingnan Chi and Changwen Hu
Controllable Assembly of Vanadium-Containing Polyoxoniobate-Based Materials and Their Electrocatalytic Activity for Selective Benzyl Alcohol Oxidation
Reprinted from: *Molecules* 2022, 27, 2862, doi:10.3390/molecules27092862 105

Hai-Yang Guo, Hui Qi, Xiao Zhang and Xiao-Bing Cui
First Organic–Inorganic Hybrid Compounds Formed by Ge-V-O Clusters and Transition Metal Complexes of Aromatic Organic Ligands
Reprinted from: *Molecules* 2022, 27, 4424, doi:10.3390/molecules27144424 119

Zhihui Ni, Hongjin Lv and Guoyu Yang
Recent Advances of Ti/Zr-Substituted Polyoxometalates: From Structural Diversity to Functional Applications
Reprinted from: *Molecules* 2022, 27, 8799, doi:10.3390/molecules27248799 135

Zheyu Wei, Jingjing Wang, Han Yu, Sheng Han and Yongge Wei
Recent Advances of Anderson-Type Polyoxometalates as Catalysts Largely for Oxidative Transformations of Organic Molecules
Reprinted from: *Molecules* **2022**, *27*, 5212, doi:10.3390/molecules27165212 . . . . **153**

Editorial

Special Issue: Research on Polyoxometalate Materials

Xiao-Bing Cui

State Key Laboratory of Inorganic Synthesis and Preparative Chemistry and College of Chemistry, Jilin University, Changchun 130021, China; cuixb@mail.jlu.edu.cn

The science of polyoxometalates (POMs) has come a long way since molybdenum blue was first described in 1778 [1]. Since then, polyoxometalates (POMs) have been showing remarkable progress and unexpected surprises in their basic principles and applications. Polyoxometalates are a special class of soluble metal oxides (intermediate state) between monomeric metal oxides and infinite metal oxides, which have amazing differences in sizes, chemical compositions, and physical properties from monomeric and infinite metal oxides. The structures of POMs are rich and complex, and their chemical compositions are mainly Mo, W, V, Nb, and Ta. Heteroatoms can be P, As, B, Al, Si, Ge, S, and other atoms, and the polyoxometalate structures can be divided into saturated and unsaturated ones. As we all know, there is a general correlation between the complexity of the structure of a compound and its displayed function. The wide variability of chemical compositions and a large number of unusual structural types make POMs exhibit a large variety of different properties, which attracts many researchers to continuously explore the synthesis strategy, structural regulation, properties, and applications of POM materials. Many of these attractive features include controllable size, composition, charge density, REDOX potential, acid strength, high solid-state thermal stability, solubility in polar/non-polar solvents, and reversible electron/proton storage.

In this context, this Special Issue aims to highlight recent results in all the fields of POMs and POM-based materials. It is composed of nine original articles, overall reporting results about the syntheses and properties of different POMs and different POM-based materials, and two review articles, one of which is about structural types, synthetic strategies, and even relevant catalytic applications of Ti/Zr-substituted POMs, and the other about the application of Anderson-type ($[XM_6O_{24}]^{n-}$) POMs with different structures in organic synthesis reactions.

Pavel A. Abramov et al. [2] studied the affinity of $[\beta\text{-}Mo_8O_{26}]^{4-}$ toward different proton sources in various conditions. It is widely known that protons are very important in the reaction of polyoxometalates. The current study reveals that the structural rearrangement of $[\beta\text{-}Mo_8O_{26}]^{4-}$ as a direct response to protonation was demonstrated. The proton transfer reaction between $(Bu_4N)_4[\beta\text{-}Mo_8O_{26}]$ and $(Bu_4N)_4H_2[V_{10}O_{28}]$ results in the formation of $[V_2Mo_4O_{19}]^{4-}$. The same type of reaction between $(Bu_4N)_4[\beta\text{-}Mo_8O_{26}]$ and $[H_4SiW_{12}O_{40}]$ leads to the formation of $[W_2Mo_4O_{19}]^{2-}$.

Yu-Fei Song et al. [3] mainly studied the conformational changes of four azobenzene covalently functionalized Keggin compounds using ion migration mass spectrometry (IMS/MS). The photo-responsive trans–cis conformational changes of azobenzene Keggin compounds were clearly revealed, which successfully opened up an important new characterization dimension for polyacids.

Sébastien Floquet et al. [4] succeeded in combining a covalently decahydro-closo-decaborate cluster $[B_{10}H_{10}]^{2-}$ with Keggin- and Dawson-type POMs through an amino-propylsilyl ligand (APTES) acting as both a linker and a spacer. Mono- and di-adduct compounds of the boron cluster were obtained with the Keggin-APTES, while only the di-adduct of the boron cluster was isolated with the Dawson-APTES. DFT studies and electrochemical studies were also conducted. Finally, electrocatalytic reduction of protons into hydrogen was evidenced in these systems.

Citation: Cui, X.-B. Special Issue: Research on Polyoxometalate Materials. *Molecules* 2023, 28, 4662. https://doi.org/10.3390/molecules28124662

Received: 2 June 2023
Revised: 6 June 2023
Accepted: 7 June 2023
Published: 9 June 2023

Copyright: © 2023 by the author. Licensee MDPI, Basel, Switzerland. This article is an open access article distributed under the terms and conditions of the Creative Commons Attribution (CC BY) license (https://creativecommons.org/licenses/by/4.0/).

Guo-Yu Yang et al. [5] synthesized three new transition metal-substituted POM compounds. [Ni$_6$(OH)$_3$(DACH)$_3$(H$_2$O)$_6$(PW$_9$O$_{34}$)]·31H$_2$O (**1**, DACH = 1,2-diami-nocyclohexane) is a Ni$_6$ cluster-substituted Keggin unit decorated with a DACH ligand. This compound is an isolated hexa-Ni-substituted Keggin unit. By introducing different organic ligands, such as rigid 5-methylisophthalate (HMIP) and flexible adipate (AP), [Ni(DACH)$_2$][Ni$_6$(OH)$_3$(DACH)$_3$(HMIP)$_2$(H$_2$O)$_2$(PW$_9$O$_{34}$)]·56H$_2$O (**2**) with a similar anionic monomeric POM cluster to compound **1** was obtained, and [Ni(DACH)$_2$][Ni$_6$(OH)$_3$(DACH)$_2$(AP)(H$_2$O)$_5$(PW$_9$O$_{34}$)]·2H$_2$O (**3**) with a novel 1-D POM cluster organic chain (POMCOF) was obtained. The synthesis of these compounds provides us with a new strategy for using chainlike dicarboxylate acid as a linker to make POMCOFs.

Bao Li et al. [6] prepared a series of triol ligand-modified Cu-centered Anderson–Evans POMs with different counterions. They combined different molybdenum sources, triol ligands, and different counter cations, such as NH^{4+}, Cu^{2+}, and Na$^+$, to systematically investigate the roles of the cations in the packing of the produced POM structures. This investigation found that the charges, sizes, and coordination manners of the countercations have an important impact on the final structures of polyanions.

Bao Li et al. [7] synthesized two new compounds of vanadomolybdates with similar unprecedented hepta-nuclear structures, which were both stabilized by triol ligands. It is known that the preparation of vanadomolybdates is relatively difficult due to their low structural stability. Therefore, the present study provides a new strategy to prepare and stabilize vanadomolybdates by using triol ligands.

Xiaoshu Qu et al. [8] successfully constructed a nanocomposite film composed of vanadium-substituted Dawson POMs and TiO$_2$ nanowires via the combination of hydrothermal and layer-by-layer self-assembly methods. Due to the unique three-dimensional core–shell nanostructure of the composite, dual-function electrochromic (EC) photomodulation and electrochemical energy storage are significantly improved. The solid electrochromic energy storage (EES) devices are prepared by using the composite films as cathodes, which were able to light up a single light-emitting diode for 20 s. Taken together, these results demonstrate that EES devices based on POMs have great potential in applications requiring multi-function supercapacitors.

Changwen Hu et al. [9] successfully synthesized two new compounds constructed from vanadium-containing Keggin-type polyoxoniobates and nickel complexes [Ni(en)]$^{2+}$ (en = ethylenediamine) by controlling and changing the hydrothermal temperature and vanadium sources. It should be noted that nickel-containing polyoxoniobates have rarely been reported previously. The selective oxidation of benzyl alcohol by the two compounds was also investigated, and the results showed that they had high catalytic activity. This study not only enriches the structural database of polyoxoniobates but also expands the catalytic applications of polyoxoniobates.

Xiao-Bing Cui et al. [10] synthesized three novel compounds based on Ge-V-O clusters by the hydrothermal method. All the previously reported Ge-V-O compounds were totally based on aliphatic organic ligands; compounds **1** and **2** are the first examples of Ge-V-O clusters containing aromatic organic ligands. The catalytic properties of these compounds for the epoxidation of styrene were also explored in this study.

In addition to nine papers on the synthesis and properties of polyoxometalates and POM-based materials, two related review articles were also published in this Special Issue.

Hongjin Lv et al. [11] mainly reviewed the structural types, synthetic strategies, and even relevant catalytic applications of Ti/Zr-substituted POMs. Transition metal-substituted POMs are a very important subclass of POMs, especially in catalytic chemistry. Common transition metal-substituted POMs are based on Cu, Co, Ni, and so on, and sometimes on lanthanide. However, Ti/Zr-substituted POMs are relatively less reported, and, to the best of my knowledge, no review about Ti/Zr-substituted POMs has been published previously. Therefore, this review gives us an overview of the Ti/Zr-substituted polyoxometalates.

The second review of this Special Issue by Yongge Wei et al. [12] reviewed the application of Anderson-type ($[XM_6O_{24}]^{n-}$) POMs with different structures in organic synthesis reactions. This will provide a new strategy for further study on the catalytic application of Anderson POMs and green catalysis.

Ultimately, it is our sincere hope that this Special Issue will serve as a reference for those who wish to learn more about POMs as an area of science, as well as help new researchers become inspired, interested, and engaged in this topic.

Funding: This research received funding from grant from Jilin Provincial Department of Science and Technology (No. 20190802027ZG).

Conflicts of Interest: The author declares no conflict of interest.

References

1. Scheele, C.W. Sämtliche Physische und Chemische Werke. Hermbstädt, D.S.F., Ed.; Martin Sändig oHG: Niederwalluf/Wiesbaden, Germany, 1971; Volume II, pp. 185–200.
2. Volchek, V.V.; Kompankov, N.B.; Sokolov, M.N.; Abramov, P.A. Proton Affinity in the Chemistry of Beta-Octamolybdate: HPLC-ICP-AES, NMR and Structural Studies. *Molecules* **2022**, *27*, 8368. [CrossRef] [PubMed]
3. Qi, B.; Jiang, L.; An, S.; Chen, W.; Song, Y.F. Detecting the Subtle Photo Responsive Conformational Bistability of Monomeric Azobenzene Functionalized Keggin Polyoxometalates by Using Ion-Mobility Mass Spectrometry. *Molecules* **2022**, *27*, 3927. [CrossRef] [PubMed]
4. Diab, M.; Mateo, A.; Cheikh, J.E.; Hajj, Z.E.; Haouas, M.; Ranjbari, A.; Guérineau, V.; Touboul, D.; Leclerc, N.; Cadot, E.; et al. Grafting of Anionic Decahydro-Closo-Decaborate Clusters on Keggin and Dawson-Type Polyoxometalates: Syntheses, Studies in Solution, DFT Calculations and Electrochemical Properties. *Molecules* **2022**, *27*, 7663. [CrossRef] [PubMed]
5. Chen, C.A.; Liu, Y.; Yang, G.Y. Designed Syntheses of Three Ni_6PW_9-Based Polyoxometalates, from Isolated Cluster to Cluster-Organic Helical Chain. *Molecules* **2022**, *27*, 4295. [CrossRef] [PubMed]
6. Wang, Y.R.; Duan, F.X.; Liu, X.T.; Li, B. Cations Modulated Assembly of Triol-Ligand Modified Cu-Centered Anderson-Evans Polyanions. *Molecules* **2022**, *27*, 2933. [CrossRef] [PubMed]
7. Chang, T.; Qu, D.; Li, B.; Wu, L.X. Organic/Inorganic Species Synergistically Supported Unprecedented Vanadomolybdates. *Molecules* **2022**, *27*, 7447. [CrossRef] [PubMed]
8. Fu, Y.; Yang, Y.Y.; Chu, D.X.; Liu, Z.F.; Zhou, L.L.; Yu, X.Y.; Qu, X.S. Vanadium-Substituted Dawson-Type Polyoxometalate-TiO_2 Nanowire Composite Film as Advanced Cathode Material for Bifunctional Electrochromic Energy-Storage Devices. *Molecules* **2022**, *27*, 4291. [CrossRef] [PubMed]
9. Li, X.X.; Zhen, N.; Liu, C.P.; Zhang, D.; Dong, J.; Chi, Y.N.; Hu, C.W. Controllable Assembly of Vanadium-Containing Polyoxoniobate-Based Materials and Their Electrocatalytic Activity for Selective Benzyl Alcohol Oxidation. *Molecules* **2022**, *27*, 2862. [CrossRef] [PubMed]
10. Guo, H.Y.; Qi, H.; Zhang, X.; Cui, X.B. First Organic-Inorganic Hybrid Compounds Formed by Ge-V-O Clusters and Transition Metal Complexes of Aromatic Organic Ligands. *Molecules* **2022**, *27*, 4424. [CrossRef] [PubMed]
11. Ni, Z.H.; Lv, H.J.; Yang, G.Y. Recent Advances of Ti/Zr-Substituted Polyoxometalates: From Structural Diversity to Functional Applications. *Molecules* **2022**, *27*, 8799. [CrossRef] [PubMed]
12. Wei, Z.Y.; Wang, J.J.; Yu, H.; Han, S.; Wei, Y.G. Recent Advances of Anderson-Type Polyoxometalates as Catalysts Largely for Oxidative Transformations of Organic Molecules. *Molecules* **2022**, *27*, 5212. [CrossRef] [PubMed]

Disclaimer/Publisher's Note: The statements, opinions and data contained in all publications are solely those of the individual author(s) and contributor(s) and not of MDPI and/or the editor(s). MDPI and/or the editor(s) disclaim responsibility for any injury to people or property resulting from any ideas, methods, instructions or products referred to in the content.

Article

Proton Affinity in the Chemistry of Beta-Octamolybdate: HPLC-ICP-AES, NMR and Structural Studies

Victoria V. Volchek [1], Nikolay B. Kompankov [1], Maxim N. Sokolov [1] and Pavel A. Abramov [1,2,*]

[1] Nikolaev Institute of Inorganic Chemistry SB RAS, 3 Akad. Lavrentiev Ave., 630090 Novosibirsk, Russia
[2] Institute of Natural Sciences and Mathematics, Ural Federal University Named after B.N. Yeltsin, 620075 Ekaterinburg, Russia
* Correspondence: abramov@niic.nsc.ru

Abstract: The affinity of $[\beta\text{-}Mo_8O_{26}]^{4-}$ toward different proton sources has been studied in various conditions. The proposed sites for proton coordination were highlighted with single crystal X-ray diffraction (SCXRD) analysis of $(Bu_4N)_3[\beta\text{-}\{Ag(py\text{-}NH_2)Mo_8O_{26}\}]$ (1) and from analysis of reported structures. Structural rearrangement of $[\beta\text{-}Mo_8O_{26}]^{4-}$ as a direct response to protonation was studied in solution with ^{95}Mo NMR and HPLC-ICP-AES techniques. A new type of proton transfer reaction between $(Bu_4N)_4[\beta\text{-}Mo_8O_{26}]$ and $(Bu_4N)_4H_2[V_{10}O_{28}]$ in DMSO results in both polyoxometalates transformation into $[V_2Mo_4O_{19}]^{4-}$, which was confirmed by the ^{95}Mo, ^{51}V NMR and HPLC-ICP-AES techniques. The same type of reaction with $[H_4SiW_{12}O_{40}]$ in DMSO leads to metal redistribution with formation of $[W_2Mo_4O_{19}]^{2-}$.

Keywords: proton transfer; octamolybdate; NMR; chromatography; structural analysis

1. Introduction

Protons play a key role in a wide range of water-associated processes, from geochemistry to biology [1]. The appearance of the Theory of Coupled Electron and Proton Transfer Reactions [2–4] opened rich prospects for chemical reactions design [5–8]. Currently, proton coupled electron transfer (PCET) processes play crucial roles in synthesis and catalysis [9,10], e.g., artificial photosynthesis systems [11–14] and PCET at interfaces [15–18].

In polyoxometalate (POM) chemistry, protonation affects the formation, stability and reactivity of polyoxoanions. Most self-assembly cascade reactions are pH driven when fast protonation-deprotonation processes provoke rapid species transformation/organization into various associates up to nanoscopic size. The study of self-assembly processes is one of the top subjects in modern chemical science [19–27]. Such a specific organization of the matter in different solutions is a research focus for a large number of research groups. For example, research groups led by T. Mak and Di Sun successfully merged polyoxometalate chemistry with that of coinage metal clusters using the self-assembly approach [28–30].

The electronic structure of polyoxoanions together with low-energy protonation makes such objects very attractive for PCET reactions. The most important catalytic process in this field is water oxidation [31,32]. Such POM catalysts as $[\{Ru^{IV}_4(OH)_2(H_2O)_4\}(\gamma\text{-}SiW_{10}O_{34})_2]^{10-}$ [33–35] and $[Co^{II}_4(H_2O)_2(B\text{-}\alpha\text{-}PW_9O_{34})_2]^{10-}$ have become classics [36,37]. Recently, $[V_6O_{13}(TRIOL^{NO_2})_2]^{2-}$ was applied to achieve concerted transfer of protons and electrons. Fully reduced clusters can induce $2e^-/2H^+$ transfer reactions from surface hydroxide ligands [38].

In the chemistry of group 6 polyoxometalates, the polyoxomolybdates are significantly more labile than the polyoxotungstates, thus making researchers favor the latter in their studies of POM chemistry. However, several studies of polyoxomolybdates' reactivity [39] and catalytic performance (electron transfer reactions) appeared [40–45]. One of the central complexes in this chemistry is $(Bu_4N)_4[\beta\text{-}Mo_8O_{26}]$ (Scheme 1), which

is a standard precursor of all reactions in organic media, leading to a huge number of materials with different properties [46–49]. Our ongoing research focuses on the use of the coordination chemistry of the [β-Mo$_8$O$_{26}$]$^{4-}$ anion in the study of silver chemistry in non-aqueous solutions [50–52]. Karoui and Ritchie used (Bu$_4$N)$_4$[β-Mo$_8$O$_{26}$] in the microwave-assisted synthesis of tris(alkoxo)molybdovanadates [V$_3$Mo$_3$O$_{16}$(O$_3$-R)]$^{2-}$ (R = C$_5$H$_8$OH or C$_4$H$_6$NH$_2$) by the reaction between [β-Mo$_8$O$_{24}$]$^{4-}$, [H$_3$V$_{10}$O$_{28}$]$^{3-}$ and pentaerythritol or tris(hydroxymethyl)aminomethane [53]. These results show the possibility of the reaction between two different types of polyoxometalates producing mixed-metal compounds based on a different structural type. Such reactions are practically unknown and can generate interesting mixed metal complexes. This is very important and can be used for various materials preparation applied in catalysis (different Mo/V oxides), photochemistry, solid-state devices (capacitors), biochemistry and biomedicine.

Scheme 1. The structure of the [β-Mo$_8$O$_{26}$]$^{4-}$ anion.

An important question is what is the trigger and the driving force of such metal redistribution reactions? In this research, we focused on the behavior of the [β-Mo$_8$O$_{26}$]$^{4-}$ anion toward protonation to answer this question. Some years ago, we suggested a straightforward hyphenated HPCL-ICP-AES technique [54] as an efficient tool to study the reaction products in different polyoxometalate systems [55–57]. In the present research, this technique helps us to have control over products' formation in different conditions.

2. Results
2.1. Structural Analysis

The structure of [β-Mo$_8$O$_{26}$]$^{4-}$ is preorganized for the coordination of different metal cations due to the presence of two trans-located lacunes (Scheme 1). During the study of complexation in the Ag$^+$/[β-Mo$_8$O$_{26}$]$^{4-}$/L (L = auxiliary ligand) systems [50,51,58], we found a large number of equilibria that can be easily shifted by the addition of different ligands. In the present case, we tested 4-aminopyridine (py-NH$_2$) as an auxiliary ligand in order to produce a 1D {-Mo$_8$-Ag-py-NH$_2$-Ag-Mo$_8$-} coordination polymer. Instead of this, the reaction gives (Bu$_4$N)$_3$[Ag(py-NH$_2$)Mo$_8$O$_{26}$] as the main product (phase purity was confirmed by XRPD, see Supplementary Materials, Figure S1). In the crystal structure (SCXRD details are collected in Supplementary Materials Table S1) Ag$^+$, [β-Mo$_8$O$_{26}$]$^{4-}$ and py-NH$_2$ combine into another type of 1D coordination polymer when [Ag(py-NH$_2$)Mo$_8$O$_{26}$]$^{3-}$ anions stack together via py-NH$_2$ . . . O=Mo interactions (Figure 1).

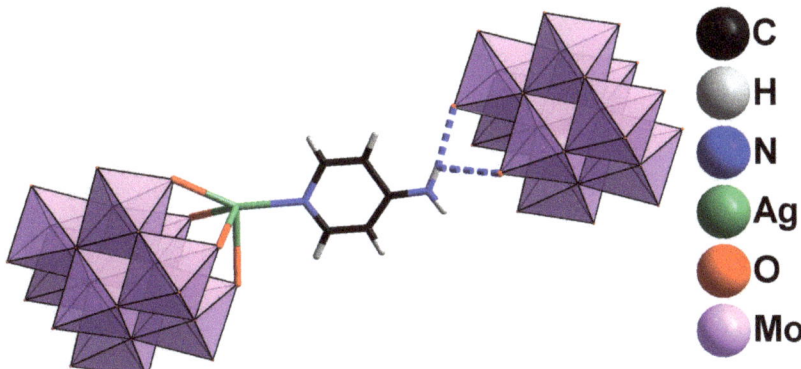

Figure 1. py-NH$_2$ ... O=Mo interactions in the crystal structure of **1**.

Three typical bonding distances surround Ag$^+$: d(Ag1-N1) = 2.27(3) Å, d(Ag1-O6) = 2.372(7) Å, d(Ag1-O2) = 2.544(6) Å and two longer contacts d(Ag1-O9) = 2.639(7) and d(Ag1-O13) = 2.689(7) Å indicate CN = 3+2 for Ag$^+$. These distances are in agreement with the previously published pyridinium complexes of this type [50]. The distances for py-NH$_2$ ... O=Mo interactions fill the interval between 2.974 and 3.285 Å. The shortest N ... O contacts 2.974 and 3.021 Å are depicted in blue in Figure 1.

The formation of this coordination polymer via NH$_2$ ... POM interactions is very interesting. We did the structural search for bonding between the oxoligands of the [β-Mo$_8$O$_{26}$]$^{4-}$ lacunes and H-atoms, and collected 11 hits (BURBOH, CASNIU, COPFIW, EWILIG, GEBYER, GISHEW, HIJSUR, MAXPUZ, MEPNIH, VEHTAF, YAGNOJ) from CCDC (ConQuest Version 2020.2.0). We will use the corresponding refcodes of deposited crystal structures as references in the description below.

The interactions between [β-Mo$_8$O$_{26}$]$^{4-}$ and Me$_2$NH(R), Me$_2$NH$_2$$^+$ and NH$_4$$^+$ in the crystal structures of YAGNOJ (a); BURBOH (b); HIJSUR (c); GISHEW (d) are shown in Figure 2.

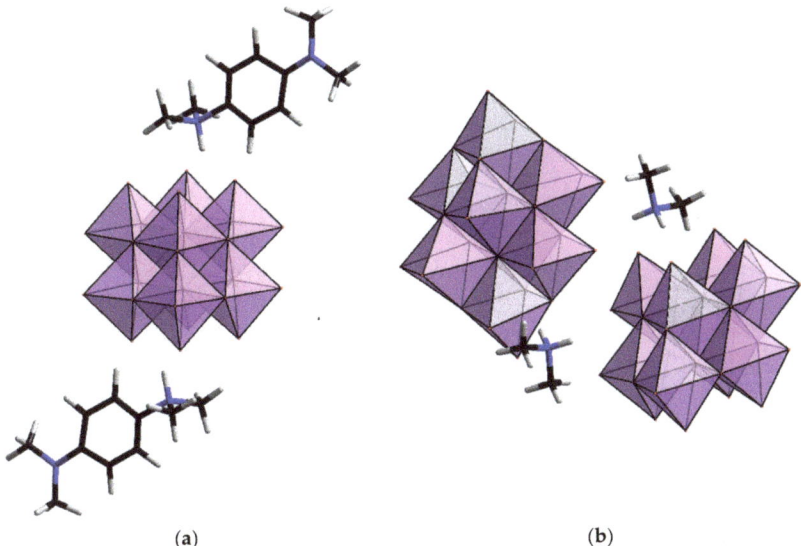

Figure 2. *Cont.*

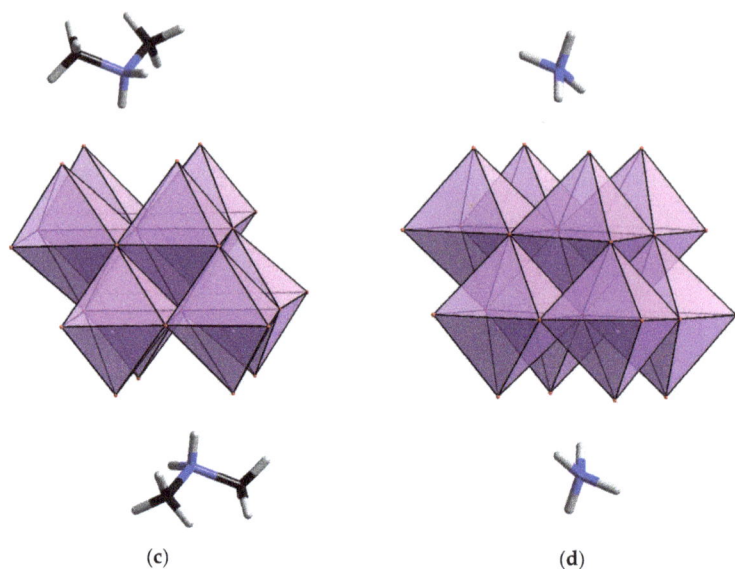

(c)   (d)

Figure 2. H...O=Mo interactions in the crystal structure of: (**a**) YAGNOJ; (**b**) BURBOH; (**c**) HIJSUR; (**d**) GISHEW.

According to the structural analysis, R_3NH^+, $R_2NH_2^+$ and NH_4^+ interact with terminal O=Mo groups of $[\beta\text{-}Mo_8O_{26}]^{4-}$ lacunes. Moreover, even Me_4N^+ can interact with the lacune (VEHTAF). In the crystal structure of GEBYER, the $[\beta\text{-}Mo_8O_{26}]^{4-}$ lacunes interact with two H_2O molecules. In the case of **1**, we detected interaction between the neutral NH_2-group protons with the O=Mo groups of polyoxomolybdate. This illustrates strong attraction between the lacune terminal oxoligands and H-atoms possessing some acidity (chiefly N–H, but also C-H in Me_4N^+). Considering this, we can suggest direct proton transfer exactly to these oxoligands-producing terminal Mo-OH group, which is highly reactive (M–O π-bonding breaking) and initiates further rearrangement of octamolybdate into hexamolybdate. The detailed mechanistic studies of this transformation are still absent. In this research, we used this channel to initiate the reaction between $[\beta\text{-}Mo_8O_{26}]^{4-}$ and different protonated polyoxometalates serving as proton source. Such direct reactions between two different polyoxometalates are poorly studied. The HPLC-ACP-AES technique was used to control the products.

2.2. Reactivity of $[\beta\text{-}Mo_8O_{26}]^{4-}$

The first candidate for this type of reaction was easily prepared $(Bu_4N)_4H_2[V_{10}O_{28}]$. The HPLC-ICP-AES chromatogram of pure $(Bu_4N)_4[\beta\text{-}Mo_8O_{26}]$ in acetonitrile shows a major molybdenum peak (t_R = 3.6 min), corresponding to the octamolybdate anion $[\beta\text{-}Mo_8O_{26}]^{4-}$, and a minor peak ($t_R$ = 4.8 min), which can be assigned as a hexamolybdate anion $[Mo_6O_{19}]^{2-}$ (Figure 3a) [59]. The profile of the major peak is asymmetric due to the presence of $[\alpha\text{-}Mo_8O_{26}]^{4-}$, according to the previous ESI-MS data, demonstrating the absence of any other molybdates in the solution [50]. The HPLC-ICP-AES chromatogram of a freshly prepared solution of $(Bu_4N)_4H_2[V_{10}O_{28}]$ shows a single peak containing vanadium (t_R = 3.0 min), which confirms the presence of individual vanadate anion $[V_{10}O_{28}]^{6-}$ in the solution (Figure 3b). Moreover, the addition of 2 eq of Bu_4NOH to the solution of $(Bu_4N)_4H_2[V_{10}O_{28}]$ does not reflect any POM transformation.

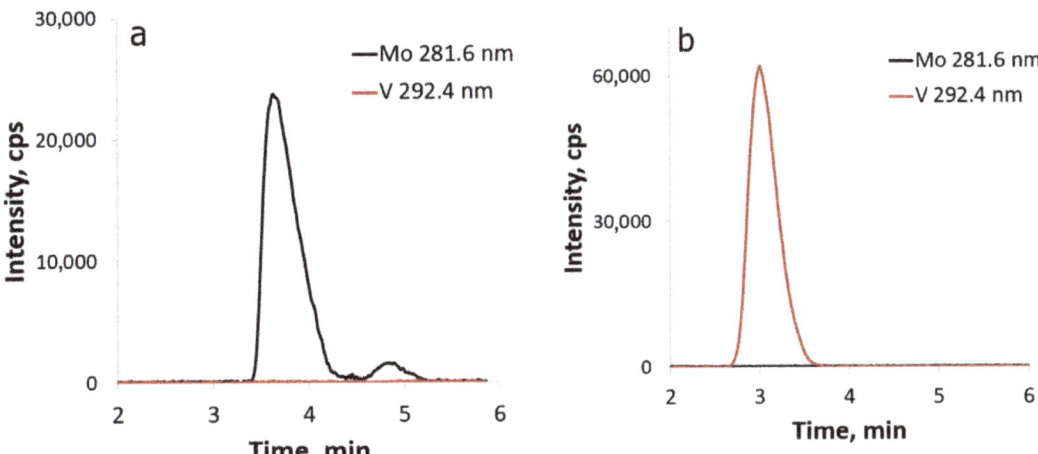

Figure 3. HPLC-ICP-AES chromatograms of (**a**) [β-Mo$_8$O$_{26}$]$^{4-}$ and (**b**) [V$_{10}$O$_{28}$]$^{6-}$ in acetonitrile.

The reaction between (Bu$_4$N)$_4$[β-Mo$_8$O$_{26}$] and (Bu$_4$N)$_4$H$_2$[V$_{10}$O$_{28}$] in DMSO does not proceed at room temperature, according to ^{51}V NMR data, which is the fastest way to check the reaction progress. The reaction mixture must be heated over 50 °C to activate the polyoxometalates' transformation. The HPLC-ICP-AES technique was used to investigate the reaction products at different molar ratios of the reagents. The reaction time was 10 min.

For molar ratio 5/1 (Mo:V = 5/1) at C_o of (Bu$_4$N)$_4$[β-Mo$_8$O$_{26}$] = 6 mM, we observed one peak with the atomic ratio Mo:V = 2.2 (t_R = 4.3 min) (Figure 4a) and a second V-free peak (t_R = 4.8 min), which may be ascribed to unreacted octamolybdate (Figure 4a). With an increase in the vanadate concentration (Mo/V = 5:2 molar ratio), the same major peak with atomic ratio Mo:V = 2.5 was observed, the intensity of which doubled (Figure 4b). In addition, a chromatogram shows a minor Mo-free peak (t_R = 3.0 min), which indicates an excess of the decavanadate anion in this case (Figure 3b).

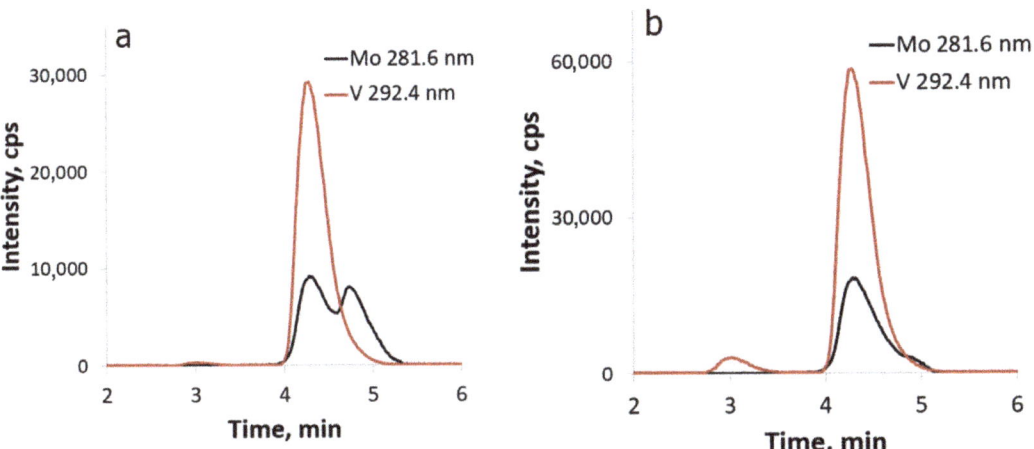

Figure 4. HPLC-ICP-AES chromatograms of the POM mixture at (**a**) the initial Mo/V ratio = 5/1 and (**b**) Mo/V ratio = 5/2.

No significant changes in the chromatograms were observed with a further increase in the concentration of vanadate. The atomic ratio Mo:V = 2.2 indicates the formation of $[V_2Mo_4O_{19}]^{4-}$ Lindqvist type anions as the reaction product. According to ^{51}V NMR, the total intensity of the other V peaks is ca. 1% of the intensity of the signal from the major product (See NMR part).

The next candidate to study the proton transfer controlled reaction with $(Bu_4N)_4[\beta\text{-}Mo_8O_{26}]$ was $[H_4SiW_{12}O_{40}]\cdot14H_2O$. Preliminary experiments showed that the reaction between $(Bu_4N)_4[\beta\text{-}Mo_8O_{26}]$ and $[H_4SiW_{12}O_{40}]\cdot14H_2O$ in acetonitrile proceeded slowly and led to the formation of a number of products in comparable amounts. Therefore, CH$_3$CN was replaced with dimethyl sulfoxide (DMSO). The HPLC-ICP-AES chromatogram of a freshly prepared solution of silicotungstic acid in DMSO shows a single peak containing tungsten (t_R = 6.2 min), which confirms the presence of individual silicotungstate anion $[SiW_{12}O_{40}]^{4-}$ in the solution (Figure 5a) (The intensities of Si lines are significantly lower than W or Mo and cannot be adequately estimated).

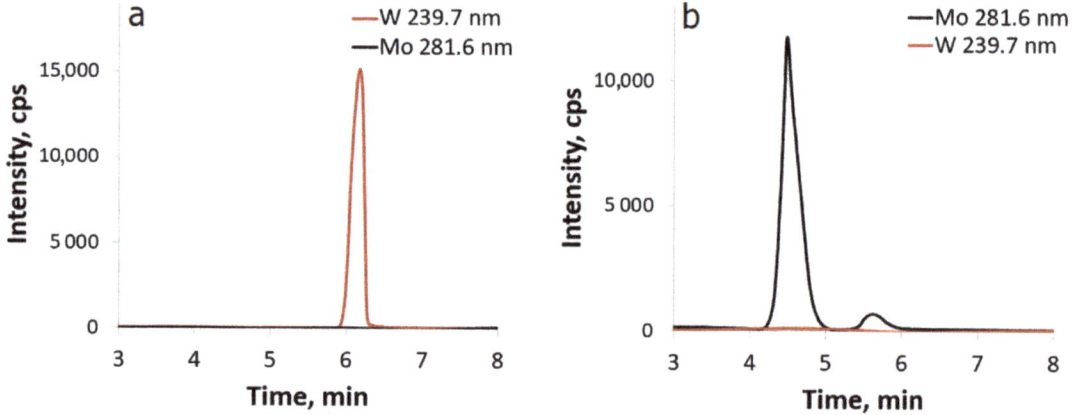

Figure 5. HPLC-ICP-AES chromatograms of (**a**) $[SiW_{12}O_{40}]^{4-}$ and (**b**) $[\beta\text{-}Mo_8O_{26}]^{4-}$ in dimethyl sulfoxide.

The HPLC-ICP-AES chromatogram of $(Bu_4N)_4[\beta\text{-}Mo_8O_{26}]$ in DMSO (Figure 5b) is similar to the chromatogram in acetonitrile (Figure 3a) and shows a major molybdenum peak (t_R = 4.5 min), corresponding to the octamolybdate anion $[\beta\text{-}Mo_8O_{26}]^{4-}$, and a minor peak of $[Mo_6O_{19}]^{2-}$ [59]. Since the viscosity of DMSO is 5 times that of acetonitrile, we were forced to reduce the concentration of the ion-pair reagent in the HPLC eluent to prevent column overpressure. Therefore, the peak retention times in DMSO increased. The HPLC-ICP-AES technique was used to investigate the reaction products between $(Bu_4N)_4[\beta\text{-}Mo_8O_{26}]$ and $[H_4SiW_{12}O_{40}]\cdot14H_2O$ at different molar ratios. For the Mo/W = 10/1 molar ratio at C_o of $(Bu_4N)_4[\beta\text{-}Mo_8O_{26}]$ = 3 mM, we observed four peaks (Figure 6a): (i) unreacted octamolybdate (t_R = 4.5 min), (ii) poorly separated peak with atomic ratio Mo:W = 2.3 (t_R = 4.7 min), (iii) hexamolybdate (t_R = 5.7 min) and (iv) Mo-free peak (t_R = 6.2 min) from unreacted silicotungstic acid. With an increase in the tungstate concentration (Mo/W = 10/2 molar ratio), the same major peak with atomic ratio Mo:W = 2.3 was observed (Figure 6b). In addition, the chromatogram shows minor W-free peaks (t_R = 4.5 min, t_R = 5.6 min) and a single peak containing tungsten (t_R = 6.2 min), which may indicate an excess of the tungstate anion. Further increase in the concentration of tungstate (Mo/W = 10/4 molar ratio) leads to the disappearance of the first molybdenum peak ($[\beta\text{-}Mo_8O_{26}]^{4-}$, t_R = 4.5 min) and an increase in the intensity of the peak of unreacted tungstate.

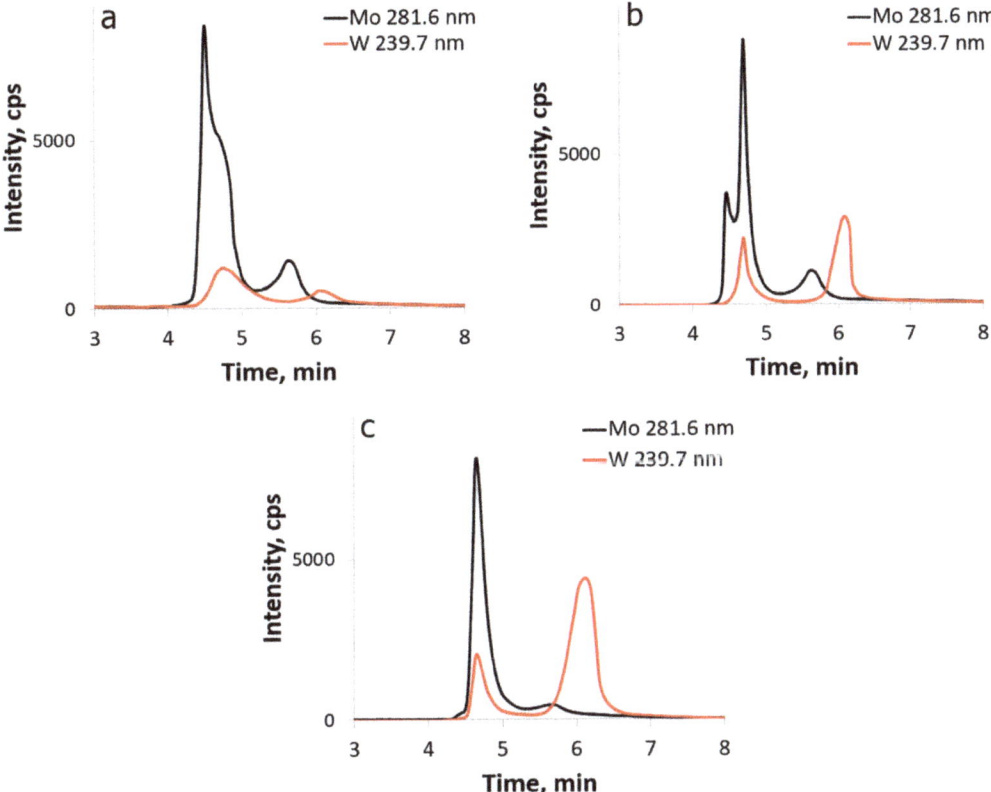

Figure 6. HPLC-ICP-AES chromatograms of the POM mixture at (**a**) the initial Mo/W ratio = 10/1, (**b**) Mo/W ratio = 10/2 and (**c**) Mo/W ratio = 10/4.

Thus, according to the HPLC-ICP-AES results, the proton transfer between the silicotungstic acid and [β-Mo$_8$O$_{26}$]$^{4-}$ triggers metal redistribution with the formation of Lindqvist type [W$_2$Mo$_4$O$_{19}$]$^{2-}$ anion as the reorganization product of [β-Mo$_8$O$_{26}$]$^{4-}$. Curiously, [α-Mo$_8$O$_{26}$]$^{4-}$ does not react in this case. The Keggin anion almost completely converts into the mixed Lindqvist at Mo/W ratio = 10/1 (Figure 6a). We reported a similar process earlier, when direct reaction of [H$_3$PW$_{12}$O$_{40}$] with [NbO(C$_2$O$_4$)$_2$]$^-$ yielded [PW$_{11}$NbO$_{40}$]$^{4-}$ [60].

The reaction between (Bu$_4$N)$_4$[β-Mo$_8$O$_{26}$] and acetic acid was investigated with the HPLC technique. The reaction was run in DMSO (C$_0$ of = 3 mM) by the addition of various concentrations of acetic acid (Figure 7).

The HPLC chromatogram of a freshly prepared solution of (Bu$_4$N)$_4$[β-Mo$_8$O$_{26}$] shows the peaks from octamolybdate [β-Mo$_8$O$_{26}$]$^{4-}$ (Figure 5, peak no. 3, t$_R$ = 4.5 min) and hexamolybdate [Mo$_6$O$_{19}$]$^{2-}$ (Figure 7, peak no. 5, t$_R$ = 5.6 min) in the ratio of 95:5. Addition of 0.001 M acetic acid decreases the octamolybdate peak intensity, while causing an increase in the hexamolybdate peak intensity and the appearance of a new peak (peak no. 4, t$_R$ = 5.1 min). Further increase in the acetic acid concentration continues to reduce the intensity of the octamolybdate peak and leads to an increase in the intensity of peak no. 4, as well as the appearance of two minor peaks (peak no. 1,2) of smaller molybdates. At an acetic acid concentration of 0.008 M, the intensity ratio of the peaks corresponding to octamolybdate (peak no. 3), the new product (peak no. 4), and hexamolydate (peak no. 5), is 1.8:3.4:1, respectively. No further changes in the ratio of species in solution was observed with an increase in the concentration of acetic acid from 0.008 M to 0.01 M; however, the intensity of all peaks decreases by 1.5, and further acidification leads to the formation of a white precipitate, which makes the HPLC analysis unapplicable.

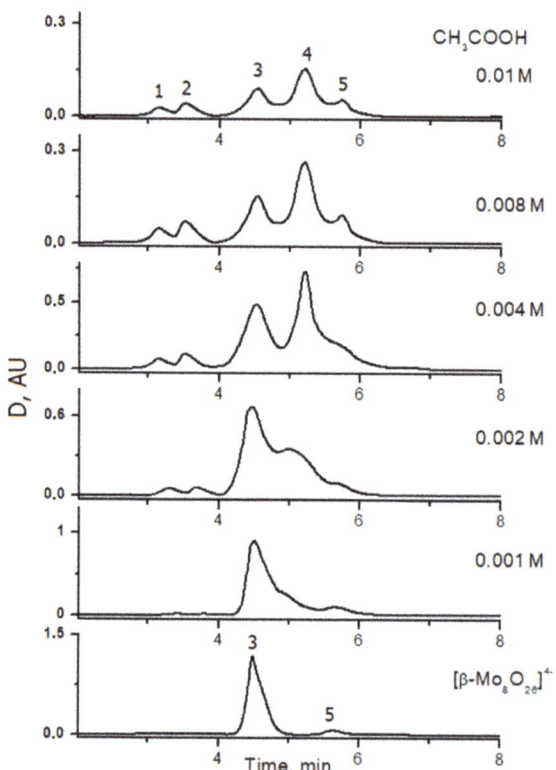

Figure 7. The HPLC-UV chromatograms of (Bu$_4$N)$_4$[β-Mo$_8$O$_{26}$] in dimethyl sulfoxide with the addition of acetic acid.

From this observation it follows that the transformation of [β-Mo$_8$O$_{26}$]$^{4-}$ into [Mo$_6$O$_{19}$]$^{2-}$ can be explained as direct dimolybdate ([Mo$_2$O$_7$]$^{2-}$) elimination, as was proposed earlier. There are two simple molybdate anions in the reaction mixture. In the literature there is a structure of K[MoO$_2$(OAc)$_3$]·HOAc [61] complex, showing the possibility of [MoO$_2$(OAc)$_3$]$^-$ existence in the solution. The new peak (Figure 7, peak no. 4) can be assigned as [Mo$_8$O$_{24}$(OAc)$_2$]$^{4-}$, with the same structure as reported for the malonate derivative ((NH$_4$)$_4$[Mo$_8$O$_{24}$(C$_3$H$_2$O$_2$)$_2$]·4H$_2$O) [62].

2.3. NMR

NMR spectroscopy was anticipated to be an informative tool to study the reaction between (Bu$_4$N)$_4$[β-Mo$_8$O$_{26}$] (**Mo8**) and (Bu$_4$N)$_4$H$_2$[V$_{10}$O$_{28}$] (**V10**) due to the presence of both ^{51}V and ^{95}Mo NMR active isotopes. We measured ^{95}Mo NMR spectra for the following solutions to study the effects of acidification of **Mo8** by Hpts (Hpts = *p*-toluenesulfonic acid) (Figure 8).

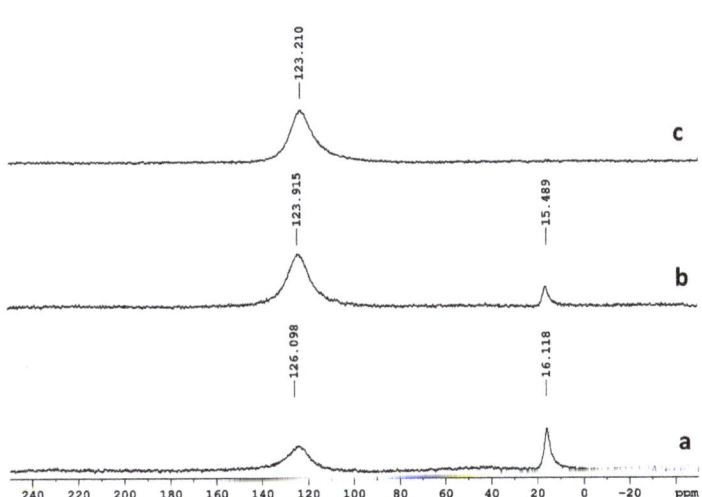

Figure 8. ^{95}Mo NMR data for **Mo8** acidification: (**a**) 70 mg **Mo8** + 2 mg of Hpts in 600 µL DMSO-d_6; (**b**) 70 mg **Mo8** + 5 mg of Hpts in 600 µL DMSO-d_6; (**c**) 70 mg **Mo8** + 10 mg of Hpts in 600 µL DMSO-d_6.

As can be seen, addition of Hpts as a non-coordinating organic acid to the solution of $(Bu_4N)_4[\beta\text{-}Mo_8O_{26}]$ leads to the disappearance of the $[\alpha\text{-}Mo_8O_{26}]^{4-}$ isomer and an increase in the amount of $[Mo_6O_{19}]^{2-}$. Simple (mononuclear or binuclear) Mo-containing complexes were not detected, most likely due to the fast exchange. The signals from such species should appear in a negative region, in comparison with the literature [63].

The reaction between **Mo8** and **V10** was studied using both ^{95}Mo and ^{51}V NMR (Figure 9).

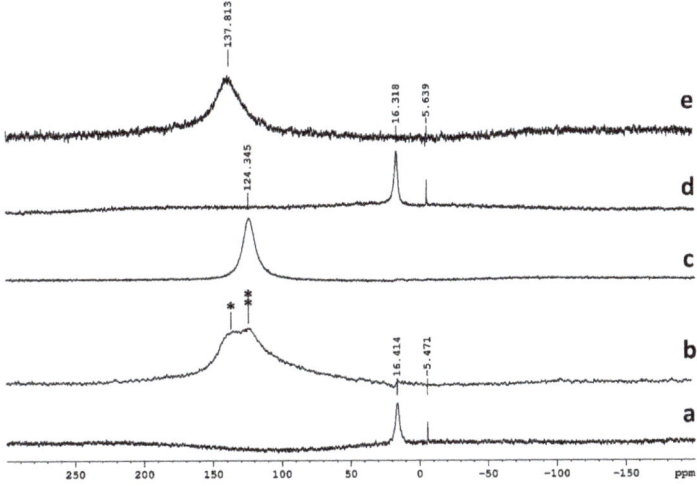

Figure 9. ^{95}Mo NMR data describing **Mo8** and **V10** reaction: (**a**) 70 mg **Mo8** (300 µL DMSO-d6) + 55 mg **V10** (300 µL DMSO-d_6) room temperature; (**b**) 70 mg **Mo8** (300 µL DMSO-d6) + 22 mg **V10** (300 µL DMSO-d_6), after 10 min at 60 °C; (**c**) 70 mg $(Bu_4N)_2[Mo_6O_{19}]$ in 600 µL DMSO-d_6; (**d**) 70 mg **Mo8** in 600 µL DMSO-d_6; (**e**) 70 mg of $(Bu_4N)_3Na[V_2Mo_4O_{19}]$ in 600 µL DMSO-d_6 * Chemical shift (CS) = 137.4301 ppm, ** CS = 123.3993 ppm.

The ^{51}V NMR spectra (Figures S2–S4) show exclusive formation of $[V_2Mo_4O_{19}]^{4-}$, to the detriment of other mixed metal Lindqvist molybdovanadates, meaning that such reactions can offer a straightforward way to this anion. Two signals in the ^{95}Mo NMR spectrum of $(Bu_4N)_4[\beta\text{-}Mo_8O_{26}]$ indicate an equilibrium between α and β isomers, as described in the literature [64]. In the case of spectrum *b* (Figure 9), the baseline correction was not as accurate, and the peaks from **Mo8** have slightly negative intensities. Moreover, due to this problem, the profile of the main signal is also not as correct. Nevertheless, we can postulate the presence of $[V_2Mo_4O_{19}]^{4-}$ and $[Mo_6O_{19}]^{2-}$ in the reaction mixture.

3. Materials and Methods

3.1. Physical Methods

$(Bu_4N)_4[\beta\text{-}Mo_8O_{26}]$, $(Bu_4N)_2[Mo_6O_{19}]$, $(Bu_4N)_2[Mo_2O_7]$, $(Bu_4N)_4H_2[V_{10}O_{28}]$ and $(Bu_4N)_3Na[V_2Mo_4O_{19}]$ were prepared according to the literature data (Inorg. Synth. Vol. 27). DMSO was distilled in vacuo over NaOH. $[H_4SiW_{12}O_{40}]\cdot 14H_2O$ was manufactured by "The Red Chemist" (Saint Petersburg, USSR) and checked with FT-IR and TGA prior to use. Other reagents were of commercial quality (Sigma Aldrich) and were used as purchased. IR spectra were recorded on a Bruker Vertex 60 FT-IR spectrometer. Elemental analysis was carried out on a MICRO Cube CHN analyzer.

Synthesis of $(Bu_4N)_3[Ag(py\text{-}NH_2)Mo_8O_{26}]$ (1): $(Bu_4N)_4[\beta\text{-}Mo_8O_{26}]$ (200 mg, mmol) was dissolved in 2 mL of DMF under sonication, and afterward, 20 mg (mmol) of py-NH$_2$ was added to the observed clear solution. Solid AgNO$_3$ (32 mg, mmol) was added to the reaction mixture under sonication. The resulting mixture was placed in Et$_2$O atmosphere at 4 °C to obtain crystalline material. The crop of large colorless crystals was isolated after 48 h. Yield was 175 mg (90% based on initial octamolybdate).

Elemental analysis. Calcd C, H, N (%) for **1**: 30.1, 5.4, 3.3; found C, H, N (%): 30.0, 5.3, 3.2.

IR (KBr, cm^{-1}): 3485 (w), 3336 (m), 3222 (w), 2959 (vs), 2929 (s), 2872 (s), 1632 (vs), 1611 (s), 1554 (w), 1522 (w), 1480 (vs), 1456 (s), 1375 (m), 1355 (w), 1333 (w), 1278 (w), 1214 (m), 1148 (w), 1100 (w), 1060 (w), 1028 (w), 1014 (m), 970 (s), 947 (vs), 928 (vs), 905 (vs), 865 (vs), 847 (vs), 825 (s), 811 (m), 700 (vs), 657 (vs), 567 (m), 551 (s), 520 (s), 472 (m), 444 (w), 407 (s).

3.2. NMR

^{51}V and ^{95}Mo NMR spectra were recorded on a Bruker Avance III 500 spectrometer (BBI detector), using NaVO$_3$ and Na$_2$MoO$_4$ as internal standards. Spectra were measured in DMSO-d_6 at room temperature using standard 5 mm NMR tubes.

3.3. X-ray Diffraction on Single Crystals

Crystallographic data and refinement details are given in Table S1 (Supplementary Materials). The diffraction data for **1** were collected on a Bruker D8 Venture diffractometer with a CMOS PHOTON III detector and IμS 3.0 source (Mo Kα radiation, λ = 0.71073 Å) at 150 K. The φ- and ω-scan techniques were employed. Absorption correction was applied by SADABS (Bruker Apex3 software suite: Apex3, SADABS-2016/2 and SAINT, version 2018.7-2; Bruker AXS Inc.: Madison, WI, USA, 2017). Structures were solved by SHELXT [65] and refined by full-matrix least-squares treatment against $|F|^2$ in anisotropic approximation with SHELX 2014/7 [66] in the ShelXle program [67]. H-atoms were refined in geometrically calculated positions.

CCDC 2215910 contains the supplementary crystallographic data. These data can be obtained free of charge via http://www.ccdc.cam.ac.uk/conts/retrieving.html, or from the Cambridge Crystallographic Data Centre, 12 Union Road, Cambridge CB2 1EZ, UK; fax: (+44) 1223-336-033; or e-mail: deposit@ccdc.cam.ac.uk.

3.4. XRPD

X-ray powder diffraction patterns were measured on a Bruker D8 Advance diffractometer using LynxEye XE T-discriminated CuKα radiation. Samples were layered on a flat plastic specimen holder.

3.5. HPLC-ICP-AES and HPLC

Separation was performed with the HPLC system Milichrom A-02 (EcoNova, Novosibirsk, Russia), equipped with a two-beam spectrophotometric detector at the wavelength range of 190–360 nm in the ion-pair mode of reversed phase chromatography (ProntoSIL 120-5-C18AQ, 2 × 75 mm), eluents: A—0.06% tetrabutylammonium hydroxide (for $(Bu_4N)_4[\beta-Mo_8O_{26}]$ and $(Bu_4N)_4H_2[V_{10}O_{28}]$), 0.02% tetrabutylammonium hydroxide (for $(Bu_4N)_4[\beta-Mo_8O_{26}]$ and $[H_4SiW_{12}O_{40}]\cdot 14H_2O$); B—acetonitrile. Gradient elution with a gradual increase in acetonitrile concentration was employed to resolve the species. ICP-AES spectrometer iCap 6500 Duo (Thermo Scientific, Waltham, MA, USA) with concentric nebulizer was applied as detector in hyphenated HPLC-ICP-AES. For the element detection Mo 281.6 nm, V 292.4 nm and W 239.7 nm, spectral lines were selected. All measurements were performed in three replicates.

The data acquisition and processing were carried out with iTEVA (Thermo Scientific, Waltham, MA, USA) software. In order to eliminate plasma quenching, we diluted the liquid coming out of the column into the spray chamber with deionized water. The steady state of the plasma and the optimal values of the analytical signals were finally achieved at the eluent flow rate of 0.25 mL min^{-1} and the eluent velocity of 3 mL min^{-1} (peristaltic pump speed—75 rpm).

4. Conclusions

This manuscript describes an affinity of $[\beta-Mo_8O_{26}]^{4-}$ lacunes for interaction with H-atoms possessing some N–H or even C–H (in Me_4N^+) acidity. We demonstrated this in the case of 1D polymeric chains formation via py-NH$_2$ and $[\beta-Mo_8O_{26}]^{4-}$ interaction. This example illustrates a general approach to the formation of soft matters based on such types of interactions. The reaction of $[\beta-Mo_8O_{26}]^{4-}$ with diluted acids generates a set of unknown complexes, according to the HPLC-ICP-AES data. Moreover, there was a new type of reactivity of $[\beta-Mo_8O_{26}]^{4-}$ combining: (i) proton transfer from another type of polyoxometalates in solution, (ii) backbone breaking and (iii) transformation into mixed Lindqvist type complexes has been demonstrated. In the case of $(Bu_4N)_4H_2[V_{10}O_{28}]$, this reaction gives $[V_2Mo_4O_{19}]^{4-}$. $[H_4SiW_{12}O_{40}]$ plays a role as a proton and W source, producing $[W_2Mo_4O_{19}]^{2-}$. The key study here is proton transfer into the lacune of $[\beta-Mo_8O_{26}]^{4-}$, generating the reactive transition state. At the current stage, it is impossible to deduce the mechanism, which is not as simple as $[Mo_2O_7]^{2-}$-elimination. In comparison with the microwave synthesis reported by Karoui and Ritchie, simple thermal activation does not need any special equipment. The addition of any triol type organic ligands into the reaction mixture will be the next step in such reactivity studies. Such an approach opens a way to new mixed functionalized complexes.

Supplementary Materials: The following supporting information can be downloaded at: https://www.mdpi.com/article/10.3390/molecules27238368/s1, Table S1: SCXRD Experimental details; Figure S1: Comparison of experimental and calculated powder diffraction patterns for **1**; Figure S2: ^{51}V NMR spectrum of $(Bu_4N)_3Na[V_2Mo_4O_{19}]$ (DMSO-d_6, r.t.); Figure S3: ^{51}V NMR spectrum of the reaction mixture containing 70 mg **Mo8** (300 µL DMSO-d_6) + 55 mg **V10** (300 µL DMSO-d_6); Figure S4: ^{51}V NMR spectrum of the reaction mixture containing 70 mg **Mo8** (300 µL DMSO-d_6) + 22 mg **V10** (300 µL DMSO-d_6), after 10 min at 60 °C.

Author Contributions: Conceptualization, P.A.A.; methodology, P.A.A.; validation, P.A.A., V.V.V. and N.B.K.; formal analysis, V.V.V. and N.B.K.; data curation, V.V.V. and N.B.K.; writing—original draft preparation, P.A.A.; writing—review and editing, P.A.A. and M.N.S.; visualization, V.V.V. and N.B.K.; supervision, P.A.A.; project administration, P.A.A.; funding acquisition, P.A.A. All authors have read and agreed to the published version of the manuscript.

Funding: This work was supported by the grant of the President of the Russian Federation for young scientists—Doctors of Sciences MD-396.2021.1.3.

Institutional Review Board Statement: Not applicable.

Informed Consent Statement: Not applicable.

Data Availability Statement: The crystallographic data have been deposited in the Cambridge Crystallographic Data Centre under the deposition codes CCDC 2215910.

Acknowledgments: The authors thank the Ministry of Science and Higher Education of the Russian Federation for access to the XRD facilities of the Nikolaev Institute of Inorganic Chemistry. The technical staff of the Institute is also thanked for their assistance.

Conflicts of Interest: The authors declare no conflict of interest. The funders had no role in the design of the study; in the collection, analyses, or interpretation of data; in the writing of the manuscript; or in the decision to publish the results.

References

1. Ishikita, H.; Saito, K. Proton transfer reactions and hydrogen-bond networks in protein environments. *J. R. Soc. Interface* **2014**, *11*, 20130518. [CrossRef] [PubMed]
2. Mayer, J.M. Proton-Coupled Electron Transfer: A Reaction Chemist's View. *Annu. Rev. Phys. Chem.* **2004**, *55*, 363–390. [CrossRef] [PubMed]
3. Hammes-Schiffer, S.; Stuchebrukhov, A.A. Theory of Coupled Electron and Proton Transfer Reactions. *Chem. Rev.* **2010**, *110*, 6939–6960. [CrossRef] [PubMed]
4. Chang, C.J.; Chang, M.C.Y.; Damrauer, N.H.; Nocera, D.G. Proton-coupled electron transfer: A unifying mechanism for biological charge transport, amino acid radical initiation and propagation, and bond making/breaking reactions of water and oxygen. *Biochim. Biophys. Acta—Bioenerg.* **2004**, *1655*, 13–28. [CrossRef] [PubMed]
5. Tyburski, R.; Liu, T.; Glover, S.D.; Hammarström, L. Proton-Coupled Electron Transfer Guidelines, Fair and Square. *J. Am. Chem. Soc.* **2021**, *143*, 560–576. [CrossRef]
6. Darcy, J.W.; Koronkiewicz, B.; Parada, G.A.; Mayer, J.M. A Continuum of Proton-Coupled Electron Transfer Reactivity. *Acc. Chem. Res.* **2018**, *51*, 2391–2399. [CrossRef]
7. Agarwal, R.G.; Coste, S.C.; Groff, B.D.; Heuer, A.M.; Noh, H.; Parada, G.A.; Wise, C.F.; Nichols, E.M.; Warren, J.J.; Mayer, J.M. Free Energies of Proton-Coupled Electron Transfer Reagents and Their Applications. *Chem. Rev.* **2022**, *122*, 1–49. [CrossRef]
8. Warren, J.J.; Tronic, T.A.; Mayer, J.M. Thermochemistry of Proton-Coupled Electron Transfer Reagents and its Implications. *Chem. Rev.* **2010**, *110*, 6961–7001. [CrossRef]
9. Siewert, I. Proton-Coupled Electron Transfer Reactions Catalysed by 3 d Metal Complexes. *Chem.—A Eur. J.* **2015**, *21*, 15078–15091. [CrossRef]
10. Cukier, R.I. Proton-Coupled Electron Transfer Reactions: Evaluation of Rate Constants. *J. Phys. Chem.* **1996**, *100*, 15428–15443. [CrossRef]
11. Huynh, M.T.; Mora, S.J.; Villalba, M.; Tejeda-Ferrari, M.E.; Liddell, P.A.; Cherry, B.R.; Teillout, A.-L.; Machan, C.W.; Kubiak, C.P.; Gust, D.; et al. Concerted One-Electron Two-Proton Transfer Processes in Models Inspired by the Tyr-His Couple of Photosystem II. *ACS Cent. Sci.* **2017**, *3*, 372–380. [CrossRef]
12. Odella, E.; Mora, S.J.; Wadsworth, B.L.; Huynh, M.T.; Goings, J.J.; Liddell, P.A.; Groy, T.L.; Gervaldo, M.; Sereno, L.E.; Gust, D.; et al. Controlling Proton-Coupled Electron Transfer in Bioinspired Artificial Photosynthetic Relays. *J. Am. Chem. Soc.* **2018**, *140*, 15450–15460. [CrossRef]
13. Odella, E.; Wadsworth, B.L.; Mora, S.J.; Goings, J.J.; Huynh, M.T.; Gust, D.; Moore, T.A.; Moore, G.F.; Hammes-Schiffer, S.; Moore, A.L. Proton-Coupled Electron Transfer Drives Long-Range Proton Translocation in Bioinspired Systems. *J. Am. Chem. Soc.* **2019**, *141*, 14057–14061. [CrossRef]
14. Odella, E.; Mora, S.J.; Wadsworth, B.L.; Goings, J.J.; Gervaldo, M.A.; Sereno, L.E.; Groy, T.L.; Gust, D.; Moore, T.A.; Moore, G.F.; et al. Proton-coupled electron transfer across benzimidazole bridges in bioinspired proton wires. *Chem. Sci.* **2020**, *11*, 3820–3828. [CrossRef]

15. Goldsmith, Z.K.; Lam, Y.C.; Soudackov, A.V.; Hammes-Schiffer, S. Proton Discharge on a Gold Electrode from Triethylammonium in Acetonitrile: Theoretical Modeling of Potential-Dependent Kinetic Isotope Effects. *J. Am. Chem. Soc.* **2019**, *141*, 1084–1090. [CrossRef]
16. Warburton, R.E.; Hutchison, P.; Jackson, M.N.; Pegis, M.L.; Surendranath, Y.; Hammes-Schiffer, S. Interfacial Field-Driven Proton-Coupled Electron Transfer at Graphite-Conjugated Organic Acids. *J. Am. Chem. Soc.* **2020**, *142*, 20855–20864. [CrossRef]
17. Lam, Y.-C.; Soudackov, A.V.; Hammes-Schiffer, S. Theory of Electrochemical Proton-Coupled Electron Transfer in Diabatic Vibronic Representation: Application to Proton Discharge on Metal Electrodes in Alkaline Solution. *J. Phys. Chem. C* **2020**, *124*, 27309–27322. [CrossRef]
18. Sarkar, S.; Maitra, A.; Lake, W.R.; Warburton, R.E.; Hammes-Schiffer, S.; Dawlaty, J.M. Mechanistic Insights about Electrochemical Proton-Coupled Electron Transfer Derived from a Vibrational Probe. *J. Am. Chem. Soc.* **2021**, *143*, 8381–8390. [CrossRef]
19. Yadav, S.; Sharma, A.K.; Kumar, P. Nanoscale Self-Assembly for Therapeutic Delivery. *Front. Bioeng. Biotechnol.* **2020**, *8*, 127. [CrossRef]
20. Tan, M.; Tian, P.; Zhang, Q.; Zhu, G.; Liu, Y.; Cheng, M.; Shi, F. Self-sorting in macroscopic supramolecular self-assembly via additive effects of capillary and magnetic forces. *Nat. Commun.* **2022**, *13*, 5201. [CrossRef]
21. Li, K.; Hu, J.-M.; Qin, W.-M.; Guo, J.; Cai, Y.-P. Precise heteroatom doping determines aqueous solubility and self-assembly behaviors for polycyclic aromatic skeletons. *Commun. Chem.* **2022**, *5*, 104. [CrossRef]
22. Cui, Z.; Jin, G.-X. Construction of a molecular prime link by interlocking two trefoil knots. *Nat. Synth.* **2022**, *1*, 635–640. [CrossRef]
23. Woods, J.F.; Gallego, L.; Pfister, P.; Maaloum, M.; Vargas Jentzsch, A.; Rickhaus, M. Shape-assisted self-assembly. *Nat. Commun.* **2022**, *13*, 3681. [CrossRef] [PubMed]
24. Jiao, Y.; Qiu, Y.; Zhang, L.; Liu, W.-G.; Mao, H.; Chen, H.; Feng, Y.; Cai, K.; Shen, D.; Song, B.; et al. Electron-catalysed molecular recognition. *Nature* **2022**, *603*, 265–270. [CrossRef]
25. Whitesides, G.M.; Boncheva, M. Beyond molecules: Self-assembly of mesoscopic and macroscopic components. *Proc. Natl. Acad. Sci. USA* **2002**, *99*, 4769–4774. [CrossRef]
26. Gartner, F.M.; Graf, I.R.; Frey, E. The time complexity of self-assembly. *Proc. Natl. Acad. Sci. USA* **2022**, *119*, e2116373119. [CrossRef]
27. Deamer, D.; Singaram, S.; Rajamani, S.; Kompanichenko, V.; Guggenheim, S. Self-assembly processes in the prebiotic environment. *Philos. Trans. R. Soc. B Biol. Sci.* **2006**, *361*, 1809–1818. [CrossRef]
28. Wang, Z.; Gupta, R.K.; Luo, G.; Sun, D. Recent Progress in Inorganic Anions Templated Silver Nanoclusters: Synthesis, Structures and Properties. *Chem. Rec.* **2020**, *20*, 389–402. [CrossRef]
29. Gao, G.-G.; Cheng, P.-S.; Mak, T.C.W. Acid-Induced Surface Functionalization of Polyoxometalate by Enclosure in a Polyhedral Silver–Alkynyl Cage. *J. Am. Chem. Soc.* **2009**, *131*, 18257–18259. [CrossRef]
30. Shi, J.-Y.; Gupta, R.K.; Deng, Y.-K.; Sun, D.; Wang, Z. Recent Advances in the Asymmetrical Templation Effect of Polyoxometalate in Silver Clusters. *Polyoxometalates* **2022**, *1*, 9140010. [CrossRef]
31. Fabre, B.; Falaise, C.; Cadot, E. Polyoxometalates-Functionalized Electrodes for (Photo)Electrocatalytic Applications: Recent Advances and Prospects. *ACS Catal.* **2022**, *12*, 12055–12091. [CrossRef]
32. Wu, Y.; Bi, L. Research Progress on Catalytic Water Splitting Based on Polyoxometalate/Semiconductor Composites. *Catalysts* **2021**, *11*, 524. [CrossRef]
33. Kuznetsov, A.E.; Geletii, Y.V.; Hill, C.L.; Morokuma, K.; Musaev, D.G. Dioxygen and Water Activation Processes on Multi-Ru-Substituted Polyoxometalates: Comparison with the "Blue-Dimer" Water Oxidation Catalyst. *J. Am. Chem. Soc.* **2009**, *131*, 6844–6854. [CrossRef]
34. Lauinger, S.M.; Piercy, B.D.; Li, W.; Yin, Q.; Collins-Wildman, D.L.; Glass, E.N.; Losego, M.D.; Wang, D.; Geletii, Y.V.; Hill, C.L. Stabilization of Polyoxometalate Water Oxidation Catalysts on Hematite by Atomic Layer Deposition. *ACS Appl. Mater. Interfaces* **2017**, *9*, 35048–35056. [CrossRef]
35. Anwar, N.; Sartorel, A.; Yaqub, M.; Wearen, K.; Laffir, F.; Armstrong, G.; Dickinson, C.; Bonchio, M.; McCormac, T. Surface Immobilization of a Tetra-Ruthenium Substituted Polyoxometalate Water Oxidation Catalyst Through the Employment of Conducting Polypyrrole and the Layer-by-Layer (LBL) Technique. *ACS Appl. Mater. Interfaces* **2014**, *6*, 8022–8031. [CrossRef]
36. Azmani, K.; Besora, M.; Soriano-López, J.; Landolsi, M.; Teillout, A.-L.; de Oliveira, P.; Mbomekallé, I.-M.; Poblet, J.M.; Galán-Mascarós, J.-R. Understanding polyoxometalates as water oxidation catalysts through iron vs. cobalt reactivity. *Chem. Sci.* **2021**, *12*, 8755–8766. [CrossRef]
37. Soriano-López, J.; Musaev, D.G.; Hill, C.L.; Galán-Mascarós, J.R.; Carbó, J.J.; Poblet, J.M. Tetracobalt-polyoxometalate catalysts for water oxidation: Key mechanistic details. *J. Catal.* **2017**, *350*, 56–63. [CrossRef]
38. Fertig, A.A.; Brennessel, W.W.; McKone, J.R.; Matson, E.M. Concerted Multiproton–Multielectron Transfer for the Reduction of O$_2$ to H$_2$O with a Polyoxovanadate Cluster. *J. Am. Chem. Soc.* **2021**, *143*, 15756–15768. [CrossRef]
39. Wang, X.-L.; Zhang, Y.; Chen, Y.-Z.; Wang, Y.; Wang, X. Two polymolybdate-directed Zn(ii) complexes tuned by a new bis-pyridine-bis-amide ligand with a diphenylketone spacer for efficient ampere sensing and dye adsorption. *CrystEngComm* **2022**, *24*, 5289–5296. [CrossRef]
40. Huang, X.; Cui, Y.; Liu, G.; Wang, H.; Ren, J.; Zhang, Y.; Shen, G.; Lv, L.; Wang, H.-W.; Chen, Y.-F. Imidazole-Dependent Assembly of Copper Polymolybdate Frameworks for One-Pot Sulfide Oxidation and C–H Activation. *Energy Fuels* **2022**, *36*, 1665–1675. [CrossRef]

41. Liu, J.; Huang, M.; Hua, Z.; Dong, Y.; Feng, Z.; Sun, T.; Chen, C. Polyoxometalate-Based Metal Organic Frameworks: Recent Advances and Challenges. *ChemistrySelect* **2022**, *7*, e202200546. [CrossRef]
42. Talib, S.H.; Yu, X.; Lu, Z.; Ahmad, K.; Yang, T.; Xiao, H.; Li, J. A polyoxometalate cluster-based single-atom catalyst for NH_3 synthesis via an enzymatic mechanism. *J. Mater. Chem. A* **2022**, *10*, 6165–6177. [CrossRef]
43. Paul, A.; Das Adhikary, S.; Kapurwan, S.; Konar, S. En route to artificial photosynthesis: The role of polyoxometalate based photocatalysts. *J. Mater. Chem. A* **2022**, *10*, 13152–13169. [CrossRef]
44. Liu, C.; Cui, C.; Dai, Y.; Liu, G.; Qiao, S.; Tao, Y.; Zhang, Y.; Shen, G.; Li, Z.; Huang, X. Two silver–containing polyoxometalate-based inorganic–organic hybrids as heterogeneous bifunctional catalysts for construction of C–C bonds and decontamination of sulfur mustard simulant. *J. Solid State Chem.* **2022**, *316*, 123547. [CrossRef]
45. Silva, D.F.; Viana, A.M.; Santos-Vieira, I.; Balula, S.S.; Cunha-Silva, L. Ionic Liquid-Based Polyoxometalate Incorporated at ZIF-8: A Sustainable Catalyst to Combine Desulfurization and Denitrogenation Processes. *Molecules* **2022**, *27*, 1711. [CrossRef]
46. Veríssimo, M.I.S.; Evtuguin, D.V.; Gomes, M.T.S.R. Polyoxometalate Functionalized Sensors: A Review. *Front. Chem.* **2022**, *10*, 840657. [CrossRef]
47. Ren, W.; Li, B.; Li, S.; Li, Y.; Gao, Z.; Chen, X.; Zang, H. Synthesis and Proton Conductivity of Two Molybdate Polymers Based on $[Mo_8O_{26}]^{4-}$ Anions. *ChemistrySelect* **2022**, *7*, e202201337. [CrossRef]
48. Hsieh, T.C.; Shaikh, S.N.; Zubieta, J. Derivatized polyoxomolybdates. Synthesis and characterization of oxomolybdate clusters containing coordinatively bound diazenido units. Crystal and molecular structure of the octanuclear oxomolybdate $(NHEt_3)_2$(n-$Bu_4N)_2[Mo_8O_{20}(NNPh)_6]$ and comparison to the structures of the parent oxomolybdate .alpha.-(n-$Bu_4N)_4[Mo_8O_{26}]$ and the tetranuclear (diazenido)oxomolybdates (n-$Bu_4N)_2[Mo_4O_{10}(OMe)_2(NNPh)_2]$ and (n-$Bu_4N)_2[Mo_4O_8(OMe)_2(NNC_6H_4NO_2)_4]$. *Inorg. Chem.* **1987**, *26*, 4079–4089.
49. Cindrić, M.; Veksli, Z.; Kamenar, B. Polyoxomolybdates and polyoxomolybdovanadates—from structure to functions: Recent results. *Croat. Chem. Acta* **2009**, *82*, 345–362.
50. Chupina, A.V.; Shayapov, V.; Novikov, A.S.; Volchek, V.V.; Benassi, E.; Abramov, P.A.; Sokolov, M.N. $[\{AgL\}_2Mo_8O_{26}]^{n-}$ complexes: A combined experimental and theoretical study. *Dalton Trans.* **2020**, *49*, 1522–1530. [CrossRef]
51. Abramov, P.A.; Komarov, V.Y.; Pischur, D.A.; Sulyaeva, V.S.; Benassi, E.; Sokolov, M.N. Solvatomorphs of $(Bu_4N)_2[\{Ag(N_2\text{-}py)\}_2Mo_8O_{26}]$: Structure, colouration and phase transition. *CrystEngComm* **2021**, *23*, 8527–8537. [CrossRef]
52. Komlyagina, V.I.; Romashev, N.F.; Kokovkin, V.V.; Gushchin, A.L.; Benassi, E.; Sokolov, M.N.; Abramov, P.A. Trapping of Ag^+ into a Perfect Six-Coordinated Environment: Structural Analysis, Quantum Chemical Calculations and Electrochemistry. *Molecules* **2022**, *27*, 6961. [CrossRef]
53. Karoui, H.; Ritchie, C. Microwave-assisted synthesis of organically functionalized hexa-molybdovanadates. *New J. Chem.* **2018**, *42*, 25–28. [CrossRef]
54. Shuvaeva, O.V.; Zhdanov, A.A.; Romanova, T.E.; Abramov, P.A.; Sokolov, M.N. Hyphenated techniques in speciation analysis of polyoxometalates: Identification of individual $[PMo_{12-x}V_xO_{40}]^{-3-x}$ (x = 1–3) in the reaction mixtures by high performance liquid chromatography and atomic emission spectrometry with inductively coupled. *Dalton Trans.* **2017**, *46*, 3541–3546. [CrossRef]
55. Mukhacheva, A.A.; Volcheck, V.V.; Sheven, D.G.; Yanshole, V.V.; Kompankov, N.B.; Haouas, M.; Abramov, P.A.; Sokolov, M.N. Coordination capacity of Keggin anions as polytopic ligands: Case study of $[VNb_{12}O_{40}]^{15-}$. *Dalton Trans.* **2021**, *50*, 7078–7084. [CrossRef]
56. Mukhacheva, A.A.; Shmakova, A.A.; Volchek, V.V.; Romanova, T.E.; Benassi, E.; Gushchin, A.L.; Yanshole, V.; Sheven, D.G.; Kompankov, N.B.; Abramov, P.A.; et al. Reactions of $[Ru(NO)Cl_5]^{2-}$ with pseudotrilacunary $\{XW_9O_{33}\}^{9-}$ (X = As III, Sb III) anions. *Dalton Trans.* **2019**, *48*, 15989–15999. [CrossRef]
57. Kuznetsova, A.A.; Volchek, V.V.; Yanshole, V.V.; Fedorenko, A.D.; Kompankov, N.B.; Kokovkin, V.V.; Gushchin, A.L.; Abramov, P.A.; Sokolov, M.N. Coordination of Pt(IV) by $\{P_8W_{48}\}$ Macrocyclic Inorganic Cavitand: Structural, Solution, and Electrochemical Studies. *Inorg. Chem.* **2022**, *61*, 14560–14567. [CrossRef]
58. Chupina, A.V.; Mukhacheva, A.A.; Abramov, P.A.; Sokolov, M.N. Complexation and Isomerization of $[\beta\text{-}Mo_8O_{26}]^{4-}$ in the Presence of Ag^+ and DMF. *J. Struct. Chem.* **2020**, *61*, 299–308. [CrossRef]
59. Pantyukhina, V.S.; Volchek, V.V.; Komarov, V.Y.; Korolkov, I.V.; Kokovkin, V.V.; Kompankov, N.B.; Abramov, P.A.; Sokolov, M.N. Tubular polyoxoanion $[(SeMo_6O_{21})_2(C_2O_4)_3]^{10-}$ and its transformations. *New J. Chem.* **2021**, *45*, 6745–6752. [CrossRef]
60. Shmakova, A.A.; Akhmetova, M.M.; Volchek, V.V.; Romanova, T.E.; Korolkov, I.; Sheven, D.G.; Adonin, S.A.; Abramov, P.A.; Sokolov, M.N. A HPLC-ICP-AES technique for the screening of $[XW_{11}NbO_{40}]^{n-}$ aqueous solutions. *New J. Chem.* **2018**, *42*, 7940–7948. [CrossRef]
61. Korpar-Čolig, B.; Cindrić, M.; Matković-Čalogović, D.; Vrdoljak, V.; Kamenar, B. Synthesis and characterization of some new acetato complexes of molybdenum(IV), (V) and (VI). *Polyhedron* **2002**, *21*, 147–153. [CrossRef]
62. Wang, S.; Mo, S.; Liu, Z.-G. Synthesis and characterization of a new octamolybdate $(NH_4)_4[Mo_8O_{24}(C_3H_2O_2)_2]\cdot 4H_2O$. *Russ. J. Inorg. Chem.* **2012**, *57*, 430–433. [CrossRef]
63. Brito, J.A.; Teruel, H.; Massou, S.; Gómez, M. ^{95}Mo NMR: A useful tool for structural studies in solution. *Magn. Reson. Chem.* **2009**, *47*, 573–577. [CrossRef] [PubMed]
64. Fedotov, M.A.; Maksimovskaya, R.I. NMR structural aspects of the chemistry of V, Mo, W polyoxometalates. *J. Struct. Chem.* **2006**, *47*, 952–978. [CrossRef]

65. Sheldrick, G.M. SHELXT—Integrated space-group and crystal-structure determination. *Acta Crystallogr. Sect. A Found. Adv.* **2015**, *71*, 3–8. [CrossRef]
66. Sheldrick, G.M. Crystal structure refinement with SHELXL. *Acta Crystallogr. Sect. C Struct. Chem.* **2015**, *71*, 3–8. [CrossRef]
67. Hübschle, C.B.; Sheldrick, G.M.; Dittrich, B. ShelXle: A Qt graphical user interface for SHELXL. *J. Appl. Crystallogr.* **2011**, *44*, 1281–1284. [CrossRef]

Communication

Detecting the Subtle Photo-Responsive Conformational Bistability of Monomeric Azobenzene Functionalized Keggin Polyoxometalates by Using Ion-Mobility Mass Spectrometry

Bo Qi, Luran Jiang, Sai An, Wei Chen * and Yu-Fei Song *

State Key Laboratory of Chemical Resource Engineering, Beijing University of Chemical Technology, Beijing 100029, China; bqi@mail.buct.edu.cn (B.Q.); lr_jiang2021@163.com (L.J.); ansai@mail.buct.edu.cn (S.A.)
* Correspondence: chenw@mail.buct.edu.cn (W.C.); songyf@mail.buct.edu.cn (Y.-F.S.)

Citation: Qi, B.; Jiang, L.; An, S.; Chen, W.; Song, Y.-F. Detecting the Subtle Photo-Responsive Conformational Bistability of Monomeric Azobenzene Functionalized Keggin Polyoxometalates by Using Ion-Mobility Mass Spectrometry. *Molecules* **2022**, *27*, 3927. https://doi.org/10.3390/molecules27123927

Academic Editor: Xiaobing Cui

Received: 25 May 2022
Accepted: 16 June 2022
Published: 19 June 2022

Publisher's Note: MDPI stays neutral with regard to jurisdictional claims in published maps and institutional affiliations.

Copyright: © 2022 by the authors. Licensee MDPI, Basel, Switzerland. This article is an open access article distributed under the terms and conditions of the Creative Commons Attribution (CC BY) license (https://creativecommons.org/licenses/by/4.0/).

Abstract: Accurately characterizing the conformational variation of novel molecular assemblies is important but often ignored due to limited characterization methods. Herein, we reported the use of ion-mobility mass spectrometry (IMS/MS) to investigate the conformational changes of four azobenzene covalently functionalized Keggin hybrids (azo-Keggins, compounds **1**–**4**). The as-prepared azo-Keggins showed the general molecular formula of $[C_{16}H_{36}N]_4[SiW_{11}O_{40}(Si(CH_2)_3NH-CO(CH_2)_nO-C_6H_4N=NC_6H_4-R)_2]$ (R = H, n = 0 (**1**); R = NO_2, n = 0 (**2**); R = H, n = 5 (**3**); R = H, n = 10 (**4**)). The resultant azo-Keggins maintained stable monomeric states in the gas phase with intact molecular structures. Furthermore, the subtle photo-responsive trans-cis conformational variations of azo-Keggins were clearly revealed by the molecular shape-related collision cross-section value difference ranging from 2.44 Å^2 to 6.91 Å^2. The longer the alkyl chains linkers were, the larger the conformational variation was. Moreover, for compounds **1** and **2**, higher stability in *trans*-conformation can be observed, while for compounds **3** and **4**, bistability can be achieved for both of them.

Keywords: polyoxometalates; azobenzene; photo-responsive; ion-mobility mass spectrometry; shape characterization

1. Introduction

Azobenzene (Azo) and its derivatives have been widely investigated as promising photochemical systems, owing to their properties of reversible *trans-cis* isomerization upon light irradiation, which further results in reversibly changed physical and optical properties such as the solubility, dipole moment, surface free energy, and mechanical actuation behaviors [1–6]. By attaching azo groups, photochemical properties could be introduced to new functional materials [7]. In recent years, polyoxometalates (POMs), a class of anionic molecular metal oxide clusters with intriguing physical properties and nanometer size [8–14], have been applied to combine with azo groups. The resultant azo-based polyoxometalates (azo-POMs) have shown fascinating photo-responsive behaviors and have been applied for to a broad area, such as lyotropic or thermotropic liquid crystal, supramolecular self-assembly, and catalysis [15–19]. Different from small molecular azo, macromolecular azo-POMs show relatively complicated physical properties because of the introduction of more parameters such as large molecular size, multiple charges, steric hindrance effect, electronic effect, etc. [20]. Although the reversible *trans-cis* conformational change of molecules was studied, the detection means were still limited to UV-Vis or NMR spectroscopy [19,21]. Moreover, the shape's variation was inevitably involved during conformational changes; however, this factor is often ignored. Therefore, it is important to develop novel characterization methods for obtaining more structural information about these azo-POM macromolecules.

The technique of ion mobility mass spectrometry (IMS/MS) has been demonstrated to be useful for larger and more complex biological molecules as early as the 1990s [22–25]. With the development of ion mobility separation technology and high-resolution electrospray ionization mass spectrometry systems, the applications of IMS/MS were expanded to the investigation of polymers, proteomics, protein digest mixture, and carbon clusters as well as the supramolecular assembly of POMs [26–32]. By coupling MS instruments that separate ions on the basis of mass-to-charge ratio and IMS instruments that separate ions based on size-to-charge ratio, a two-dimensional mobility-mass spectrum could be obtained with information on size, charge, shape, and mass dimensions. In particular, the isomers or conformers have different shapes while very similar mass information can be separated in the mobility space [18,29]. During the past five years, the conformational variation of azobenzene was studied by using IMS/MS. However, most related studies were focused on the organic azo-based small molecules with CCS values less than 277 Å^2 [33,34]. In contrast, few works were reported for investigating the azo-based large macromolecules, which may bring a more complex while fascinating isomerization behavior. Therefore, it should be interesting to study azo-based POMs by using IMS/MS, due to their 1–10 nm molecular sizes, multiple negative charges, and redox properties, which were quite different with small azo molecules.

To best of our knowledge, the only work for studying the azo-POMs by IMS/MS was reported by Song, Cronin, and co-workers [18]. Taking the advantage of IMS/MS, the high CCS value of 600 Å^2 and the bistability conformational variations of azo-modified Mn-Anderson POMs (denoted as azo-Andersons) were observed. A significant difference in the collision cross-section (CCS) value of the oligomeric compounds was observed between the *trans*- and *cis*-conformation of the azo groups. However, these oligomeric structures involved the influence of intermolecular arrangement, which may overestimate shape differences caused by the isomerization of the single molecule itself. Therefore, applying the novel azo-POM systems with relatively stable monomeric molecules is crucial for detecting the bistability nature of azo-based macromolecules with the more subtle and intriguing reversible variation. Moreover, it was also intriguing to develop the substituents effect on the stability of *trans*- and *cis*-isomers.

2. Results and Discussion

In this work, four azo ligands with different lengths of alkyl linkers and substituent groups were covalently grafted onto the lacunary Keggin of $[SiW_{11}O_{39}]^{8-}$ (SiW_{11}) (denoted as azo-Keggins). The as-prepared azo-Keggins showed the general molecular formula of $[C_{16}H_{36}N]_4[SiW_{11}O_{40}(Si(CH_2)_3-NH-CO(CH_2)_nO-C_6H_4N=NC_6H_4-R)_2]$ (R = H, n = 0 (**1**); R = NO$_2$, n = 0 (**2**); R = H, n = 5 (**3**); R = H, n = 10 (**4**)). Different from the previously reported azo-Andersons, in which two azo ligands were attached to the opposite sides of $MnMo_6O_{18}$, the resultant azo-Keggins contained two azo ligands on the same sides of SiW_{11} (Figure 1). Considering strong intermolecular interactions was the prerequisite for the formation of large aggregates in the solution or gas phase; these azo-Keggins were expected to have weak intermolecular interactions due to hindrance from Keggin clusters with large molecular size and strong electrostatic repulsion.

The azo-POMs can be detected by electrospray ionization mass spectrometry (ESI-MS), as the aggregates commonly yielded multiple species with similar m/z resulting in overlapping envelopes in the mass spectra [18]. The ESI-MS spectra of compounds **1–4** were acquired by directly using their acetonitrile solutions. In each case, a series of notable peaks was observed and can be well assigned to the corresponding fragment ions, demonstrating that these hybrid structures were successfully prepared and remained intact both in the solution and gas phase. As shown in Figure 2, compounds **1–4** showed MS signals at m/z 1898.1, 1943.0, 1968.1, and 2038.2, which can be assigned to the fragment ions of $[X_{1-4}+2TBA]^{2-}$ (X_{1-4} = the anionic part of compounds **1–4**; TBA = tetrabutylammonium). All four peaks provided the unambiguous isotopic distribution envelopes without any similar overlapping fashion, suggesting the monomeric state of azo-Keggins. Full spectra and a list of identified peaks are provided in the Figures S1–S4.

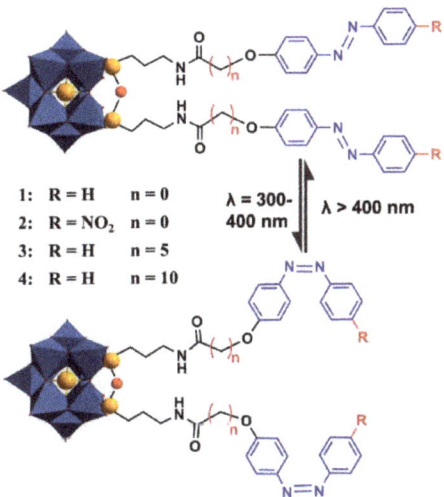

Figure 1. The lacunary Keggin cluster $[SiW_{11}O_{39}]^{8-}$ functionalized by photo-responsive azobenzene groups with different lengths of linkers and substituent groups (Compound **1–4**). Upon UV irradiation, the azo bond switches from *trans-* to *cis-*conformation. Under visible light irradiation, the bond switches back again.

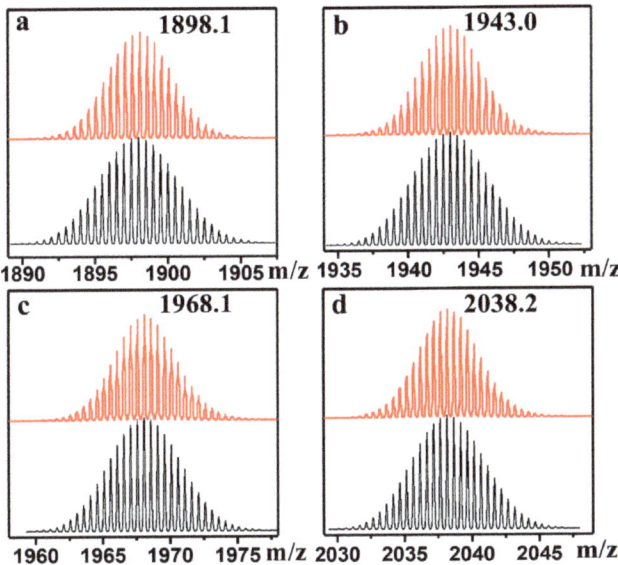

Figure 2. ESI-MS spectra of the compound **1–4** (red peaks in **a–d**) showing the *m/z* towards fragment ions of $[X_{1-4} + 2TBA]^{2-}$ (X_{1-4} = anionic part of **1–4**). The corresponding simulated isotopic patterns of the $[X_{1-4} + 2TBA]^{2-}$ were indicated as the grey peaks in **a–d**.

After confirming the monomeric state of azo-Keggins with intact molecular structure, IMS/MS measurements were applied to analyze their bistability conformation and reversible dynamics, considering that MS was unable to provide the information on shape.

A commercially available travelling-wave ion-mobility spectrometer (TWIMS) was coupled with mass measurements to separate the ions according to their mobility and to derive the collision cross-section (CCS) data of the ions. Figure 3 showed the 2D IMS/MS spectrum of compound **1**, with the peak intensity displayed by a color-coded logarithmic scale. The x-axis represented the m/z range from MS, and the y-axis represented the drift time from IMS. As we might expect, the spectrum of compound **1** showed a clearer situation than that of azo-Andersons, where larger oligomeric structures or the higher charge states were minor. Same results were observed in the 2D IMS/MS spectra of **2–4** both in their *trans*- and *cis*-conformations (Figures S5–S12). The intense (yellow) line of the peaks could be easily assigned to the individual cluster ions as observed in ESI-MS. All these peak envelopes could be assigned. The intensive peak at m/z = 1989.1 and drift time = 10.8 ms was assigned to the monomeric Keggin hybrids: $[X_1 + 2TBA]^{2-}$ (X_1 = anionic part of **1**). The corresponding higher charge peaks were assigned to minor aggregates: $[2X_1 + 5TBA]^{3-}$, $[3X_1 + 8TBA]^{4-}$, and $[4X_1 + 11TBA]^{5-}$ (Figure 3).

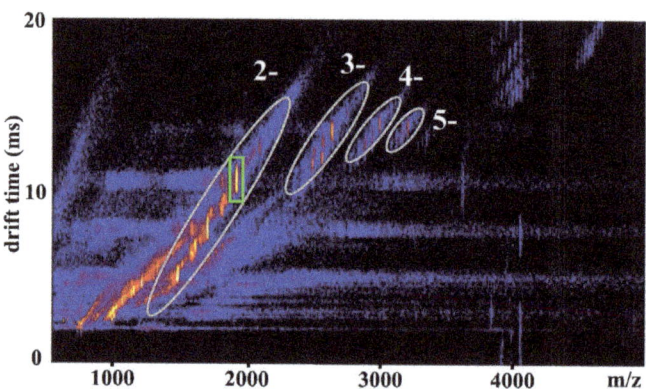

Figure 3. The 2D IMS/MS spectrum of compound **1**. The diagonal lines of similarly charged species were encircled by ellipsoids and the charges of these species were given in white. The peak in green rectangle was assigned to $[X_1 + 2TBA]^{2-}$.

With the further inspection of the IMS/MS spectrum of compound **1**, the conformational variation of these azo functionalized POM macromolecules could be observed. It was well known that the conformation of the azo bond can change from *trans* to *cis* upon UV irradiation at 365 nm and reversibly recovered back to *trans* conformation when the sample was exposed to visible light. The two conformations were relatively stable under the constant light condition, which enabled the detection of the bistability of azo-Keggins and the reversible transformation process. Additional UV-Vis and ^{1}H NMR experiments were conducted before and after UV irradiation at 365 nm. Moreover, the kinetic curve of the UV absorbance with time increasing was provided to follow the isomerization process (Figure S20). The results showed that almost all *trans*-isomers were changed to *cis*-isomers after 5 h UV light irradiation. The *cis*-isomer samples were immediately measured by IMS/MS under the exclusion of light. Figure 4 showed the drift time spectra for the main MS peaks of anionic part of compounds **1–4** with two TBA cations $[X_{1-4} + 2TBA]^{2-}$. Before and after UV irradiation, the drift time peaks (as indicated by green rectangle in Figure 3) manifested a clear shift, suggesting that the shape of compounds was changed. In IMS, the drift velocity of the ions was proportional to the electric field, and proportionality constant K was related to the CCS of ions [23]. On the basis of the CCS value that can be measured directly from the drift time, information about the chemical structure and 3D conformation of the ions was further provided (Tables S1–S4 Supplementary Materials).

Furthermore, the stability of specific isomers can be reflected in the intensity of drift time peaks. As shown in Figure 4, **1-UV** showed a lower peak intensity than its isomer

1-Vis, suggesting the lower stability of *cis* conformation than that of *trans* conformation. The decreased stability was attributed to the linkage of the Keggin POM cluster, which was considered a strong electron-withdrawing group [35]. Comparing the peak intensity of **2-UV** and **2-Vis**, the terminal substituent of electron-withdrawing NO_2 groups further decreased the stability of *cis* conformation [36]. Similar peak intensities were observed between two conformational isomers toward **3** and **4**, suggesting that the effect of substituent on the stability of isomers was diminishing with the increased length of alkyl chains between POMs and azo groups. Therefore, the conformation bistability of compounds **3** and **4** could be achieved.

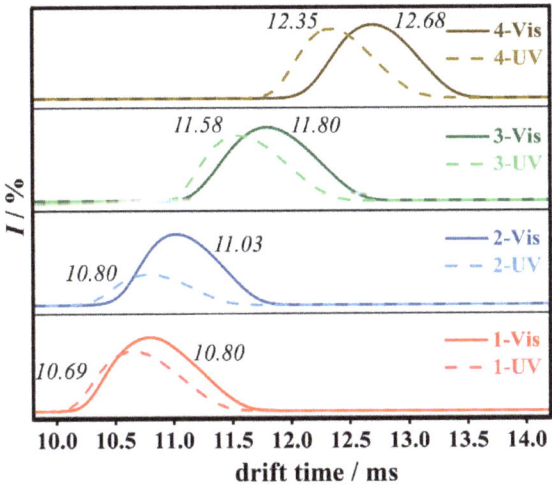

Figure 4. Comparison of the drift time graphs of the main MS peaks of compounds **1–4** with visible (solid lines) and UV (dash lines) irradiation. The spectra were consistent for all three repeated cycles.

As shown in Figure 5, the CCS value between the *trans* and *cis* conformations of compounds **1–4** showed the difference ranging from 6.91 Å^2 for the linker with the longest alkyl chain to 2.44 Å^2 for the smallest compound. Compared with the dimeric Anderson azo-POMs (CCS difference from 26 Å^2 to 13 Å^2), monomeric azo-Keggins exhibited a higher resolution difference, which could more precisely reflect the subtle conformational change of macromolecules derived from the azo group itself. The CCS differences for compounds **1–2** were 2.44 Å^2 and 5.07 Å^2, respectively, which were consistent with both of the roughly calculated conformational difference of the azo or nitro-azo groups by ChemBio3D Software and 2.1 Å^2 CCS difference of the conformational variation of the protonated azo reported in the literature [35]. As for compounds **3–4**, the greater CCS difference value was attributed to the more significant shape variation caused by flexible alkane chains. Therefore, for macromolecular azo-Keggins, the conformational change originated mainly from the transformation of organic azo ligands, while the influence from bulky POM clusters was minor. The IMS/MS measurement toward compounds **1–4** with reversible UV/Vis irradiation was repeated three times, and the resulting spectra were consistent through all repetitions of the isomerization, suggesting reproducibility.

It is worthwhile to note that our previous work reported the investigation of the IMSMS study of azo-Anderson assemblies [18]. In contrast, the azo-Keggin assemblies reported herein showed very different results: (1) The resultant azo-Keggins could stabilize the monomeric state, while azo-Anderson favored to form aggregates (dimer, trimer, and tetramer, etc.). Moreover, the conformational variation of azo-Keggins was more precisely reflected in IMS/MS results with a smaller CCS difference (2.44 Å^2 to 6.91 Å^2) than that of azo-Anderson aggregates with a large CCS difference (13 Å^2 to 26 Å^2), which was due to the influence of the intermolecular arrangement. (2) The azo-Andersons with substituents of electron donor groups (such as alkoxyl chains) showed a higher stability of *cis*-conformation compared with that of a *trans*-conformation. As complementary, azo-Keggins with a substituent of electron-withdrawing groups exhibited a higher stability of *trans*-conformation than that of *cis*-conformation. When long alkyl chains were present in azo-Keggin assemblies (-C5 and -C10 alkyl chains were linked between azo and POMs), the bistability of both *trans*- and *cis*- isomers can be achieved.

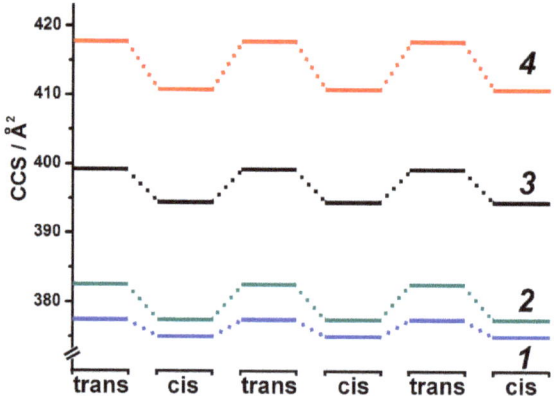

Figure 5. The changes in collision cross-section of compounds **1–4** when switching between *trans* and *cis* conformations of the azo bond. The formula of the azo-Keggin monomers was $[X_{1-4} + 2TBA]^{2-}$.

3. Materials and Methods

3.1. General Materials

All materials and reagents were purchased from commercial sources and used without further purification. $K_8SiW_{11}O_{39} \cdot 13H_2O$ and azo-functionalized compounds **1**, **3** and **4** were prepared according to the literature [15]. Fourier transform infrared (FT-IR) spectra were recorded on JASCO FT-IR 410 spectrometeror or a JASCO FT-IR 4100 spectrometer (JASCO, Tokyo, Japan). ^1H NMR spectroscopy was performed on a Bruker DPX 400 spectrometer (Bruker, Zurich, Switzerland) using the solvent's signal as an internal standard. Quantitative elemental analyses of C, H, and N were determined by microanalysis services within the College of Chemistry, Beijing University of Chemical Technology.

3.2. IMS/MS Studies and Determination of Collision Cross-Sections

All measurements were performed on a Synapt™ G2 HDMS™ from Waters (Massachusetts, USA) with equipment of a Quadrupole and Time-of-flight (Q/ToF) module for MS analysis. The IMS section is a travelling-wave IMS, which is located between the Q- and ToF-sections consisting of a trap cell, an ion-mobility cell, and a transfer cell. All compounds were directly dissolved in HPLC-grade acetonitrile at a concentration of 10^{-5} M. The solutions were filtered through a syringe filter (0.2 μm) before being injected into the spectrometer via a Harvard syringe pump at a flow rate of 5 μL·min^{-1}. The parameters for IMS/MS measurement of all compounds have been set up with the following: ESI capillary voltage: 1.93 kV; sample cone voltage: 10 V; extraction cone voltage: 4.6 V; source temperature: 80 °C; desolvation temperature: 120 °C; cone gas flow: 15 L·h^{-1} (N$_2$); desolvation gas flow: 500 L·h^{-1} (N$_2$); source gas flow: 0 mL·min^{-1}; trap gas flow: 2.5 mL·min^{-1}; helium cell gas flow: 200 mL·min^{-1}; IMS gas flow: 90.00 mL·min^{-1}; IMS DC entrance: 25.0; helium cell DC: 35.0; helium exit: −5; IMS bias: 30; IMS DC exit: 0; IMS wave velocity: 700 m·s^{-1}; IMS wave height: 40 V.

Collision cross-sections (CCSs) were estimated following calibration with Equine Cytochrome C and T10 olgiothymidine to determine instrument-dependent parameters

A and B from published CCSs data [5], as previously described in the literature [24]. The analysis of MS spectra was carried out using Mass Lynx V4.1 Software supplied by Waters. Driftscope V2.1 was used for the analysis of IMS/MS spectra and calibration of the drift cell for the determination of CCSs. IMS experiments were performed in -ve mode. All drift times and cross-sections are quoted for the intact 2- fragment (i.e., $[X_{1-4} + 2TBA]^{2-}$) for the following reasons: (1) $[X_{1-4} + 2TBA]^{2-}$ forms were the largest molecular fragment in ESI-MS spectra. (2) $[X_{1-4} + 2TBA]^{2-}$ forms had relatively intact molecular structure instead of partial structural fragments. (3) As the characteristic peak, $[X_{1-4} + 2TBA]^{2-}$ existed in all IMS/MS spectra of compounds 1–4.

3.3. Synthesis of $(Bu_4N)_4\{SiW_{11}O_{39}[O(SiCH_2CH_2CH_2NHCOR)_2]\}$ ($R = -OC_6H_4NNC_6H_5NO_2$, Compound 2)

When (E)-4-((4-nitrophenyl)diazenyl)phenyl (3-(triethoxysilyl)propyl)carbamate (0.77 g, 1.57 mmol) was dissolved in a solution of H_2O/CH_3CN ($v/v = 20/60$ mL) at room temperature, a turbid solution was formed to which $K_8SiW_{11}O_{39}\cdot 13H_2O$ (1.95 g, 0.60 mmol) was added. The pH of the resulting reaction mixture was slowly adjusted to 0.7 with 1M HCl, and the clear solution was stirred overnight. Then, CH_3CN was evaporated and NBu_4Br (4.20 g, 13 mmol) was added to the aqueous solution to precipitate the expected product. The precipitate was collected and washed with deionized water (100 mL), ethanol (100 mL), and diethyl ether (100 mL) before being dried in a vacuum. Yield: 2.25 g (85.8%). IR (KBr, cm^{-1}): ν = 3418 (m), 2961 (m), 2873 (m), 1742 (m), 1525 (m), 1485 (m), 1344 (m), 1205 (m), 1141 (w), 1044 (m), 964 (s), 947 (s), 904 (vs.), 856 (m), 804 (s), 535 (w). ^{1}H NMR (CD$_3$CN-d_3, 400 MHz, ppm) δ = 8.32 (d, ArH, 4H), 7.97 (d, ArH, 4H), 7.97 (d, ArH, 4H), 7.92 (d, ArH, 4H), 7.31 (d, ArH, 4H), 3.29 (t, CH$_2$, 4H), 3.12 (t, CH$_2$, 32H), 1.86 (s, CH$_2$, 4H), 1.62 (q, CH$_2$, 32H), 1.37 (m, CH$_2$, 32H), 0.97 (t, CH$_3$, 48H), 0.76 (t, CH$_2$, 4H). ^{13}C NMR (CD$_3$CN-d_3, 100 MHz, ppm) δ = 156.6, 155.8, 155.0, 150.2, 149.7, 125.8, 125.3, 124.2, 123.7, 59.2, 44.1, 24.3, 24.0, 20.3, 13.9, 10.8. ESI-MS (negative mode, CH$_3$CN): m/z: 1943 $[X_2 + 2TBA]^{2-}$. Anal. Calcd. for $C_{96}H_{174}N_{12}O_{48}Si_3W_{11}$ (4370.96): C, 26.38; H, 4.01; N, 3.85; Found: C, 25.91; H, 4.02; N, 3.61.

4. Conclusions

In conclusion, four azo-Keggins macromolecules with hybrid structures in which the photo-responsive azo moieties with different lengths of alkyl linkers and substituents were grafted onto the SiW$_{11}$ clusters were applied to investigate their photo-responsive conformational variation properties. ESI-MS of compounds 1–4 showed that all fragment ions can be well assigned without observing any similar overlapping fashion in each mass envelopes, suggesting that the monomeric state of these azo-Keggins was intact both in the solution and in the gas phase. The 2D IMS/MS spectra further indicated the high stability of the monomeric state which presented as the main peak envelops. Moreover, with the reversible UV and Vis photoirradiation, the subtle conformational change of these azo-Keggins resulted in different CCS values with a higher resolution of 2.44 Å^2. Compared with the previously reported data for large assemblies of azo-Andersons, the conformational variation results of monomeric azo-Keggins could more precisely indicate the effect of each group on the shape and stability of isomers: (1) The longer the linker's length was, the greater the conformational variation. (2) Conformational bistability was achieved for compounds 3 and 4, while for compounds 1 and 2 with electron-withdrawing substituents, higher stability in the *trans* conformation was observed. This work opened up a crucial new characterization dimension for which information on size, shape, charge, mass, and variation can be revealed. Providing deep insight into the shape changes of POMs into areas of self-assembly, catalysis, and responsive behavior from external stimulation such as photo, thermal, magnetic, etc., was helpful.

Supplementary Materials: The following supporting information can be downloaded at: https://www.mdpi.com/article/10.3390/molecules27123927/s1, Figures S1–S4: ESI-MS spectrum of compounds 1–4 and notable peaks assignment. Figures S5, S7, S9, and S11: 2D IMS/MS spectrum of compounds 1–4 before UV-irradiation. Figures S6, S8, S10 and S12: 2D IMS/MS spectrum of compound **1**–**4** after UV-irradiation. Figures S13–S16: Drift time of compounds $[X_{1-4} + 2TBA]^{2-}$ before and after UV-irradiation. Tables S1–S4: Change in drift time and CCS of monomeric compound $[X_{1-4} + 2TBA]^{2-}$ when switching between the *trans* and the *cis* conformation of the azo bond. Synthesis method of organic compounds used for the synthesis of compounds **1** and **2**. Figures S17–S19: FT-IR, ^{1}H NMR, and ^{13}C NMR spectra of compounds **1** and **2**. Figure S20: (a) ^{1}H NMR spectra of compound **2** in d_3-CH$_3$CN before (bottom) and after (top) UV irradiation at 365 nm. (b) UV/Vis spectra of compound **2** in CH$_3$CN. (c) The kinetic curve of UV absorbance with increasing time. (d) UV absorbance at 343 nm upon alternating UV irradiation at 365 nm and visible light for five cycles. Figure S21: The models of the azo and nitro-azo groups with *trans*- and *cis*-conformations. The models were constructed by ChemBio3D Software, and the shape difference was roughly calculated by atom distance.

Author Contributions: Conceptualization, W.C. and Y.-F.S.; methodology B.Q. and L.J.; validation, B.Q. and S.A.; formal analysis, S.A. and L.J.; investigation, W.C. and B.Q.; writing—original draft preparation, B.Q.; writing—review and editing, Y.-F.S.; visualization, B.Q. and Y.-F.S.; supervision, W.C. and Y.-F.S.; project administration, W.C. and Y.-F.S.; funding acquisition, Y.-F.S. All authors have read and agreed to the published version of the manuscript.

Funding: This research was funded by Beijing Natural Science Foundation (2202039), the National Nature Science Foundation of China (22101017, 22178019, 21625101, 21808011, and 21521005), the National Key Research and Development Program of China (2017YFB0307303), and the Fundamental Research Funds for the Central Universities (XK1802-6, XK1803-05, XK1902, and 12060093063).

Institutional Review Board Statement: Not applicable.

Informed Consent Statement: Not applicable.

Data Availability Statement: Not applicable.

Conflicts of Interest: The authors declare no conflict of interest.

Sample Availability: Samples of the compounds are available from the authors.

References

1. Bandara, H.M.D.; Burdette, S.C. Photoisomerization in different classes of azobenzene. *Chem. Soc. Rev.* **2012**, *41*, 1809–1825. [CrossRef] [PubMed]
2. Benkhaya, S.; M'Rabet, S.; El Harfi, A. Classifications, properties, recent synthesis and applications of azo dyes. *Heliyon* **2020**, *6*, e03271. [CrossRef] [PubMed]
3. Dattler, D.; Fuks, G.; Heiser, J.; Moulin, E.; Perrot, A.; Yao, X.; Giuseppone, N. Design of collective motions from synthetic molecular switches, rotors, and motors. *Chem. Rev.* **2020**, *120*, 310–433. [CrossRef] [PubMed]
4. Tamai, N.; Miyasaka, H. Ultrafast dynamics of photochromic systems. *Chem. Rev.* **2000**, *100*, 1875–1890. [CrossRef] [PubMed]
5. Xu, W.-C.; Sun, S.; Wu, S. Photoinduced reversible solid-to-liquid transitions for photoswitchable materials. *Angew. Chem. Int. Ed.* **2019**, *58*, 9712–9740. [CrossRef]
6. Zhu, F.; Tan, S.; Dhinakaran, M.K.; Cheng, J.; Li, H. The light-driven macroscopic directional motion of a water droplet on an azobenzene–calix[4]arene modified surface. *Chem. Commun.* **2020**, *56*, 10922–10925. [CrossRef]
7. Pardo, R.; Zayat, M.; Levy, D. Photochromic organic–inorganic hybrid materials. *Chem. Soc. Rev.* **2011**, *40*, 672–687. [CrossRef]
8. Cameron, J.M.; Guillemot, G.; Galambos, T.; Amin, S.S.; Hampson, E.; Mall Haidaraly, K.; Newton, G.N.; Izzet, G. Supramolecular assemblies of organo-functionalised hybrid polyoxometalates: From functional building blocks to hierarchical nanomaterials. *Chem. Soc. Rev.* **2022**, *51*, 293–328. [CrossRef]
9. Liu, J.; Shi, W.; Ni, B.; Yang, Y.; Li, S.; Zhuang, J.; Wang, X. Incorporation of clusters within inorganic materials through their addition during nucleation steps. *Nat. Chem.* **2019**, *11*, 839–845. [CrossRef]
10. Qi, B.; An, S.; Luo, J.; Liu, T.; Song, Y.-F. Enhanced macroanion recognition of superchaotropic Keggin clusters achieved by synergy of anion–π and anion–cation interactions. *Chem. Eur. J.* **2020**, *26*, 16802–16810. [CrossRef]
11. Yin, P.; Li, D.; Liu, T. Solution behaviors and self-assembly of polyoxometalates as models of macroions and amphiphilic polyoxometalate–organic hybrids as novel surfactants. *Chem. Soc. Rev.* **2012**, *41*, 7368–7383. [CrossRef] [PubMed]
12. Zhang, T.-T.; Li, G.; Cui, X.-B. Three new polyoxoniobates functioning as different oxidation catalysts. *Cryst. Growth Des.* **2021**, *21*, 3191–3201. [CrossRef]
13. Zheng, S.-T.; Yang, G.-Y. Recent advances in paramagnetic-TM-substituted polyoxometalates (TM = Mn, Fe, Co, Ni, Cu). *Chem. Soc. Rev.* **2012**, *41*, 7623–7646. [CrossRef] [PubMed]
14. Luo, J.; Chen, K.; Yin, P.; Li, T.; Wan, G.; Zhang, J.; Ye, S.; Bi, X.; Pang, Y.; Wei, Y.; et al. Effect of cation-π interaction on macroionic self-assembly. *Angew. Chem. Int. Ed.* **2018**, *57*, 4067–4072. [CrossRef]

15. Chen, W.; Ma, D.; Yan, J.; Boyd, T.; Cronin, L.; Long, D.-L.; Song, Y.-F. 0D to 1D switching of hybrid polyoxometalate assemblies at the nanoscale by using molecular control. *ChemPlusChem* **2013**, *78*, 1226–1229. [CrossRef]
16. Han, S.; Cheng, Y.; Liu, S.; Tao, C.; Wang, A.; Wei, W.; Yu, H.; Wei, Y. Selective oxidation of anilines to azobenzenes and azoxybenzenes by a molecular Mo oxide catalyst. *Angew. Chem. Int. Ed.* **2021**, *60*, 6382–6385. [CrossRef]
17. Lin, C.-G.; Chen, W.; Omwoma, S.; Song, Y.-F. Covalently grafting nonmesogenic moieties onto polyoxometalate for fabrication of thermotropic liquid-crystalline nanomaterials. *J. Mater. Chem. C* **2015**, *3*, 15–18. [CrossRef]
18. Thiel, J.; Yang, D.; Rosnes, M.H.; Liu, X.; Yvon, C.; Kelly, S.E.; Song, Y.-F.; Long, D.-L.; Cronin, L. Observing the hierarchical self-assembly and architectural bistability of hybrid molecular metal oxides using ion-mobility mass spectrometry. *Angew. Chem. Int. Ed.* **2011**, *50*, 8871–8875. [CrossRef]
19. Guo, Y.; Gong, Y.; Yu, Z.; Gao, Y.; Yu, L. Rational design of photo-responsive supramolecular nanostructures based on an azobenzene-derived surfactant-encapsulated polyoxometalate complex. *RSC Adv.* **2016**, *6*, 14468–14473. [CrossRef]
20. Kumar, G.S.; Neckers, D.C. Photochemistry of azobenzene-containing polymers. *Chem. Rev.* **1989**, *89*, 1915–1925. [CrossRef]
21. Pieroni, O.; Fissi, A.; Angelini, N.; Lenci, F. Photoresponsive polypeptides. *Acc. Chem. Res.* **2001**, *34*, 9–17. [CrossRef] [PubMed]
22. Baumbach, J.I.; Eiceman, G.A. Ion mobility spectrometry: Arriving on site and moving beyond a low profile. *Appl. Spectrosc.* **1999**, *53*, 338A–355A. [CrossRef] [PubMed]
23. Kanu, A.B.; Dwivedi, P.; Tam, M.; Matz, L.; Hill, H.H., Jr. Ion mobility–mass spectrometry. *J. Mass Spectrom.* **2008**, *43*, 1–22. [CrossRef]
24. Merenbloom, S.I.; Flick, T.G.; Williams, E.R. How hot are your ions in twave ion mobility spectrometry? *J. Am. Soc. Mass. Spectrom.* **2012**, *23*, 553–562. [CrossRef]
25. Mukhopadhyay, R. IMS/MS: Its time has come. *Anal. Chem.* **2008**, *80*, 7918–7920. [CrossRef] [PubMed]
26. Barrère, C.; Maire, F.; Afonso, C.; Giusti, P. Atmospheric solid analysis probe–ion mobility mass spectrometry of polypropylene. *Anal. Chem.* **2012**, *84*, 9349–9354. [CrossRef] [PubMed]
27. Canterbury, J.D.; Yi, X.; Hoopmann, M.R.; MacCoss, M.J. Assessing the dynamic range and peak capacity of nanoflow LC–AIMS–MS on an ion trap mass spectrometer for proteomics. *Anal. Chem.* **2008**, *80*, 6888–6897. [CrossRef]
28. Hoskins, J.N.; Trimpin, S.; Grayson, S.M. Architectural differentiation of linear and cyclic polymeric isomers by ion mobility spectrometry-mass spectrometry. *Macromolecules* **2011**, *44*, 6915–6918. [CrossRef]
29. Robbins, P.J.; Surman, A.J.; Thiel, J.; Long, D.-L.; Cronin, L. Use of ion-mobility mass spectrometry (IMS-MS) to map polyoxometalate Keplerate clusters and their supramolecular assemblies. *Chem. Commun.* **2013**, *49*, 1909–1911. [CrossRef]
30. Scott, C.D.; Ugarov, M.; Hauge, R.H.; Sosa, E.D.; Arepalli, S.; Schultz, J.A.; Yowell, L. Characterization of large fullerenes in single-wall carbon nanotube production by ion mobility mass spectrometry. *J. Phys. Chem. C* **2007**, *111*, 36–44. [CrossRef]
31. Tang, K.; Li, F.; Shvartsburg, A.A.; Strittmatter, E.F.; Smith, R.D. Two-dimensional gas-phase separations coupled to mass spectrometry for analysis of complex mixtures. *Anal. Chem.* **2005**, *77*, 6381–6388. [CrossRef] [PubMed]
32. Trimpin, S.; Inutan, E.D.; Karki, S.; Elia, E.A.; Zhang, W.-J.; Weidner, S.M.; Marshall, D.D.; Hoang, K.; Lee, C.; Davis, E.T.J.; et al. Fundamental studies of new ionization technologies and insights from IMS-MS. *J. Am. Soc. Mass. Spectrom.* **2019**, *30*, 1133–1147. [CrossRef] [PubMed]
33. Scholz, M.S.; Bull, J.N.; Coughlan, N.J.A.; Carrascosa, E.; Adamson, B.D.; Bieske, E.J. Photoisomerization of protonated azobenzenes in the gas phase. *J. Phys. Chem. A* **2017**, *121*, 6413–6419. [CrossRef] [PubMed]
34. Galanti, A.; Santoro, J.; Mannancherry, R.; Duez, Q.; Diez-Cabanes, V.; Valášek, M.; De Winter, J.; Cornil, J.; Gerbaux, P.; Mayor, M.; et al. A new class of rigid multi(azobenzene) switches featuring electronic decoupling: Unravelling the isomerization in individual photochromes. *J. Am. Chem. Soc.* **2019**, *141*, 9273–9283. [CrossRef] [PubMed]
35. Lachkar, D.; Vilona, D.; Dumont, E.; Lelli, M.; Lacôte, E. Grafting of secondary diolamides onto $[P_2W_{15}V_3O_{62}]^{9-}$ generates hybrid heteropoly acids. *Angew. Chem. Int. Ed.* **2016**, *55*, 5961–5965. [CrossRef] [PubMed]
36. Diana, R.; Caruso, U.; Piotto, S.; Concilio, S.; Shikler, R.; Panunzi, B. Spectroscopic behaviour of two novel azobenzene fluorescent dyes and their polymeric blends. *Molecules* **2020**, *25*, 1368. [CrossRef] [PubMed]

Article

Grafting of Anionic Decahydro-*Closo*-Decaborate Clusters on Keggin and Dawson-Type Polyoxometalates: Syntheses, Studies in Solution, DFT Calculations and Electrochemical Properties

Manal Diab [1,2], Ana Mateo [3], Joumada El Cheikh [4], Zeinab El Hajj [1,2], Mohamed Haouas [1], Alireza Ranjbari [5], Vincent Guérineau [6], David Touboul [6], Nathalie Leclerc [1], Emmanuel Cadot [1], Daoud Naoufal [2,*], Carles Bo [2,*] and Sébastien Floquet [1,*]

1. Institut Lavoisier de Versailles, CNRS, UVSQ, Université Paris-Saclay, 78035 Versailles, France
2. Laboratory of Organometallic and Coordination Chemistry, LCIO, Faculty of Sciences I, Lebanese University, Hadath 6573, Lebanon
3. Institute of Chemical Research of Catalonia (ICIQ), The Barcelona Institute of Science and Technology, 43007 Tarragona, Spain
4. Equipe de Recherche et Innovation en Electrochimie pour l'énergie (ERIEE), Institut de Chimie Moléculaire et des Matériaux d'Orsay (ICMMO), UMR CNRS 8182, Université Paris-Sud, Université Paris-Saclay, 91405 Orsay, France
5. Institut de Chimie Physique, CNRS, UMR 8000, Université Paris-Saclay, 91405 Orsay, France
6. Institut de Chimie des Substances Naturelles, CNRS UPR2301, Université Paris-Sud, Université Paris-Saclay, 91198 Gif-sur-Yvette, France
* Correspondence: dnaoufal@ul.edu.lb (D.N.); cbo@iciq.cat (C.B.); sebastien.floquet@uvsq.fr (S.F.)

Abstract: Herein we report the synthesis of a new class of compounds associating Keggin and Dawson-type Polyoxometalates (POMs) with a derivative of the anionic decahydro-*closo*-decaborate cluster $[B_{10}H_{10}]^{2-}$ through aminopropylsilyl ligand (APTES) acting as both a linker and a spacer between the two negatively charged species. Three new adducts were isolated and fully characterized by various NMR techniques and MALDI-TOF mass spectrometry, notably revealing the isolation of an unprecedented monofunctionalized SiW_{10} derivative stabilized through intramolecular H-H dihydrogen contacts. DFT as well as electrochemical studies allowed studying the electronic effect of grafting the decaborate cluster on the POM moiety and its consequences on the hydrogen evolution reaction (HER) properties.

Keywords: polyoxometalate; hybrid; decaborate; DFT; NMR; hydrogen evolution reaction

1. Introduction

Polyoxometalates (POMs) and POM-based materials constitute a highly versatile class of compounds rich in more than several thousand inorganic compounds, which can be finely tuned at the molecular level. Because of their stunning compositions, diversified architectures and their rich electrochemical redox behaviors, they are known to display numerous properties or applications in many domains such as supramolecular chemistry [1–3], catalysis [4–6], electro-catalysis [7–9], and medicine, especially when POMs are functionalized with organic groups or complexes [10–13].

On their side, hydroborates represent a wide family of anionic clusters, for which many reports demonstrated their interest in different areas, especially in the biomedical domain [14–18]. This property thus makes the studies of borane derivatives of a great interest. In particular, the $[B_{10}H_{10}]^{2-}$ cluster offers the possibility of various selective functionalizations [19,20] leading for example to *closo*-decaborate-triethoxysilane precursor, which can be coordinated to luminescent dye doped silica nanoparticles, hence facilitating the tracing of the *closo*-decaborate drug pathway in BNCT (Boron Neutron Capture Therapy) [21,22].

Driven by the synthetic challenge that constitutes the association of two anionic species with two complementary redox characters, reductive for hydroborates and oxidative for POMs, and by the biomedical applications which could be reached by associating these two families of compounds, this study aims to find the right strategy to design such POM-borate adducts and to study their chemical properties.

In a previous paper, we demonstrated that it is possible to covalently graft decaborate clusters to an Anderson-type polyoxometalate functionalized with the well-known TRIS ligand (TRIS = tris(hydroxymethyl)aminomethane), namely $[Mn^{III}Mo_6O_{18}(TRIS)_2]^{3-}$ [23]. Nevertheless, the compound $[Mn^{III}Mo_6O_{18}(TRIS-B_{10})_2]^{7-}$ resulting from the coupling between both components revealed to be fragile, probably because of the rigidity of the linker and the close proximity of both anionic components. This weakness is confirmed by DFT calculations indicating an athermic or slightly exothermic process for the formation of the adducts with Anderson-TRIS hybrid POMs.

In the field of hybrid POMs, the organosilyl derivatives of vacant polyoxotungstates as $[PW_9O_{34}]^{9-}$, $[SiW_{10}O_{36}]^{8-}$, $[PW_{11}O_{39}]^{7-}$, $[SiW_{11}O_{39}]^{6-}$, or $[P_2W_{17}O_{61}]^{10-}$ offer large diversities of compounds exhibiting a wide panel of applications [24,25]. Among them, the divacant POM Keggin $[SiW_{10}O_{36}]^{8-}$ (noted hereafter SiW$_{10}$) and the monovacant POM Dawson $[P_2W_{17}O_{61}]^{10-}$ (noted hereafter P$_2$W$_{17}$) derivatives are probably the most used because of their stability, their topology and the richness of their electrochemical properties in reduction. In particular, by reacting with aminopropyltri(ethoxy)silane (called APTES) they can provide two very useful platforms, noted respectively **SiW$_{10}$-APTES** and **P$_2$W$_{17}$-APTES** (see Figure 1), for elaborating functional hybrid molecular architectures.

The aim of this study is to use these two different platforms to prepare new hybrid compounds associating an anionic decaborate boron cluster (denoted hereafter B$_{10}$) with Keggin and Dawson POM derivatives. The choice of polyoxotungstate moieties rather than Mo-based POMs is based on its stability towards reduction. The employment of a long and flexible linker as APTES is essential to tackle the challenge of combining efficiently a reduced anionic boron cluster with an anionic oxidized polyoxometalate. The use of APTES linker should limit the repulsion between the two components, while its flexibility allows more easily accommodating the two entities. Finally, as shown in Figure 1, due to monovacant and divacant characters of P$_2$W$_{17}$ and SiW$_{10}$, respectively, it is worth noting that the relative conformations of the chains are different. For SiW$_{10}$-APTES, the two alkyl chains are oriented nearly in parallel, whereas the monovacancy of P$_2$W$_{17}$ imposes divergent directions for the two alkyl chains. This topology is well adapted for designing triangular or square molecular species as evidenced by Izzet et al. [2,26], and in our case, we expect that these two kinds of conformation could lead to different types of adducts incorporating B$_{10}$ clusters. In this study, we thus report the synthesis, the full characterization in solution by various NMR techniques, the electronic, the electrochemical and the electrocatalytic properties of three new hybrid POMs. In the absence of XRD structures, DFT studies provide a fine structural description of these hybrids and rationalization of their properties.

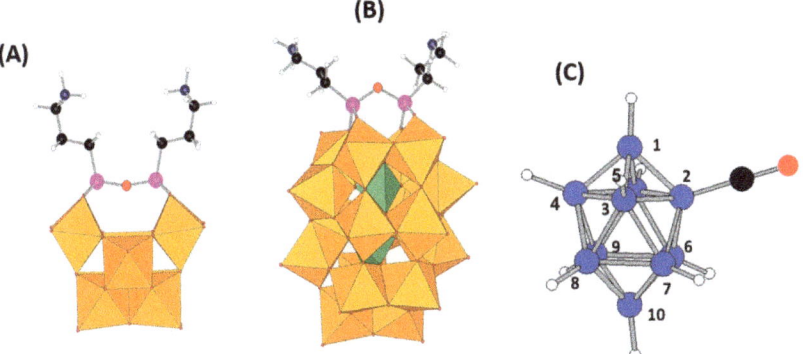

Figure 1. Molecular structures (DFT-optimized geometry) of (**A**) **SiW$_{10}$-APTES** and (**B**) **P$_2$W$_{17}$-APTES** platforms highlighting the two different topologies of the APTES linker, and of (**C**) [B$_{10}$H$_9$CO]$^-$ (X-ray diffraction structure from reference [27]). Legend: C in black, H in white, N in dark blue, Si in pink, O in red, B in blue, WO$_6$ octahedra in orange and PO$_4$ tetrahedra in green.

2. Results and Discussion

2.1. Syntheses

The synthesis of hybrid POMs can be achieved through different strategies. In the present study, the best synthetic procedure to get the targeted hybrid POMs has been to react first the lacunary POMs "SiW$_{10}$" and "P$_2$W$_{17}$" with two aminopropyltri(ethoxy)silane molecules (APTES) to give the two POM-APTES precursors (see Figure 2) of formulas (TBA)$_3$H[(SiW$_{10}$O$_{36}$)(Si(CH$_2$)$_3$NH$_2$)$_2$O]·3H$_2$O (denoted hereafter **SiW$_{10}$-APTES**) and (TBA)$_5$H[P$_2$W$_{17}$O$_{61}$(Si(CH$_2$)$_3$NH$_2$)$_2$O]·6H$_2$O, denoted hereafter **P$_2$W$_{17}$-APTES**. The syntheses of these two precursors were adapted from Mayer at al. [28] by reaction of k$_8$(γ-SiW$_{10}$O$_{36}$)·12H$_2$O or K$_{10}$α–P$_2$W$_{17}$O$_{61}$·20H$_2$O with 3-aminopropyltriethoxy silane in presence of TBABr in H$_2$O/CH$_3$CN medium acidified by concentrated HCl (for more details see experimental section in Supplementary Materials). Note that for each, the proton usually written as counter-cation is in fact probably an ammonium arm R-NH$_3^+$.

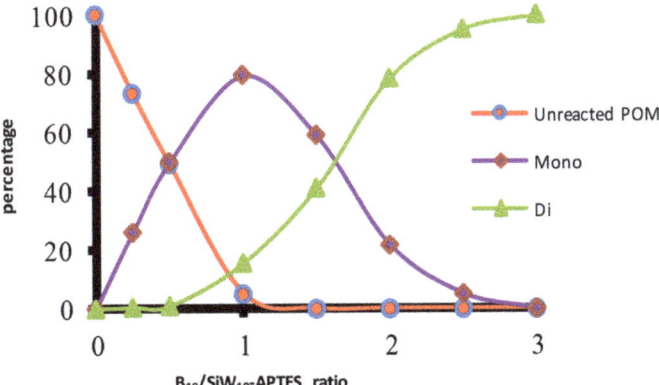

Figure 2. Evolution of the proportions of the products in the system **SiW$_{10}$-APTES**/B$_{10}$H$_9$CO/DIPEA as a function of B$_{10}$H$_9$CO/**SiW$_{10}$-APTES** ratio at fixed DIPEA/B$_{10}$H$_9$CO ratio of 2. The proportion of each species are determined by integration of the ^{29}Si NMR signals. Reproduced with permission from the doctoral thesis manuscript of Dr Manal Diab, University Paris Saclay/Lebanese University, May 2018.

The synthetic strategy to get POM-borate adducts is then to combine the amines of these POM-APTES precursors with the reactive carbonyl of the decaborate cluster $[B_{10}H_9CO]^-$ (Figure 1C) to give an amide function connecting both components. Since the boron cluster can react with water for giving a carboxylic acid and since heating the synthetic mixture above 40–50 °C led to some degradation products or to some reduction in the Dawson derivative by the hydrodecaborate cluster, reactions have been conducted at room temperature and under nitrogen atmosphere. Furthermore, the coupling reaction needs the presence of a base both to help the deprotonation of the ammonium arm(s) of the POM-APTES precursors and to trap the proton produced by the coupling reaction. A moderate and a bulky organic base, diisopropylethylamine (DIPEA), was thus used to avoid the competition with APTES for the coupling reaction with $[B_{10}H_9CO]^-$.

To quickly circumscribe the optimal conditions for the synthesis of the POM-borate adducts, ^{29}Si, ^{31}P and ^{1}H NMR titrations were conducted by varying the ratios of the three reactants $[B_{10}H_9CO]^-$/POM-APTES/DIPEA (all details are given in the Supplementary Materials).

For the $[B_{10}H_9CO]^-$/**SiW$_{10}$-APTES**/DIPEA system, the ^{29}Si NMR studies in solution reveal that it is possible to modulate the coupling reaction between $[B_{10}H_9CO]^-$ and POM-APTES precursors by playing on the amounts of DIPEA and of $[B_{10}H_9CO]^-$. For this tri-reactants system, the successive formation of two POM-borate species identified as mono- and di-adduct compounds was demonstrated thanks to their molecular symmetries (C_s versus C_{2v}). Besides, the crucial role of DIPEA in the reaction of $[B_{10}H_9CO]^-$ with POM-APTES precursors was clearly evidenced. No reaction occurs when no base is used. NMR titration studies allowed establishing that the optimal quantity of base was two equivalents for one equivalent of $[B_{10}H_9CO]^-$. The Figure 2 shows for instance the proportions of SiW$_{10}$-derivatives determined by the integration of the different peaks obtained by ^{29}Si NMR in the system **SiW$_{10}$-APTES**/$[B_{10}H_9CO]^-$/DIPEA as a function of $[B_{10}H_9CO]^-$/**SiW$_{10}$-APTES** ratio at fixed DIPEA/$[B_{10}H_9CO]^-$ ratio of 2.

Starting from **SiW$_{10}$-APTES**, it evidences first the formation of a mono-adduct, which predominates for ration B_{10}/**SiW$_{10}$-APTES** = 1, before being converted into a di-adduct. The NMR titrations studies allowed establishing that using proportions **SiW$_{10}$-APTES**/$B_{10}H_9CO$/DIPEA = 1/3/6 lead to the pure di-adduct denoted **SiW$_{10}$-diB$_{10}$**, while using 1/1/2 ratios lead to around 80% of mono-adduct mixed with some unreacted starting POM and the di-adduct. The separation of compounds has not been possible but considering the effect of the added DIPEA amounts, we succeeded to reduce the formation of the di-adduct and thus to get the mono-adduct compound denoted **SiW$_{10}$-monoB$_{10}$** with a good purity by decreasing the quantity of DIPEA in the proportions **SiW$_{10}$-APTES**/$B_{10}H_9CO$/DIPEA = 1/1/1.5. The Figure 3 summarizes the experimental conditions used to isolate POM-borate adducts.

Similar NMR studies were also performed in solution with the Dawson derivative **P$_2$W$_{17}$-APTES** (see Supplementary Materials). In contrast to SiW$_{10}$ derivatives, the formation of mono- and di-adduct of the Dawson derivative are not so separated as for SiW$_{10}$. Therefore, we failed to isolate the mono-adduct as pure product. Nevertheless, we can obtain quantitatively the di-adduct compound in the reaction mixture when ratios **P$_2$W$_{17}$-APTES**/$B_{10}H_9CO$/DIPEA = 1/3/6 are used.

To summarize, the multistep coupling reactions have successfully been monitored by ^{29}Si and ^{31}P NMR, fully described in the Supplementary Materials, revealing that intermediate products can be followed and isolated. From these results, we established the experimental conditions allowing to selectively synthesize with good yields the mono adduct of SiW$_{10}$ POM and the di-adducts of both POMs as mixed TBA$^+$ and DIPEAH$^+$ salts, namely (TBA)$_3$(DIPEAH)$_3$[(SiW$_{10}$O$_{36}$)(B$_{10}$H$_9$CONHC$_3$H$_6$Si)(NH$_2$C$_3$H$_6$Si)O]·3H$_2$O denoted **SiW$_{10}$-monoB$_{10}$**, (TBA)$_{6.5}$(DIPEAH)$_{1.5}$[(SiW$_{10}$O$_{36}$)(B$_{10}$H$_9$CONHC$_3$H$_6$Si)$_2$O]·2H$_2$O denoted **SiW$_{10}$-diB$_{10}$**, and (TBA)$_6$(DIPEAH)$_4$[(P$_2$W$_{17}$O$_{61}$)(B$_{10}$H$_9$CONHC$_3$H$_6$Si)$_2$O]·3H$_2$O, denoted **P$_2$W$_{17}$-diB$_{10}$** (See Experimental Section in Supplementary Materials for more details). All adducts were isolated as powders and were characterized by FT-IR, TGA, elemental analysis, MALDI-TOF and NMR techniques. It should be noted that to our knowledge, **SiW$_{10}$-**

monoB$_{10}$ is the first example of a POM-APTES monoadduct isolated so far from the direct synthesis. All studies in the literature usually reported di-adducts with such types of hybrid POMs [29–31].

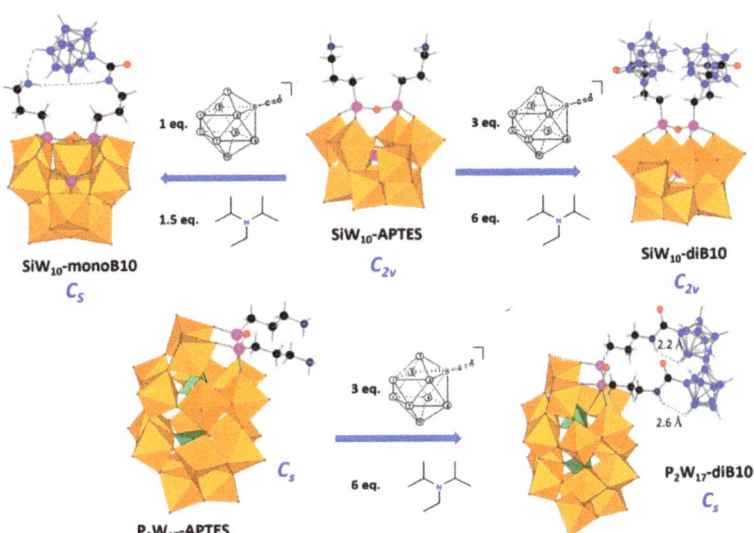

Figure 3. Scheme of syntheses of POM-borates adducts. The optimal quantities of reactants were determined by NMR titration studies. The reactions are performed in dry acetonitrile, at room temperature under inert atmosphere. Molecular structures are optimized geometry obtained by DFT. Legend: C in black, H in white, N in dark blue, Si in pink, O in red, B in blue, WO$_6$ octahedra in orange and PO$_4$ tetrahedra in green.

2.2. FT-IR Spectroscopy

FT-IR spectra are given in Figures S11 and S12 in Supplementary Materials. The FT-IR spectra of **SiW$_{10}$-monoB$_{10}$**, **SiW$_{10}$-diB$_{10}$** and **P$_2$W$_{17}$-diB$_{10}$** evidence that the integrity of the POM part is maintained compared to the POM-APTES precursors. Furthermore, the association of the [B$_{10}$H$_9$CO]$^-$ cluster is demonstrated by the disappearance of the carbonyl CO band at 2098 cm^{-1} in the B$_{10}$H$_9$CO$^-$ cluster, while the broad band located at 2464–2470 cm^{-1} typical for B-H vibration bands of the decaborate moiety within the three compounds **SiW$_{10}$-monoB$_{10}$**, **SiW$_{10}$-diB$_{10}$** and **P$_2$W$_{17}$-diB$_{10}$** is significantly shifted from that observed at 2517 cm^{-1} for the [B$_{10}$H$_9$CO]$^-$ precursor [22,23].

2.3. Characterizations by MALDI-TOF Mass Spectrometry

Mass spectrometry (MS) is a very efficient technique for the characterization of polyoxometalates in solution. In our case, we did not succeed in getting mass spectra with reasonable signal-to-noise ratio and exploitable data by the usual electrospray ESI-MS technique. On the contrary, Matrix-Assisted Laser Desorption/Ionization coupled to a Time-of-Flight mass spectrometer (MALDI-TOF) revealed to be an effective technique for hybrid POMs characterization, as shown for example by Mayer and coworkers on "SiW$_{10}$" and "P$_2$W$_{17}$" organosilyl derivatives [28,32]. MALDI-TOF technique is applied on samples which are diluted in a matrix solution (DCTB in our case, DCTB = Trans-2-[3-(4-ter-Butylphenyl)-2-propenylidene] malonitrile) and then co-crystallized on a conductive target. Thanks to a laser irradiation, it allows producing singly charged species (cationic or anionic) and presenting the great advantage to strongly limit the number of peaks in comparison with ESI-MS spectra, where multiply charged species are generated. In the present study, the experiments were performed in both negative and positive modes (see Figure S14 in the

Supplementary Materials for the example of SiW_{10}-diB_{10}). According to previous works in this field, the best results were obtained in the positive mode, although the anionic character of the POM [28,32]. Indeed, as seen in the Supplementary Materials for SiW_{10}-diB_{10}, the intensity reached in the negative mode appears lower, but the number of peaks is higher as there are more degradation species. Even thought our systems are polyanionic, they are more efficiently analyzed as monocationic species resulting from adducts between POMs and counter cations such as TBA^+ and H^+ in our case (H^+ coming notably from $DIPEAH^+$ cations or protonated amines). Furthermore, the monocationic character of the species is confirmed in all cases by the shift between peaks in the isotopic massifs.

The precursor **SiW_{10}-APTES** and the compounds **SiW_{10}-monoB_{10}**, **SiW_{10}-diB_{10}** and **P_2W_{17}-diB_{10}**, were thus analyzed by this technique in the positive mode. The results are gathered in Table S1 (see Supplementary Materials). The full spectra and a zoom on the target compounds with a spectrum simulated with IsoPro3 software are shown in Figure 4 for SiW_{10} derivatives and in Figures S15 and S16 for P_2W_{17} ones.

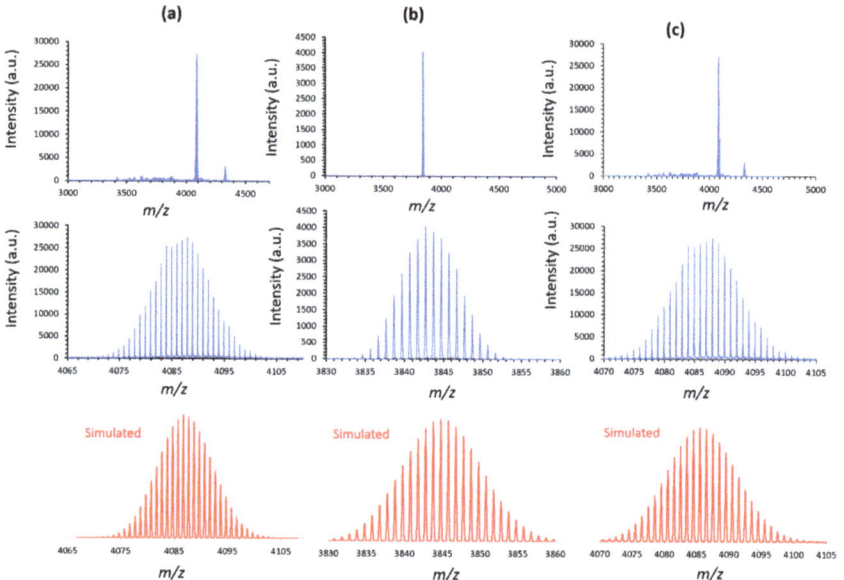

Figure 4. Reflector positive ion MALDI-TOF spectra of (**a**) **SiW_{10}-APTES**, (**b**) **SiW_{10}-monoB_{10}**, and (**c**) **SiW_{10}-diB_{10}**. Zooms of major peaks in the 3000–5000 m/z range are displayed with their respective simulated spectra.

As shown in Figure 4 the spectrum of the precursor **SiW_{10}-APTES** (Figure 4a) displays a major peak centered at m/z 4087.3 and a minor peak at m/z 4328.3. The first peak is assigned to the monocationic species $\{(TBA)_3H_2[(SiW_{10}O_{36})O(SiC_3H_6NH_2)_2](CH_3CN)_2(H_2O)_8(DCTB)_2\}^+$ (calculated m/z 4087.3), while the second peak is attributed to the species $\{(TBA)_4H[(SiW_{10}O_{36})O(SiC_3H_6NH_2)_2](CH_3CN)_2(H_2O)_8(DCTB)_2\}^+$ (calculated m/z 4328.7). The two peaks correspond to the expected hybrid POM associated with some TBA^+ and H^+ cations, some solvates and two molecules of the DCTB matrix. Note that the presence of amines on the APTES part of the POM could probably favor the formation of intermolecular interactions with solvates and DCTB molecules. Such an adduct with DCTB is also observed with the precursor **P_2W_{17}-APTES** (Figure S15, Supplementary Materials) but not seen with the other POMs functionalized with B_{10} clusters. The MALDI-TOF spectrum of **P_2W_{17}-APTES** indeed exhibits a major peak corresponding to the expected precursor associated with one molecule of the DCTB matrix at m/z 6058.7 (calculated m/z 6057.9 for $(TBA)_6H[(P_2W_{17}O_{61})O(SiC_3H_6NH_2)_2](DCTB)\}^+$) and a minor peak at m/z 6300.0

(calculated m/z 6299.4 for $(TBA)_7[(P_2W_{17}O_{61})O(SiC_3H_6NH_2)_2](DCTB)\}^+$). The attribution of the peaks is definitely confirmed thanks to the fitting of the isotopic distribution massifs. The latter are mainly due to the isotopic distribution of the 10 or 17 tungsten atoms of the POMs, which appears consistent with the experimental spectrum (see Figure 4a and Figure S15, respectively).

The spectrum of **SiW$_{10}$-monoB$_{10}$** depicted in Figure 4b shows only one experimental peak at m/z 3843.2 which is perfectly consistent with the calculated mass for the monocationic product $\{(TBA)_4H_3[(SiW_{10}O_{36})O(SiC_3H_6NH_2)(SiC_3H_6NHCOB_{10}H_9)](CH_3CN)(H_2O)_3\}^+$ (calculated m/z 3843.1). It evidences the formation of the expected adduct **SiW$_{10}$-monoB$_{10}$** and thus indirectly the grafting of one $(B_{10}H_9CO)^-$ cluster to **SiW$_{10}$-APTES**. The simulated spectrum agrees well with the experimental data, which supports this assumption although the presence of one B_{10} cluster does not modify significantly the isotopic massif.

The MALDI-TOF spectrum of **SiW$_{10}$-diB$_{10}$** shown in Figure 4c displays a major peak centered at m/z 4085.8, which fully agrees with the expected di-grafted compound $\{(TBA)_4H_5[(SiW_{10}O_{36})O(SiC_3H_6NHCOB_{10}H_9)_2](CH_3CN)_2(H_2O)_6\}^+$ (m/z calculated 4084.4) and a minor peak at m/z = 4330.2 consistent with the species $\{(TBA)_5H_4[(SiW_{10}O_{36})O(SiC_3H_6NHCOB_{10}H_9)_2](CH_3CN)_3(H_2O)_4\}^+$ (m/z calculated 4330.8). This result confirms the formation of the expected di-grafted compound.

Finally, the case of **P$_2$W$_{17}$-diB$_{10}$**, appears more complicated, certainly due to a higher charge of the hybrid POM (10-) and a larger surface, which both favor intermolecular interactions with solvent molecules and cations. For technical reasons, the MALDI-TOF spectrum shown in Figure S16 (see Supplementary Materials) was recorded in linear mode, which does not favor the high resolution in contrast with other compounds. The spectrum displays an intense and broad experimental peak centered at m/z 6048.1, while four smaller peaks are found, respectively, at m/z 6289.6, 6431.7, 6672.5 and 6813.8. All these peaks are consistent with di-grafted species of general formula $\{(TBA)_xH_y[(P_2W_{17}O_{61})O(SiC_3H_6NHCOB_{10}H_9)_2](CH_3CN)_z(H_2O)_t\}^+$ ($x + y = 11$, $z = 0–3$ and $t = 5–6$). Regarding the main peak, the latter appears much broader than expected for only one species. Moreover, the resolution of the isotopic massif is lost. In fact, the experimental spectrum likely corresponds to a spectra superimposition of monocationic species of general formula $\{(TBA)_5H_6[(P_2W_{17}O_{61})O(SiC_3H_6NHCOB_{10}H_9)_2](CH_3CN)_x(H_2O)_y\}^+$ with x ranging from 1 to 5 and y from 0 to 8 (m/z in the range 6035.70 to 6076.75). Some simulated spectra are given in Figure S16 in Supplementary Materials.

2.4. NMR Studies in Solution

In the absence of crystallographic data, the three obtained hybrid systems have been thoroughly characterized by multinuclear NMR spectroscopy in order to verify their structures in solution. ^1H, ^{11}B, ^{13}C, ^{15}N, ^{29}Si, ^{31}P, and ^{183}W NMR spectra were recorded in CD$_3$CN at room temperature. The data are gathered in Table S2 (Supplementary Materials), while selected spectra are given in Figures 5 and 6 and in Figures S17–S32 (Supplementary Materials).

As shown in Figure 5a and in Figures S17–S20 (Supplementary Materials), $^{11}B\{^1H\}$ NMR spectrum of $[B_{10}H_9CO]^-$ undergoes a significant change upon coupling with **SiW$_{10}$-APTES** or **P$_2$W$_{17}$-APTES**. In particular, the signal at -44.4 ppm specific for the equatorial boron atom bearing the substituent CO in $[B_{10}H_9CO]^-$ (B2 atom, see Figure 1c) is strongly shifted to ca. -25 ppm in the spectra of **SiW$_{10}$-monoB$_{10}$**, **SiW$_{10}$-diB$_{10}$** and **P$_2$W$_{17}$-APTES** in agreement with the grafting of the cluster on the POM.

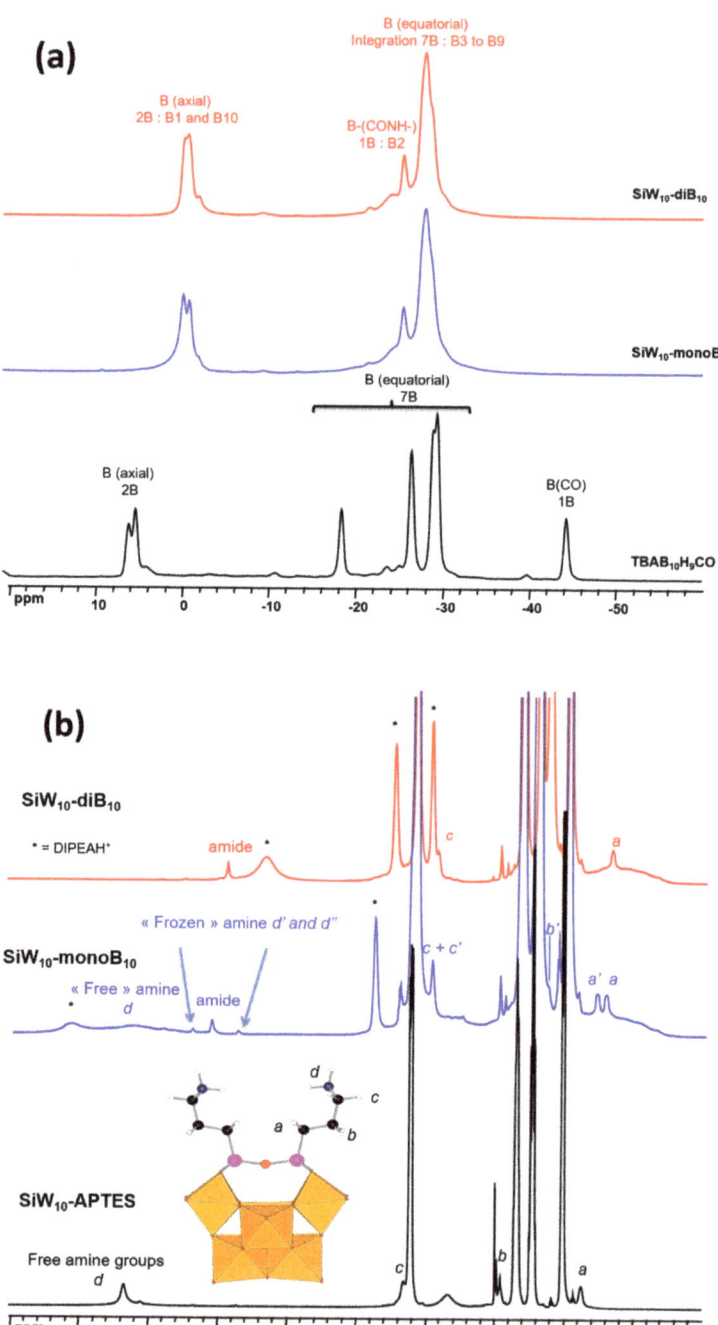

Figure 5. (a) $^{11}B\{^{1}H\}$ NMR spectra of **SiW$_{10}$-monoB$_{10}$**, **SiW$_{10}$-diB$_{10}$** and TBA[B$_{10}$H$_9$CO] in CD$_3$CN. (b) ^{1}H NMR spectra of **SiW$_{10}$-monoB$_{10}$**, **SiW$_{10}$-diB$_{10}$** and **SiW$_{10}$-APTES** in CD$_3$CN. * indicates the signal of the protonated amine DIPEAH+ present as a counter-cation. Reproduced with permission from the doctoral thesis manuscript of Dr Manal Diab, University Paris Saclay/Lebanese University, May 2018.

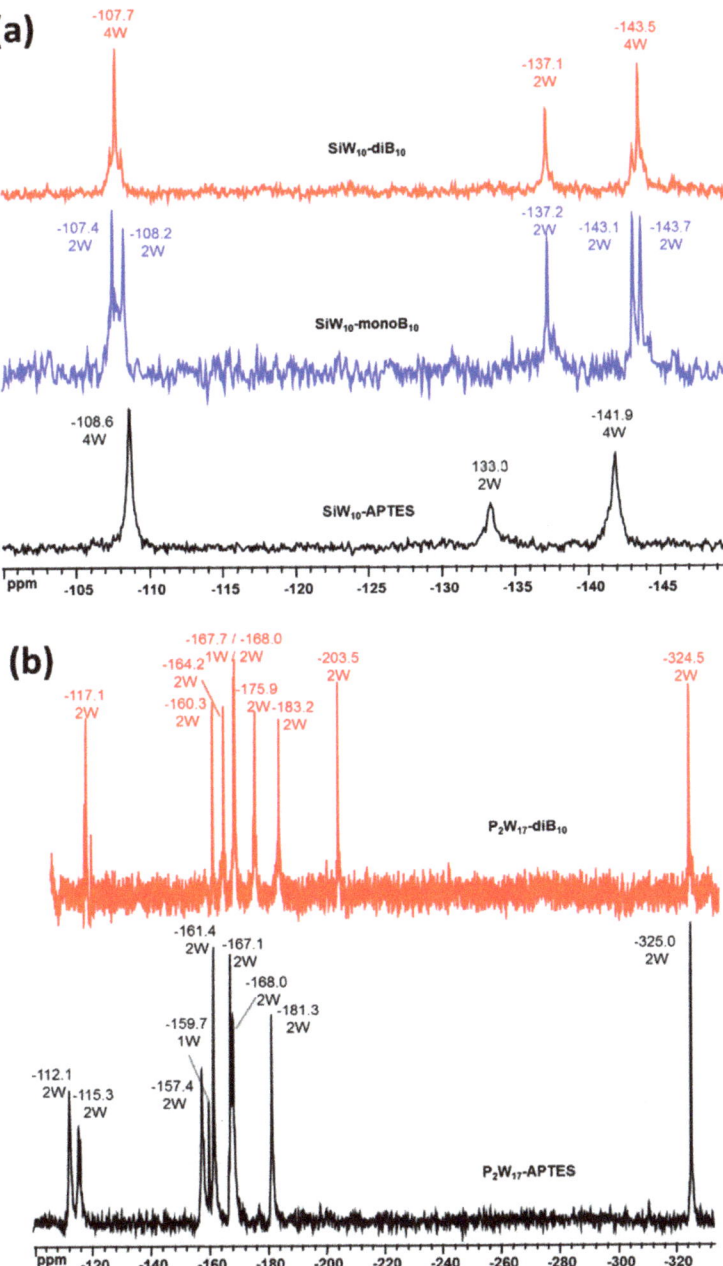

Figure 6. (a) ^{183}W NMR spectra of **SiW$_{10}$-monoB$_{10}$**, **SiW$_{10}$-diB$_{10}$** and **SiW$_{10}$-APTES** in CD$_3$CN. (b) ^{183}W NMR spectra of **P$_2$W$_{17}$-diB$_{10}$** and **P$_2$W$_{17}$-APTES** in CD$_3$CN.

Concomitantly, the ^{1}H NMR spectrum of the mono adduct **SiW$_{10}$-monoB$_{10}$**, exhibits a splitting of the signals for the three methylene groups –CH$_2$- of the APTES linker, denoted a, b, c (see Figure 5b), because of the lowering of the symmetry of **SiW$_{10}$-APTES** from C$_{2v}$ to C$_s$. In addition, a new peak at 6.12 ppm assigned to an amide function is observed.

For the remaining amine function, a broad signal is observed at 7.4 ppm (d), but together with two other broad signals at 5.70 and 6.33 ppm (d' and d''), attributed to the amine function in a frozen configuration in which the interaction with B_{10} cluster generates two inequivalent protons as depicted in Figure 7a (DFT optimized structure). These assumptions are confirmed by ^1H-^{15}N HMBC (Heteronuclear Multiple Bond Correlation) NMR spectrum (Figure S25) revealing two ^{15}N signals at -251 ppm (amide) correlated to the proton signal at 6.12 ppm and at -272 ppm (free amine) correlated to the two protons peaks at 5.70 and 6.33 ppm.

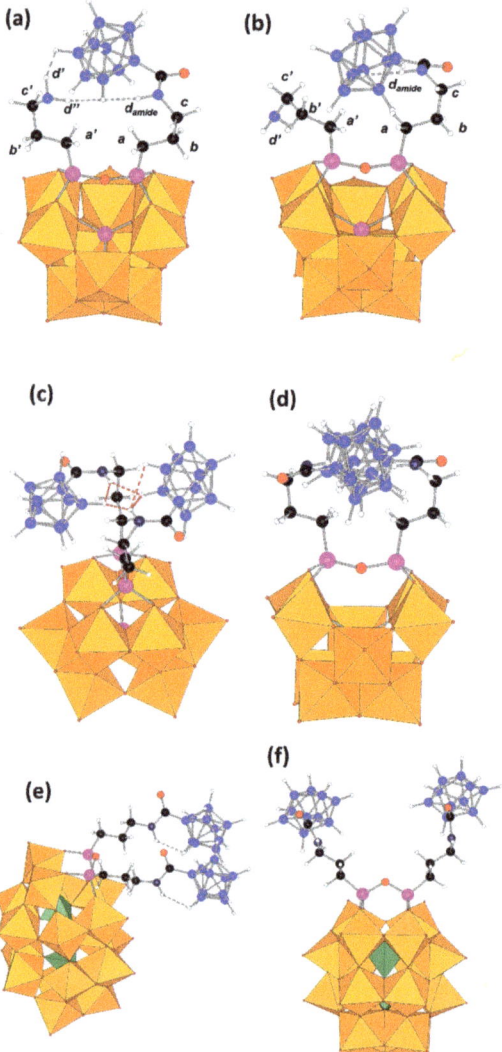

Figure 7. Optimized molecular structures of the POM-borates derivatives. **SiW$_{10}$-monoB$_{10}$** (**a**) in «closed» form and (**b**) in «open» form; (**c**,**d**) two views of the most stable configuration of **SiW$_{10}$-diB$_{10}$**; (**e**,**f**) two views of the most stable configuration of **P$_2$W$_{17}$-diB$_{10}$**. Dashed lines are given for shortest H-H contacts. Legend: C in black, H in white, B in blue, N in dark blue, Si in pink, WO$_6$ octahedra in orange and PO$_4$ tetrahedra in green.

Grafting a second $[B_{10}H_9CO]^-$ group on the **SiW$_{10}$-APTES** platform allows recovering the C_{2v} symmetry and thus one set of peaks was observed for the linker and especially the protons "a", in addition to an amide peak at 5.94 ppm (Figure S24, Supplementary Materials). The signals of the amine at 7.4, 5.7, and 6.3 ppm disappear in agreement with the reaction of $[B_{10}H_9CO]^-$ groups with this function. Similarly, the ^{1}H NMR spectrum of **P$_2$W$_{17}$-diB$_{10}$** compared to that of **P$_2$W$_{17}$-APTES** (Figure S27, Supplementary Materials) evidences the appearance of a sharp peak at 5.94 ppm assigned to an amide function, while the signal of the free amine at 7.03 ppm in the precursor **P$_2$W$_{17}$-APTES** disappears.

To further confirm our assignments of the signal of the amide function, ^{1}H-^{1}H ROESY (Rotating frame Overhause Effect SpectroscopY) and ^{13}C NMR experiments were performed on **SiW$_{10}$-diB$_{10}$** and **P$_2$W$_{17}$-diB$_{10}$** (Figures S28–S32, Supplementary Materials). Cross REO peaks involving the amide proton (5.94 ppm in both compounds) and some equatorial B-H protons of the B$_{10}$ cluster at 0.4 ppm and the protons "c" of the APTES chains can be seen in Figures S28 and S29 (Supplementary Materials). This demonstrates the spatial proximity between these protons that interact between each other through dipolar contacts. ^{13}C NMR spectra of **SiW$_{10}$-monoB$_{10}$**, **SiW$_{10}$-diB$_{10}$**, and **P$_2$W$_{17}$-diB$_{10}$** (Figures S30–S32) notably exhibits a signal at 203 ppm assigned to a carbon atom from an amide group, which is confirmed by 2D ^{1}H-^{13}C HMBC NMR spectrum of **SiW$_{10}$-monoB$_{10}$** evidencing a correlation between this ^{13}C signal at 203 ppm and the ^{1}H amide signal at 6.12 ppm. Besides, in both cases of **SiW$_{10}$-monoB$_{10}$** and **SiW$_{10}$-diB$_{10}$** this ^{13}C signal appears as a poorly resolved quadruplet with a coupling constant of 95 Hz consistent with a $^1J_{^{13}C-^{11}B}$ coupling.

Therefore, ^{1}H, ^{1}H-^{15}N HMBC, ^{1}H-^{1}H ROESY, ^{13}C and ^{1}H-^{13}C HMBC NMR experiments unambiguously confirm the formation of an amide group in our three adducts by reaction of the amines of POM-APTES precursors with the carbonyl group of the cluster $[B_{10}H_9CO]^-$. The modification of the $^{11}B\{^1H\}$ NMR spectra of the boron cluster after its reaction with the POM-APTES precursors further confirms such results.

^{29}Si, ^{31}P and ^{183}W NMR probe the POM part in compounds **SiW$_{10}$-APTES**, **SiW$_{10}$-monoB$_{10}$**, **SiW$_{10}$-diB$_{10}$**, **P$_2$W$_{17}$-APTES** and **P$_2$W$_{17}$-diB$_{10}$** (see Figure 6 and Figures S21–S23 in Supplementary Materials). The unsymmetrical environment in the mono adduct **SiW$_{10}$-monoB$_{10}$** is clearly confirmed through the appearance of two peaks for Si of the different linker arms at −61.9 and −63.3 ppm, while only one signal was observed at −62.3 ppm for the symmetrical di adduct **SiW$_{10}$-diB$_{10}$** with only a small shift from the initial **SiW$_{10}$-APTES** precursor (−62.5 ppm). In addition, for all the compounds, a single peak is observed for the Si atom in the central cavity of the SiW$_{10}$ POM moiety which is almost not affected by the grafting of the boron clusters and the resulting changes of symmetry of the adducts (Figure S21, Supplementary Materials). In case of **P$_2$W$_{17}$-APTES** and **P$_2$W$_{17}$-diB$_{10}$**, both compounds exhibit only one signal assigned to the two equivalent Si atoms of the APTES linker (Figure S22, Supplementary Materials).

The ^{183}W NMR spectra of precursors and adducts are given in Figure 6. For **SiW$_{10}$-monoB$_{10}$** the ^{183}W NMR spectrum displays five peaks of intensities 2:2:2:2:2 in agreement with the expected low C_s symmetry, while three resonances are observed for the di adduct **SiW$_{10}$-diB$_{10}$** and the initial precursor **SiW$_{10}$-APTES** of intensities (4:2:4) consistent with their C_{2v} symmetry (Figure 6a). Figure 6b shows the ^{183}W NMR spectrum of **P$_2$W$_{17}$-diB$_{10}$** which differs significantly from its precursor. Both compounds exhibit nine NMR lines of integration 2:2:2:1:2:2:2:2:2 in agreement with the expected C_s symmetry, but their positions are slightly changed. This is due to the modification of the P$_2$W$_{17}$ moiety induced by the grafting of the two $[B_{10}H_9CO]^-$ clusters. Additionally, ^{31}P NMR spectra of **P$_2$W$_{17}$-APTES** and **P$_2$W$_{17}$-diB$_{10}$** display two signals (Figure S23, Supplementary Materials), wherein one of them showed a common chemical shift, while the second exhibited a small shift from −13.4 ppm in **P$_2$W$_{17}$-APTES** to −13.6 ppm in **P$_2$W$_{17}$-diB$_{10}$**.

In conclusion, these experiments focused on the POM part fully agree in terms of molecular symmetries with the formation of the expected mono- or di-adducts with B$_{10}$ clusters.

2.5. Computational Studies

The molecular geometries of **SiW$_{10}$-APTES**, **SiW$_{10}$-monoB$_{10}$**, and **SiW$_{10}$-diB$_{10}$**, as well as those of P$_2$W$_{17}$-APTES, P$_2$W$_{17}$-monoB$_{10}$ and P$_2$W$_{17}$-diB$_{10}$ were fully optimized at a DFT level including implicit solvent effects (see Figure 7 and Supplementary Materials for computational details). We considered the most relevant plausible conformers. Firstly, regarding **SiW$_{10}$-APTES**, it exhibits two main conformers as defined by the orientation of the two amine organic arms, which we called them open and closed forms. The small difference in their relative energy, less than 1 kcal·mol^{-1} in favor of the closed form (represented in Figure 1a), forecasted that further substitution would easily overcome any initial geometric preference in the reactants. Indeed, upon B$_{10}$ incorporation a much more complex situation arises. For **SiW$_{10}$-monoB$_{10}$** we characterized five conformers, two arising from the closed reactant and three species from the open reactant form. In the most stable conformer (Figure 7a), which arise from the closed form, the decaborate moiety interacts favorably with the amine hydrogens (d' and d") of the unreacted arm through strong dihydrogen contacts. In the most stable open form (Figure 7b), although interaction between arms is almost neglected, the H amide atom develops other interactions. Overall, the most stable conformer given in Figure 7a is 11 kcal·mol^{-1} below the second one (Figure 7b). All five conformers lie in a narrow 20 kcal·mol^{-1} range. For the double substituted **SiW$_{10}$-diB$_{10}$**, since the additional repulsion arising from the negatively charged B$_{10}$ groups, we could only characterize two forms: an open (not shown) and a closed one (two views on Figure 7c,d). The energy difference between both species was computed to be just only 5 kcal·mol^{-1}. We highlight, as dashed lines in Figure 7, those hydrogen atoms of the organic arm and the B$_{10}$ moiety that lie close in three-dimensional space.

Due to the monovacant character of P$_2$W$_{17}$, the Si-O-Si angle of the APTES moiety grafted to the POM strongly differs from that observed for the divacant SiW$_{10}$ POM (see Figure 1). The topology of the two arms, and thus the connectivity of the two POM-APTES derivatives, strongly differs. Therefore, as seen in Figure 1, closed form is not possible for P$_2$W$_{17}$-APTES. Only one geometry could thus be considered. Then, for the mono-substituted P$_2$W$_{17}$-monoB$_{10}$, two conformers were characterized, one open and one folded, the folded one being more stable by only 2.2 kcal·mol^{-1}. For the di-substituted Dawson derivative, only one open shaped product could be characterized (see Figure 7e,f).

Hydrogen atoms of the decaborate moieties possess a hydride character. Consequently, they can establish hydrogen-hydrogen contacts with protic solvent or with functional groups like amines or amides. In the present structures, many H-H dihydrogen contacts between the amine organic arms from APTES moiety and hydrogen atoms from decaborate clusters were observed. For instance, for **SiW$_{10}$-monoB$_{10}$** (Figure 7a), the hydrogen atoms d' and d" from the «free» amino group are found 2.21 and 1.99 Å far from an H atom belonging to the B$_{10}$ cluster, which are quite short distances. This fact agrees with the couplings observed in the NMR experiments.

The thermodynamics of the formation of the mono- and di-adducts of the Keggin and the Dawson species is computed exergonic in all cases as seen in Figure 8. For **SiW$_{10}$-APTES**, both the formation of the mono- and bi-derivative were computed exergonic, 22.1 kcal·mol^{-1} for the **SiW$_{10}$-monoB$_{10}$**, and 13.0 kcal·mol^{-1} for the **SiW$_{10}$-diB$_{10}$**. The formation of **SiW$_{10}$-monoB$_{10}$** is clearly favored and the strong dihydrogen contacts between amino group and the grafted B$_{10}$ cluster undoubtedly strongly stabilize such a species compared to the di-adduct **SiW$_{10}$-diB$_{10}$**. For P$_2$W$_{17}$-APTES, also the two substitutions are favorable, 6.8 kcal·mol^{-1} for the first, and 7.9 kcal·mol^{-1} for the second.

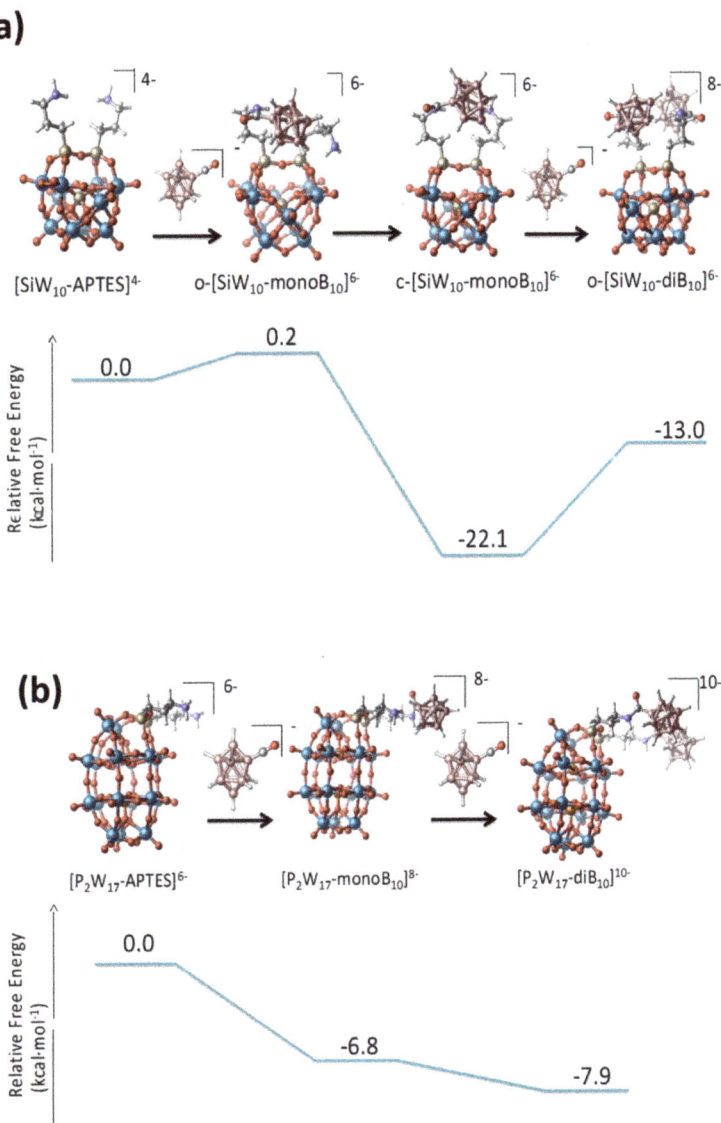

Figure 8. Energetic profiles of the formation of mono- and di-adduct from the starting precursors in CD$_3$CN. (**a**) **SiW$_{10}$-APTES**, **SiW$_{10}$-monoB$_{10}$** (open and closed isomers), and **SiW$_{10}$-diB$_{10}$**; (**b**) **P$_2$W$_{17}$-APTES**, **P$_2$W$_{17}$-monoB$_{10}$**, and **P$_2$W$_{17}$-diB$_{10}$**.

The computed reaction free energies for the **SiW$_{10}$-APTES** and P$_2$W$_{17}$-APTES derivatives are fully consistent with the experimental findings. For the keggin derivatives, the strong stabilization of the mono-adduct allows isolating both mono and di-adduct thanks to the formation of strong H-H contacts. In contrast, for the Dawson derivatives, the small difference of energies between mono- and bi-adducts (1.1 eV only) does not permit isolating the mono-adduct. Besides, an excess of [B$_{10}$H$_9$CO]$^-$ (3 equivalents/POM instead of 2) is needed to get the pure di-adduct compound to avoid the formation of a mixture between mono and di-grafted adducts. A similar situation was previously obtained with

Anderson-type derivatives since the mono and di-adduct of B_{10} with $[MnMo_6(Tris)_2]^{3-}$ are only separated by 6 eV and it was not possible to get mono-adduct [22]. This result highlights the role of the topology of the POM-APTES compounds and their faculty to stabilize species thanks to intramolecular interactions.

Finally, DFT studies provided the frontier orbitals for each compound (see Figure 9 and Figure S33 in Supplementary Materials). The results obtained for Keggin and Dawson derivatives exhibit the same feature. For POM-APTES the HOMO is located on one (for **SiW$_{10}$-APTES**) or two (**P$_2$W$_{17}$-APTES**) amines of the APTES part, while the LUMO are localized on the W atoms of the POM part. By grafting the B_{10} clusters, the LUMO levels are slightly affected. LUMO remains localized on W atoms and only minor changes in energy are observed.

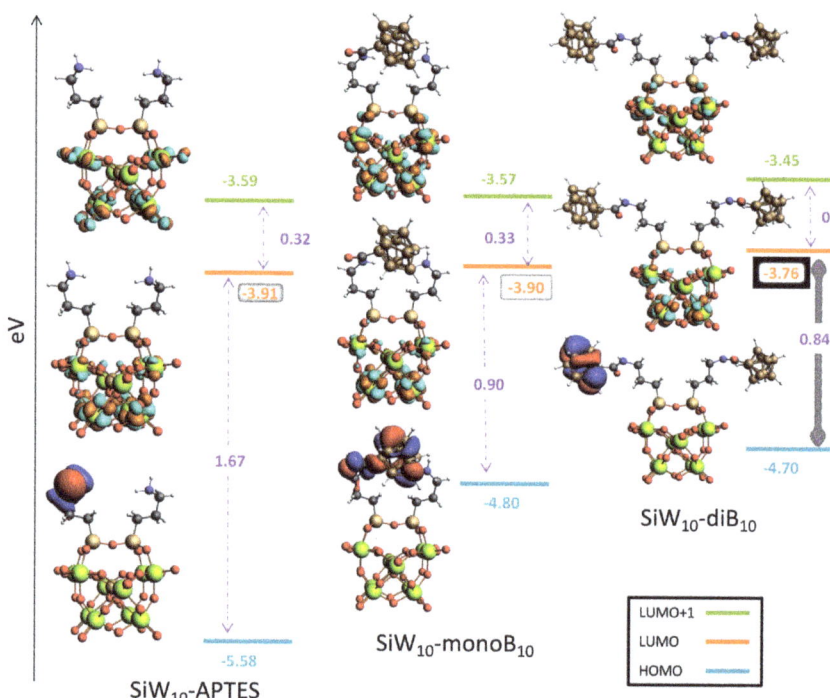

Figure 9. Frontier orbitals energies and energy gaps (eV) for the **SiW$_{10}$-APTES**, **SiW$_{10}$-monoB$_{10}$** and **SiW$_{10}$-diB$_{10}$** species. Color code: W green, O red, Si light brown, B dark brown, C grey, N blue, H white; HOMO: red/blue; LUMO: orange/cyan. MO surfaces plotted at a 0.03 isovalue.

Conversely, the HOMO levels are drastically modified by the introduction of B_{10} clusters. Electrons of the HOMO orbitals are now mainly localized on one grafted B_{10} cluster. Interestingly, for **SiW$_{10}$-monoB$_{10}$** the HOMO is delocalized between one B_{10} cluster and the amine of the second arm, which strongly interacts with the B_{10} through H-H contacts. The HOMO energy level increases in all cases within the range 0.78 to 0.92 eV. The HOMO-LUMO gaps, are thus significantly reduced upon the B_{10} grafting. Indeed, for **SiW$_{10}$-APTES**, the gap decreases from 1.67 to 0.90 eV for the first substitution, and to 0.94 eV for the second. For **P$_2$W$_{17}$-APTES**, the gap evolves from 1.56 to 0.77 eV for the first substitution, and to 0.76 eV for the second.

2.6. Electrochemical Properties

The electronic spectra of compounds **SiW$_{10}$-monoB$_{10}$**, **SiW$_{10}$-diB$_{10}$** and **P$_2$W$_{17}$-diB$_{10}$** recorded in CH_3CN containing 0.1 mol.L^{-1} TBAClO$_4$ (TBAP, tetrabutylammonium perchlorate) at room temperature and 2.10^{-4} mol.L^{-1} concentration are depicted in Figures S35 and S36 (Supplementary Materials).

The electronic spectra of precursors **SiW$_{10}$-APTES** and **P$_2$W$_{17}$-APTES** display absorption bands in ultraviolet region corresponding to transition between p-orbitals of the oxo ligands and d-type orbitals centered on tungsten [33,34], while the cluster [B$_{10}$H$_9$CO]$^-$ exhibits weak absorption band between 300 and 200 nm notably assigned to $\pi-\pi^*$ transitions [35]. Considering that the main contribution of the spectra comes from the LMCT band involving the W atoms and that the LUMO band centered on tungsten atoms are only slightly modified upon grafting of B$_{10}$ cluster, no drastic changes are expected in the POM-B$_{10}$ adducts. Indeed, the electronic spectra of **SiW$_{10}$-monoB$_{10}$** and **P$_2$W$_{17}$-diB$_{10}$** match well with the sum of the spectra of **SiW$_{10}$-APTES** or **P$_2$W$_{17}$-APTES** and one or two times that of [B$_{10}$H$_9$CO]$^-$, respectively. The spectrum of **SiW$_{10}$-diB$_{10}$** slightly differs from the sum of the component's spectra probably due to a larger variation of LUMO energy level from **SiW$_{10}$-APTES** to **SiW$_{10}$-diB$_{10}$** and additional constraints due to the vicinity of the two boron clusters.

Since no evolution of the spectra were observed within 24 h in such a medium, the compounds appear chemically stable in these experimental conditions. The cyclic voltammograms (CVs) were thus recorded for all the SiW$_{10}$ and P$_2$W$_{17}$ derivatives and are given in Figure 10 and in Figures S37 and S38 (Supplementary Materials), while the anodic and cathodic potentials are gathered in Table S3 (Supplementary Materials).

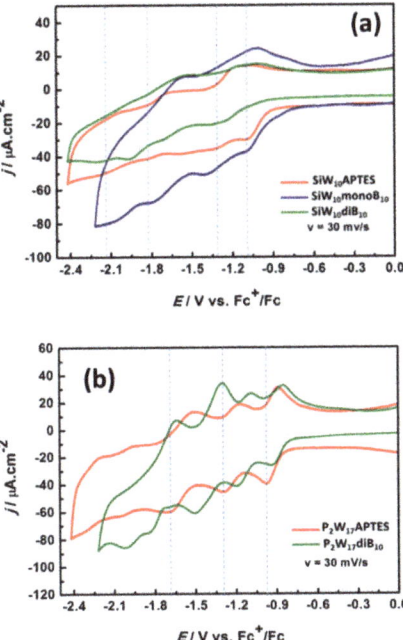

Figure 10. Comparison of cyclic voltammograms (**a**) for the three compounds **SiW$_{10}$-APTES**, **SiW$_{10}$-monoB$_{10}$** and **SiW$_{10}$-diB$_{10}$**, and (**b**) for **P$_2$W$_{17}$-APTES** and **P$_2$W$_{17}$-diB$_{10}$** in the reduction part. The electrolyte was CH$_3$CN + 0.1 M TBAClO$_4$. Dashed lines are only guide for eyes.

As depicted in Figure 10a and Figure S37, the CV of **SiW$_{10}$-APTES** is poorly resolved and it is difficult to identify confidently all the reduction processes corresponding to the successive reduction in W^{VI} centers into W^V, well-known to be monoelectronic in non-aqueous solvents [36]. These waves seem nevertheless reversible with processes better resolved in oxidation. Besides, an irreversible process attributed to the oxidation of amine function of the APTES linker is also observed in oxidation around +0.452 V vs. Fc$^+$/Fc (see Supplementary Materials). As for their parent precursor, CVs of **SiW$_{10}$-monoB$_{10}$** and **SiW$_{10}$-diB$_{10}$** display poorly resolved reversible electronic transfers, which appear shifted towards more negative potentials and one irreversible oxidation process around +0.452 V vs. Fc$^+$/Fc assigned to the oxidation of the remaining amine function and/or of the B$_{10}$ cluster (see Figure S39, Supplementary Materials). This behavior agrees with the increase in the charge from 4- in **SiW$_{10}$-APTES** to 6- in **SiW$_{10}$-monoB$_{10}$** and 8- in **SiW$_{10}$-diB$_{10}$** and the electron donating character of the boron cluster [37] but does not evidence a strong electronic effect of the boron cluster on the POMs electronic properties.

The CVs of **P$_2$W$_{17}$-APTES** and **P$_2$W$_{17}$-diB$_{10}$** are given in Figure 10b and in Figure S38 (Supplementary Materials) and appears much more resolved than those of SiW$_{10}$ derivatives. The CV of **P$_2$W$_{17}$-APTES** displays four reversible electronic transfers with cathodic peak potentials assigned to successive mono- or bi-electronic reductions of W^{VI} centers into W(+V) [38] and two irreversible oxidation processes at E_{pa} = +0.452 and +0.759 V, assigned to the oxidation of the terminal amine groups of the APTES linkers. Conversely to the di-adduct compound **SiW$_{10}$-diB$_{10}$**, the CVs of the Dawson derivative **P$_2$W$_{17}$-diB$_{10}$** give four reversible reduction processes significantly shifted towards the more positive potential compared to **P$_2$W$_{17}$-APTES** and one irreversible reduction process at E_{pc} = −2.030 V vs. Fc$^+$/Fc, which was not observed in the precursor. The opposite effect was expected. This effect probably results from a combination of a charge effect, the presence of protons (in DIPEAH+ cations) and of an electronic effect of boron cluster on P$_2$W$_{17}$ moiety but at this stage it is difficult to have a clear explanation of the contribution of all these effects which can be antagonist.

Although reduction waves in the Dawson derivatives are not very well resolved, it can be observed that the di-substituted species (green line in Figure 10) is reduced at lower potentials than **P$_2$W$_{17}$-APTES**, in agreement with the fact that the LUMO and LUMO+1 raise in energy upon B$_{10}$ attachment. Also, the successive reduction waves seem just shifted left, which would conform with the almost constant difference in the LUMO and LUMO+1 energies along the series.

2.7. Electrocatalytic Properties for the Reduction in Protons into Hydrogen (HER)

Many POMs are known to catalyze protons reduction into hydrogen in aqueous or in non-aqueous conditions [7,39,40]. We verified by UV-Vis spectroscopy that [B$_{10}$H$_9$CO]$^-$ and its adducts with POMs are stable in CH$_3$CN in the presence of excess acetic acid (20 equivalents). In these conditions, it was interesting to study the reactivity of these compounds in regard to the electro-catalytic reduction in protons into hydrogen. The experiments were performed in CH$_3$CN + 0.1 M TBAP by using acetic acid as a source of protons, and as a weak acid in such a medium (pKa = 22.3) [41].

Figure 11 and Figure S40 (Supplementary Materials) show the evolution of CVs upon stepwise addition of acetic acid up to 20 equivalents of acid/POM for all P$_2$W$_{17}$ and SiW$_{10}$ derivatives, respectively. For all the compounds, the addition of acetic acid, gives a new irreversible reduction wave, which grows gradually with the amount of acid, expressed as γ = [acid]/[POM]. As shown in Figure 11 and Figure S40, at a given potential of −2.2 V vs Fc$^+$/Fc, a linear dependence of the catalytic current versus γ is obtained, a behavior featuring the electro-catalytic reduction of protons. However, the effect of the addition of acetic acid in the solution appears stronger for P$_2$W$_{17}$ derivatives than for SiW$_{10}$ based compounds.

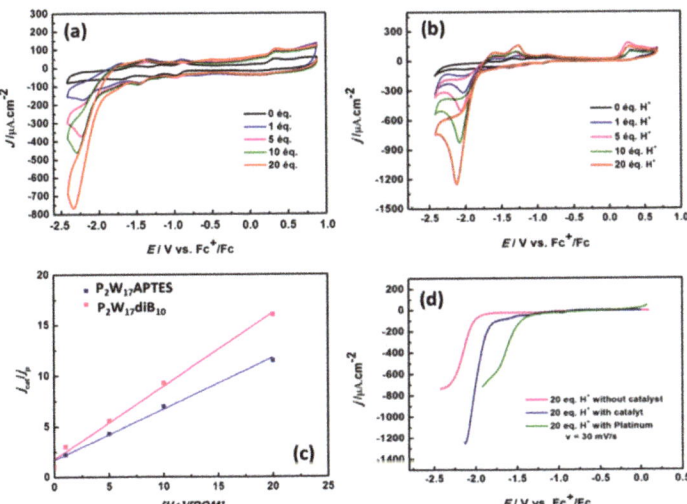

Figure 11. Cyclic voltammograms of (a) P_2W_{17}-APTES and (b) P_2W_{17}-diB10 after addition of variable amounts of acetic acid. (c) Plots of the cathodic currents measured at −2.2 V vs Fc+/Fc as a function of the ratio [acid]/[POM] for P_2W_{17}-APTES and P_2W_{17}-diB10. (d) Comparison of HER with and without catalyst P_2W_{17}-diB10, and with platinum after addition of an excess of acetic acid corresponding to the quantity added for a ratio [acid]/[POM] = 20. In all cases, the electrolyte was CH_3CN + 0.1 M $TBAClO_4$. The reference electrode was a saturated calomel electrode (SCE). Reproduced with permission from the doctoral thesis manuscript of Dr Manal Diab, University Paris Saclay/Lebanese University, May 2018.

To evidence the electrocatalytic process, linear voltammetry of P_2W_{17}-diB$_{10}$ in the presence of 20 equivalents of acetic acid was performed and compared to similar experiments performed without catalyst or on platinum electrode (Figure 11d). We notice that in the presence of the catalyst P_2W_{17}-diB$_{10}$, the current density is almost doubled and there is a 250-mV overvoltage decrease compared to the solution without catalyst. Indeed, the proton reduction with respect to platinum starts at −1.400 V vs. Fc+/Fc, while it starts at −1.750 V vs. Fc+/Fc with catalyst and at −2.000 V vs. Fc+/Fc without catalyst. Finally, the formation of hydrogen is unambiguously demonstrated by gas chromatography analysis during electrolysis performed at −2.200 V vs Fc+/Fc during 4.5 h (Figure S40, Supplementary Materials).

To compare the efficiency of all compounds, the catalytic efficiency (CAT) can be estimated using Equation (1):

$$CAT = \frac{100 * (J_{(POM+20\ eq.\ CH3COOH)} - J_{(POM\ alone)})}{J_{(POM\ alone)}} \quad (1)$$

Table 1 summarizes the *CAT* values measured for our products at −2.2 V vs Fc+/Fc. Interestingly, the two precursors **SiW$_{10}$-APTES** and **P$_2$W$_{17}$-APTES** exhibit similar efficiency. Also, the efficiency of **SiW$_{10}$-monoB$_{10}$** and **SiW$_{10}$-diB$_{10}$** adducts are lower than that of **SiW$_{10}$-APTES**, while it is the opposite for **P$_2$W$_{17}$-diB$_{10}$**, which appears much more efficient than its parent precursor, in agreement with cyclic voltammetry experiments. Indeed, a less negative reduction potential of the POM part should facilitate the electro-catalytic reduction in protons.

Table 1. Electrocatalytic efficiency for the reduction of protons into hydrogen at E = −2.2 V vs. Fc+/Fc for 20 equivalents of CH$_3$COOH added in CH$_3$CN.

Compound	Catalytic Efficiency (%)
SiW$_{10}$-APTES	827
SiW$_{10}$-monoB$_{10}$	524
SiW$_{10}$-diB$_{10}$	548
P$_2$W$_{17}$-APTES	854
P$_2$W$_{17}$-diB$_{10}$	1340

In terms of mechanism, three key steps have to be considered: the protonation, the reduction in the catalyst and the transfer of electron towards the protons to give dihydrogen. For protonation step, since catalysis is observed in all compounds, it must occur on the most basic sites, either on the oxo groups of the POM moiety, on the free amine groups in **SiW$_{10}$-APTES** and **P$_2$W$_{17}$-APTES** or on boron clusters for **SiW$_{10}$-diB$_{10}$** and **P$_2$W$_{17}$-diB$_{10}$**. DFT calculations evidence that the most nucleophilic sites are found on the oxo ligands of the POM parts which are consequently the preferential sites for protonation (see Figure 12 and Figure S34 in Supplementary Materials).

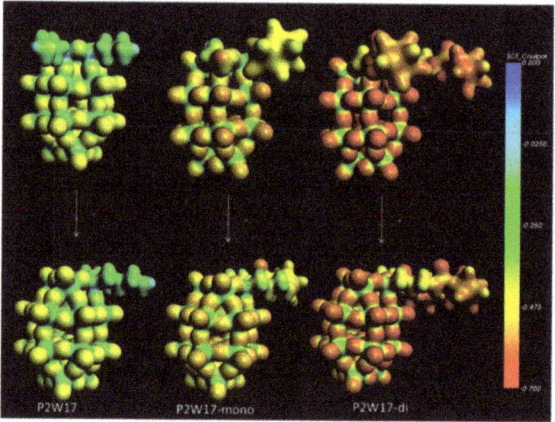

Figure 12. Two views of the molecular electrostatic potential in atomic units (a.u.) projected onto an electron density isosurface (0.03 e·au^{-3}) for **P$_2$W$_{17}$-APTES**, **P$_2$W$_{17}$-monoB$_{10}$** and **P$_2$W$_{17}$-diB$_{10}$** species.

For the reduction step, as seen in Figure S41c,d in Supplementary Materials, during electrolysis, the P$_2$W$_{17}$ derivatives turned to blue as expected for the reduction in such species before returning back colorless when the current is stopped indicating that the reduced POM probably transfers electrons to protons to produce hydrogen. We understand well that if this reduction occurs at higher potential, it should favor the process. **P$_2$W$_{17}$-diB$_{10}$** is thus logically the most efficient compound.

To sum up, even if the decaborate cluster is probably not directly involved in the HER process, it plays two indirect roles: (1) the covalent grafting on POMs increases the electronic density on the POM which should facilitate the protonation step, and (2) the covalent grafting can modifies the reduction potential of the POM moieties in POM-borate adducts, which favors the reduction step of the POM species when shifted towards more positive potentials as observed in **P$_2$W$_{17}$-diB$_{10}$**.

3. Conclusions

In this work, we succeeded in combining covalently the anionic [B$_{10}$H$_9$CO]$^-$ cluster with anionic SiW$_{10}$ and P$_2$W$_{17}$ derivatives using functionalized silyl derivatives (APTES) as a linker. The coupling between the two families of anionic parts appears much stronger than that with Anderson-type POMs we previously reported [23] and detailed NMR study allowed establishing the optimized conditions for the synthesis of target compounds. Hence, the selective isolation of mono- and di-adduct compounds of boron cluster with **SiW$_{10}$-APTES**, namely [(SiW$_{10}$O$_{36}$)(B$_{10}$H$_9$CONHC$_3$H$_6$Si)(NH$_2$C$_3$H$_6$Si)O]$^{6-}$ and [(SiW$_{10}$O$_{36}$)(B$_{10}$H$_9$CONHC$_3$H$_6$Si)$_2$O]$^{8-}$ was successfully achieved, while only the di-adduct [(P$_2$W$_{17}$O$_{61}$)(B$_{10}$H$_9$CONHC$_3$H$_6$Si)$_2$O]$^{10-}$ was isolated with **P$_2$W$_{17}$-APTES**. To the best of our knowledge, it is the first time that a mono-adduct can be isolated directly from the synthesis by functionalization of the **SiW$_{10}$-APTES** precursor. DFT studies supported by experimental NMR data evidenced that the formation of intramolecular H-H dihydrogen contact is the driving force for the preferred formation of the mono-adduct species and such a synthetic strategy could open the route toward the formation of hybrid POMs with two different functional groups.

All these compounds were fully characterized by multi-NMR techniques including ^{1}H, ^{11}B, ^{13}C, ^{15}N, ^{29}Si, ^{183}W and ^{31}P as well as multi-dimensional correlations such as COSY, HMBC (^{1}H-^{13}C and ^{1}H-^{15}N) and ROESY NMR allowing focusing on each part of the adducts, i.e., POM, linker and boron cluster. These characterizations demonstrated unambiguously the formation of the targeted adducts and were also consistent with FT IR and MALDI-TOF spectrometry data. DFT studies permitted to get optimized structures for all compounds consistent with the NMR data.

The electrochemical studies allowed studying the electronic effects of the grafting of the reducing boron cluster on some oxidized POMs with probable antagonist effect between charge effect and the variation of frontiers orbitals levels upon grafting of B$_{10}$ cluster. Finally, electro-catalytic reduction in protons into hydrogen was evidenced for these systems, the best efficiency being obtained with **P$_2$W$_{17}$-diB$_{10}$**. The process appears mainly effective on the POM part while the boron cluster participates only indirectly to the process.

Supplementary Materials: The following supporting information can be downloaded at: https://www.mdpi.com/article/10.3390/molecules27227663/s1. Experimental section including general methods, the ^{29}Si, ^{31}P and ^{1}H NMR titration studies (Figures S1–S10) and the syntheses of compounds used in this study; FT-IR spectra (Figures S11 and S12); MALDI-TOF data (Table S1) and spectra (Figures S13–S16); a summary of NMR data (Table S2), ^{11}B NMR studies (Figures S17–S20), ^{29}Si NMR spectra (Figures S21 and S22), ^{31}P NMR spectra (Figure S23), ^{1}H NMR studies including COSY, ROESY and ^{1}H-^{15}N HMBC experiments (Figures S24–S29); ^{13}C NMR studies (Figures S30–S32); a DFT part including computational details, frontiers orbitals of P$_2$W$_{17}$ derivatives (Figure S33) and electron density maps on SiW$_{10}$ derivatives (Figure S34); electronic spectra (Figures S35 and S36), additional cyclovoltammograms (Figures S37–S39) and table of potentials (Table S3); Evidence of the production of hydrogen and picture of the reduced POM during the HER process (Figures S40 and S41).

Author Contributions: Synthesis and solution studies, M.D. and Z.E.H.; elemental analyses, N.L.; supervision of NMR studies, M.H.; MALDI-TOF experiments, V.G. and D.T.; DFT Calculation, A.M. and C.B.; electrochemical studies, J.E.C. and A.R.; supervision of the work and funding acquisition, E.C., D.N. and S.F. All authors have read and agreed to the published version of the manuscript.

Funding: This research was funded by a public grant overseen by the French National Research Agency as part of the «Investissements d'Avenir» program (Labex Charm3at, ANR-11-LABX-0039-grat), by the Paris Ile-de-France Region-DIM "Respore", and by a mobility program PHC CEDRE (project POMBORON n°42237UG). M.D. thanks AZM association and IUF for financial support and Z.E.H. thanks CampusFrance for Eiffel doctoral grant. We also thank CERCA Program of the Generalitat de Catalunya, the ICIQ Foundation, and the Spanish Ministerio de Ciencia e Innovación through project PID2020-112806RB-I00 and through the Severo Ochoa Excellence Accreditation 2020–2023 (CEX2019-000925-S, MCI/AEI).

Institutional Review Board Statement: Not applicable.

Informed Consent Statement: Not applicable.

Data Availability Statement: A data set collection of computational results is available in the ioChem-BD repository and can be accessed via https://dx.doi.org/10.19061/iochem-bd-1-217 (accessed on 1 November 2022).

Acknowledgments: We acknowledge the Centre National de la Recherche Scientifique (CNRS) and the Ministère de l'Education Nationale de l'Enseignement Supérieur, de la Recherche et de l'Innovation (MESRI) for their financial support. This study results from an international collaboration supported by IRN-CNRS 2019–2023.

Conflicts of Interest: The authors declare no conflict of interest.

References

1. Lin, C.G.; Fura, G.D.; Long, Y.; Xuan, W.M.; Song, Y.F. Polyoxometalate-based supramolecular hydrogels constructed through host–guest interactions. *Inorg. Chem. Front.* **2017**, *4*, 789–794. [CrossRef]
2. Izzet, G.; Abecassis, B.; Brouri, D.; Piot, M.; Matt, B.; Serapian, S.A.; Bo, C.; Proust, A. Hierarchical Self-Assembly of Polyoxometalate-Based Hybrids Driven by Metal Coordination and Electrostatic Interactions: From Discrete Supramolecular Species to Dense Monodisperse Nanoparticles. *J. Am. Chem. Soc.* **2016**, *138*, 5093–5099. [CrossRef] [PubMed]
3. Zhu, Y.; Yin, P.C.; Xiao, F.P.; Li, D.; Bitterlich, E.; Xiao, Z.C.; Zhang, J.; Hao, J.T.; Liu, B.; Wang, Y.; et al. Bottom-Up Construction of POM-Based Macrostructures: Coordination Assembled Paddle-Wheel Macroclusters and Their Vesicle-like Supramolecular Aggregation in Solution. *J. Am. Chem. Soc.* **2013**, *135*, 17155–17160. [CrossRef] [PubMed]
4. Vickers, J.W.; Lv, H.J.; Sumliner, J.M.; Zhu, G.B.; Luo, Z.; Musaev, D.G.; Geletii, Y.V.; Hill, C.L. Differentiating Homogeneous and Heterogeneous Water Oxidation Catalysis: Confirmation that $[Co_4(H_2O)_2(\alpha-PW_9O_{34})_2]^{10-}$ Is a Molecular Water Oxidation Catalyst. *J. Am. Chem. Soc.* **2013**, *135*, 14110–14118. [CrossRef] [PubMed]
5. Sarma, B.B.; Neumann, R. Polyoxometalate-mediated electron transfer–oxygen transfer oxidation of cellulose and hemicellulose to synthesis gas. *Nat. Commun.* **2014**, *5*, 4621. [CrossRef] [PubMed]
6. Natali, M.; Orlandi, M.; Berardi, S.; Campagna, S.; Bonchio, M.; Sartorel, A.; Scandola, F. Photoinduced Water Oxidation by a Tetraruthenium Polyoxometalate Catalyst: Ion-pairing and Primary Processes with Ru(bpy)$_3$$^{2+}$ Photosensitizer. *Inorg. Chem.* **2012**, *51*, 7324–7331. [CrossRef]
7. Keita, B.; Nadjo, L. Polyoxometalate-based homogeneous catalysis of electrode reactions: Recent achievements. *J. Mol. Catal. A Chem.* **2007**, *262*, 190–215. [CrossRef]
8. Freire, C.; Fernandes, D.M.; Nunes, M.; Abdelkader, V.K. POM & MOF-based Electrocatalysts for Energy-related Reactions. *ChemCatChem* **2018**, *10*, 1703–1730.
9. Zhao, J.-W.; Li, Y.-Z.; Chen, L.-J.; Yang, G.-Y. Research progress on polyoxometalate-based transition-metal–rare-earth heterometallic derived materials: Synthetic strategies, structural overview and functional applications. *Chem. Commun.* **2016**, *52*, 4418–4445. [CrossRef]
10. Hasenknopf, B. Polyoxometalates: Introduction to a class of inorganic compounds and their biomedical applications. *Front. Biosci.* **2005**, *10*, 275–287. [CrossRef]
11. Yamase, T. Anti-tumor, -viral, and -bacterial activities of polyoxometalates for realizing an inorganic drug. *J. Mater. Chem.* **2005**, *15*, 4773–4782. [CrossRef]
12. Bijelic, A.; Aureliano, M.; Rompel, A. The antibacterial activity of polyoxometalates: Structures, antibiotic effects and future perspectives. *Chem. Commun.* **2018**, *54*, 1153–1169. [CrossRef] [PubMed]
13. Bijelic, A.; Aureliano, M.; Rompel, A. Polyoxometalates as Potential Next-Generation Metallodrugs in the Combat Against Cancer. *Angew. Chem. Int. Ed.* **2019**, *58*, 2980–2999. [CrossRef] [PubMed]
14. Sivaev, I.B.; Bregadze, V.I.; Kuznetsov, N.T. Derivatives of the closo-dodecaborate anion and their application in medicine. *Russ. Chem. Bull.* **2002**, *51*, 1362–1374. [CrossRef]
15. Zhu, Y.; Hosmane, N.S. Nanostructured boron compounds for cancer therapy. *Pure Appl. Chem.* **2018**, *90*, 653–663. [CrossRef]
16. Hey-Hawkins, E.; Viñas-Teixidor, C. *Boron-Based Compounds: Potential and Emerging Applications in Medicine*; John Wiley&Sons Ltd.: Hoboken, NJ, USA, 2018.
17. Hosmane, N.S. *Boron Science: New Technologies and Applications*; CRC Press: Boca Raton, FL, USA, 2012.
18. Zhu, Y.; Hosmane, N.S. Applications and perspectives of boron-enriched nanocomposites in cancer therapy. *Future Med. Chem.* **2013**, *5*, 705–714.
19. Sivaev, I.B.; Prikaznov, A.V.; Naoufal, D. Fifty years of the closo-decaborate anion chemistry. *Collect. Czechoslov. Chem. Commun.* **2010**, *75*, 1149–1199. [CrossRef]
20. Mahfouz, N.; Abi Ghaida, F.; El Hajj, Z.; Diab, M.; Floquet, S.; Mehdi, A.; Naoufal, D. Recent Achievements on Functionalization within closo-Decahydrodecaborate $[B_{10}H_{10}]^{2-}$ Clusters. *Chem. Sel.* **2022**, *7*, e202200770.
21. Abi-Ghaida, F.; Clement, S.; Safa, A.; Naoufal, D.; Mehdi, A. Multifunctional Silica Nanoparticles Modified via Silylated-Decaborate Precursors. *J. Nanomater.* **2015**, *2015*, 608432. [CrossRef]
22. Abi-Ghaida, F.; Laila, Z.; Ibrahim, G.; Naoufal, D.; Mehdi, A. New triethoxysilylated 10-vertex closo-decaborate clusters. Synthesis and controlled immobilization into mesoporous silica. *Dalton Trans.* **2014**, *43*, 13087–13095. [CrossRef]
23. Diab, M.; Mateo, A.; Al Cheikh, J.; Haouas, M.; Ranjbari, A.; Bourdreux, F.; Naoufal, D.; Cadot, E.; Bo, C.; Floquet, S. Unprecedented coupling reaction between two anionic species of a closo-decahydrodecaborate cluster and an Anderson-type polyoxometalate. *Dalton Trans.* **2020**, *49*, 4685–4689. [CrossRef] [PubMed]
24. Izzet, G.; Volatron, F.; Proust, A. Tailor-made Covalent Organic-Inorganic Polyoxometalate Hybrids: Versatile Platforms for the Elaboration of Functional Molecular Architectures. *Chem. Rec.* **2017**, *17*, 250–266. [CrossRef] [PubMed]

25. Proust, A.; Matt, B.; Villanneau, R.; Guillemot, G.; Gouzerh, P.; Izzet, G. Functionalization and post-functionalization: A step towards polyoxometalate-based materials. *Chem. Soc. Rev.* **2012**, *41*, 7605–7622. [CrossRef] [PubMed]
26. Piot, M.; Hupin, S.; Lavanant, H.; Afonso, C.; Bouteiller, L.; Proust, A.; Izzet, G. Charge Effect on the Formation of Polyoxometalate-Based Supramolecular Polygons Driven by Metal Coordination. *Inorg. Chem.* **2017**, *56*, 8490–8496. [CrossRef]
27. Shelly, K.; Knobler, C.B.; Hawthorne, M.F. Synthesis of monosubstituted derivatives of closo decahydrododecaborate(2-). X-ray crystal structures of $[closo\text{-}2\text{-}B_{10}H_9CO]^-$ and $[closo\text{-}2\text{-}B_{10}H_9NCO]^{2-}$. *Inorg. Chem.* **1992**, *31*, 2889–2892. [CrossRef]
28. Mayer, C.R.; Fournier, I.; Thouvenot, R. Bis- and Tetrakis(organosilyl) Decatungstosilicate, $[\gamma\text{-}SiW_{10}O_{36}(RSi)_2O]^{4-}$ and $[\gamma\text{-}SiW_{10}O_{36}(RSiO)_4]^{4-}$: Synthesis and Structural Determination by Multinuclear NMR Spectroscopy and Matrix-Assisted Laser Desorption/Desorption Time-of-Flight Mass Spectrometry. *R. Chem. Eur. J.* **2000**, *6*, 105–110. [CrossRef]
29. Bonchio, M.; Carraro, M.; Scorrano, G.; Bagno, A. Photooxidation in Water by New Hybrid Molecular Photocatalysts Integrating an Organic Sensitizer with a Polyoxometalate Core. *Adv. Synth. Catal.* **2004**, *346*, 648–654. [CrossRef]
30. Modugno, G.; Monney, A.; Bonchio, M.; Albrecht, M.; Carraro, M. Transfer Hydrogenation Catalysis by a N-Heterocyclic Carbene (NHC) Iridium Complex on a Polyoxometalate Platform. *Eur. J. Inorg. Chem.* **2014**, *14*, 2356–2360. [CrossRef]
31. Carraro, M.; Modugno, G.; Fiorani, G.; Maccato, C.; Sartorel, A.; Bonchio, M. Organic-Inorganic Molecular Nano-Sensors: A Bis-Dansylated Tweezer-Like Fluoroionophore Integrating a Polyoxometalate Core. *Eur. J. Org. Chem.* **2012**, 281–289. [CrossRef]
32. Mayer, C.R.; Roch-Marchal, C.; Lavanant, H.; Thouvenot, R.; Sellier, N.; Blais, J.C.; Sécheresse, F. New Organosilyl Derivatives of the Dawson Polyoxometalate $[\alpha_2\text{-}P_2W_{17}O_{61}(RSi)_2O]^{6-}$: Synthesis and Mass Spectrometric Investigation. *Chem. Eur. J.* **2004**, *10*, 5517–5523. [CrossRef]
33. Maestre, J.M.; Lopez, X.; Bo, C.; Poblet, J.-M.; Daul, C. A DFT Study of the Electronic Spectrum of the α-Keggin Anion $[Co^{II}W_{12}O_{40}]^{6-}$. *Inorg. Chem.* **2002**, *41*, 1883–1888. [CrossRef] [PubMed]
34. Ravelli, D.; Dondi, D.; Fagnoni, M.; Albini, A.; Bagno, A. Predicting the UV spectrum of polyoxometalates by TD-DFT. *J. Comput. Chem.* **2011**, *32*, 2983–2987. [CrossRef] [PubMed]
35. Pogula, L.P.; Ramakrishna, D.S. Charge Distribution and UV Absorption Spectra of Liquid Crystals with Structural Element $[closo\text{-}B_{10}H_{10}]^{2-}$: A Theoretical Approach. *IARJSET* **2015**, *2*, 92–95.
36. Kemmegne-Mbouguen, J.C.; Floquet, S.; Zang, D.; Bonnefont, A.; Ruhlmann, L.; Simonnet-Jégat, C.; López, X.; Haouas, M.; Cadot, E. Electrochemical properties of the $[SiW10O36(M2O2E2)]6-$ polyoxometalate series (M = Mo(v) or W(v); E = S or O). *New J. Chem.* **2019**, *43*, 1146–1155.
37. Sivaev, I.B.; Prikaznov, A.V.; Anufriev, S.A. On relative electronic effects of polyhedral boron hydrides. *J. Organomet. Chem.* **2013**, *747*, 254–256. [CrossRef]
38. Boujtita, M.; Boixel, J.; Blart, E.; Mayer, C.R.; Odobel, F. Redox properties of hybrid Dawson type polyoxometalates disubstituted with organo-silyl or organo-phosphoryl moieties. *Polyhedron* **2008**, *27*, 688–692. [CrossRef]
39. Liu, R.J.; Zhang, G.J.; Cao, H.B.; Zhang, S.J.; Xie, Y.B.; Haider, A.; Kortz, U.; Chen, B.H.; Dalal, N.S.; Zhao, Y.S.; et al. Enhanced proton and electron reservoir abilities of polyoxometalate grafted on graphene for high-performance hydrogen evolution. *Energy Environ. Sci.* **2016**, *9*, 1012–1023. [CrossRef]
40. Keita, B.; Kortz, U.; Holzle, L.R.B.; Brown, S.; Nadjo, L. Efficient Hydrogen-Evolving Cathodes Based on Proton and Electron Reservoir Behaviors of the Phosphotungstate $[H_7P_8W_{48}O_{184}]^{33-}$ and the Co(II)-Containing Silicotungstates $[Co_6(H_2O)_{30}\{Co_9Cl_2(OH)_3(H_2O)_9(\beta\text{-}SiW_8O_{31})_3\}]^{5-}$ and $[\{Co_3(B\text{-}\beta\text{-}SiW_9O_{33}(OH))(B\text{-}\beta\text{-}SiW_8O_{29}OH)_2\}_2]^{22-}$. *Langmuir* **2007**, *23*, 9531–9953.
41. Fourmond, V.; Jacques, P.A.; Fontecave, M.; Artero, V. H$_2$ evolution and molecular electrocatalysts: Determination of overpotentials and effect of homoconjugation. *Inorg. Chem.* **2010**, *49*, 10338–10347. [CrossRef]

Article

Designed Syntheses of Three {Ni$_6$PW$_9$}-Based Polyoxometalates, from Isolated Cluster to Cluster-Organic Helical Chain

Chong-An Chen, Yan Liu and Guo-Yu Yang *

MOE Key Laboratory of Cluster Science, School of Chemistry and Chemical Engineering,
Beijing Institute of Technology, Beijing 102488, China; cca@bit.edu.cn (C.-A.C.); liuyanik@163.com (Y.L.)
* Correspondence: ygy@bit.edu.cn or ygy@fjirsm.ac.cn

Abstract: Three new hexa-Ni-substituted Keggin-type polyoxometalates (POMs), [Ni$_6$(OH)$_3$-(DACH)$_3$(H$_2$O)$_6$(PW$_9$O$_{34}$)]·31H$_2$O (**1**), [Ni(DACH)$_2$][Ni$_6$(OH)$_3$(DACH)$_3$(HMIP)$_2$(H$_2$O)$_2$(PW$_9$O$_{34}$)]·56 H$_2$O (**2**), and [Ni(DACH)$_2$][Ni$_6$(OH)$_3$(DACH)$_2$(AP)(H$_2$O)$_5$(PW$_9$O$_{34}$)]·2H$_2$O (**3**) (DACH = 1,2-Diami- nocyclohexane, MIP = 5-Methylisophthalate, AP = Adipate) were successfully made in the presence of DACH under hydrothermal conditions. **1** is an isolated hexa-Ni-substituted Keggin unit decorated by DACH. In order to further construct POM cluster-organic frameworks (POMCOFs) on the basis of **1**, by analyzing the steric hindrances and orientations of the POM units, the rigid HMIP and flexible AP ligands were successively incorporated, and another anionic monomeric POM **2** and the new 1D POM cluster organic chain (POMCOC) **3** were obtained. HMIP ligand still acts as a decorating group on the Ni$_6$ core of **2** but results in the different spatial arrangement of the {Ni$_6$PW$_9$} units. AP ligands in **3** successfully bridge adjacent isolated POM cluster units to 1D POMCOC with left-hand helices. The AP in **3** is the longest aliphatic carboxylic acid ligand in POMs, and the 1D POM cluster-AP helical chain represents the first 1D POMCOC with a helical feature.

Keywords: polyoxometalates; hydrothermal syntheses; cluster-organic frameworks; helical chain

1. Introduction

In the past century, polyoxometalates (POMs) have been widely researched for their abundant structures and applications in catalytic [1–3], magnetic [4], and electrical fields [5,6]. In order to enrich POMs' structural chemistry and further expand or optimize their applications, researchers have started to design and construct POM cluster organic frameworks (POMCOFs) [7–9] which is a new and promising branch of cluster organic frameworks (COFs) [10–12]. Since the POMCOF was reported [13], considerable efforts have been made in building POMCOFs with Keggin-/Anderson-/Lindqvist-POM secondary building units (SBUs) and rigid aromatic organic linkers [7,8,14,15]. However, compared with the traditional MOFs, the designed syntheses of POMCOFs are still facing huge challenges for the following two reasons: (1) POM clusters have large negative charges and oxygen-rich surfaces, which facilitate their bonding to metal cations, rather than the O-/N-donors from organic linkers. (2) POMs are rigid and stable clusters, therefore, the steric hindrance effects of POM SBUs and linkers need to be well-matched during assembly. Hence, how to choose proper POM SBUs and organic linkers is the key to constructing POMCOFs.

Among seven typical types of POMs, only Anderson-/Lindqvist-/Keggin-types have been successfully applied as SBUs in POMCOFs. Since 2016, the first Anderson-type POM-based heterometallic cluster organic framework was made; Anderson-type POMs have become the popular choice for SBUs [8]. The combination of Anderson-type SBU and rigid bifunctional tris(alkoxo) ligand with a pyridyl group opens up the gate of Anderson-type POMCOFs' world. Lindqvist-type POMs are important members of the POMs family. Though five different elements can all produce the Lindqvist-type [M$_6$O$_{19}$]$^{n−}$ (M = V^V, Nb^V, Ta^V, Mo^{VI}, W^{VI}) cluster, only polyoxovanadates have been successfully applied as

SBUs in Lindqvist-type POMCOFs [15,16]. So far, most of the reported POMCOFs are made with Keggin-type POM SBUs [7,13,14,17–20]. In these POMCOFs, most of the SBUs are saturated {ε-$M_4PMo_{12}O_{40}$} (M=La, Zn) [13,14,17–19], of which, the incorporation of M (M=Zn^{2+}, La^{3+}) provide the easier bonding sites than the saturated {PMo_{12}} units for organic linkers. Our group has long been devoted to transition metal substituted POMs (TMSPs) based on the trilacunary Keggin fragments under hydrothermal conditions. From our perspective, the trilacunary sites of the [XW_9O_{34}] (P, W, Ge) unit can act as structure-directing agents (SDAs) to induce transition metal ions' aggregation to cluster, on which the terminal end of water molecules may facilitate the substitutions of organic linkers in constructing Keggin-type POMCOFs. Since the first hexa-Ni^{II} substituted TMSP based on trilacunary Keggin fragments was made [4], we have been working on POMs structural chemistry based on hexa-Ni^{II}-substituted POMs and have already mastered the synthetic conditions of hexa-Ni^{II} substituted Keggin POMs. By using {Ni_6PW_9} SBUs and rigid aromatic carboxylate ligands, we have built a series of novel Keggin-type POMCOFs [7,20]. Hence, we believe that some other intriguing POMCOFs can be made by using {Ni_6PW_9} SBUs with proper organic linkers.

Rigid and semi-rigid aromatic carboxylate ligands are the common linkers being used in making POMCOFs [7,20–22]; their rigid structures are favorable for the stabilization of the frameworks. However, the large steric hindrance effects of POMs and rigid ligands sometimes cannot match to form POMCOFs. To overcome this difficulty, aliphatic dicarboxylic acid may be a potential candidate due to its smaller steric hindrances and better flexibilities, which may produce some intriguing frameworks with helical or interpenetrating features that cannot be obtained with rigid aromatic ligands. However, little relevant research has been made, including two typical examples containing aliphatic dicarboxylic acid-bridges for a 2D POMCOF and a tetramer [23,24]. Hence, in this work, we first made an isolated hexa-Ni-substituted Keggin-type POM [$Ni_6(OH)_3$-$(DACH)_3(H_2O)_6(PW_9O_{34})$]·$31H_2O$ (1) under hydrothermal conditions. The abundant terminal water molecules on the Ni_6 cores are potential substitution sites for organic linkers, which help us to further construct POMCOFs. When we first applied the rigid carboxylate ligand MIP, another hexa-Ni-substituted Keggin-type monomer [$Ni(DACH)_2$]-[$Ni_6(OH)_3(DACH)_3(HMIP)_2$-$(H_2O)_2(PW_9O_{34})$]·$56H_2O$ (2) was obtained, HMIP ligands still decorate on the Ni_6 cores, failing to bridge the POM clusters. By analyzing the structure of 2, we used the aliphatic dicarboxylate AP ligands as a linker and a new 1D POMCOC with helical chains [$Ni(DACH)_2$][$Ni_6(OH)_3(DACH)_2(AP)(H_2O)_5(PW_9O_{34})$]·$2H_2O$ (3) was made. To the best of our knowledge, the AP in 3 is the longest aliphatic dicarboxylic acid being incorporated in POMs. Moreover, the 1D helical chain of 3 is the first 1D POMCOC with helical features.

2. Experimental Section

2.1. General Procedure

All the reagents were analytical grade and used without any further purification. Na_{10}[A-α-PW_9O_{34}]·$7H_2O$ was prepared by a method from the literature [25]. Meso-form DACH was used in the syntheses. The powder X-Ray diffraction (PXRD) patterns of the three compounds were collected on a Bruker D8 Advance X-ray diffractometer (Bruker, Karlsruhe, Germany) with Cu Kα radiation (λ = 1.54056 Å) and 2θ scanning from 5–50°. UV-Vis absorption spectra were obtained on a Shimadzu UV3600 spectrometer (Shimadzu, Kyoto, Japan) with wavelengths from 190 to 800 nm. IR spectra were recorded on a Nicolet iS10 FT-IR spectrometer (Thermo Fisher Scientific, Waltham, MA, USA) with the wavenumbers ranging from 4000 to 400 cm^{-1}. Thermogravimetric analyses were conducted on a Mettler Toledo TGA/DSC 1100 analyzer (Mettler Toledo, Zurich, Switzerland) heating up from 25–1000 °C (heating rate: 10 °C/h) under an air atmosphere. Elemental analyses proceeded on the EuroEA3000 elemental analyzer (EuroVector, Pavia, Italy).

2.2. Syntheses

2.2.1. Synthesis of 1

Na$_9$[A-α-PW$_9$O$_{34}$]·7H$_2$O (0.320 g, 0.125 mmol) and NiCl$_2$·6H$_2$O (0.820 g, 3.44 mmol) were stirred in 9 mL 0.5 mol/L sodium acetate buffer (pH = 4.8) for 10 min; then, 3 mL DACH (Scheme 1a) was slowly dropped in and continually stirred for 30 min. The resulting solution was sealed in a 25 mL Teflon-lined stainless-steel autoclave and heated at 170 °C for 5 days. After cooling down to room temperature and washing with distilled water, green rod-like crystals were obtained with a yield of 34% (based on Na$_9$[A-α-PW$_9$O$_{34}$]·7H$_2$O). Elemental analysis calcd (%): C, 5.93; H, 3.26; and N, 2.30 (based on [Ni$_6$(OH)$_3$(DACH)$_3$(H$_2$O)$_6$(PW$_9$O$_{34}$)]·31H$_2$O). Found: C, 7.10; H, 2.17; N, 2.81. IR(KBr, cm^{-1}): 3428(s), 3332(w), 3280(w), 2929(w), 2856(w), 1628(w), 1588(w), 1449(w), 1377(w), 1231(w), 1119(w), 1039(s), 941(vs), 842(vs), 796(vs), and 716(vs).

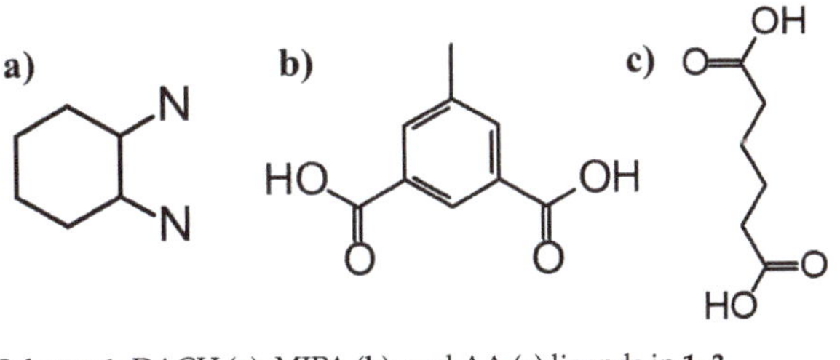

Scheme 1. DACH (a), MIPA (b), and AA (c) ligands in **1–3**.

2.2.2. Syntheses of 2 and 3

The synthetic procedures of **2** and **3** were the same as **1**, except for the adding of MIPA (5-Methylisophthalic Acid, Scheme 1b) (0.200 g, 1.11 mol) and AA (Adipic Acid, Scheme 1c) ligands (0.200 g, 1.37 mol) for **2** and **3**, with the yield of 36% and 28%, respectively. Based on Na$_9$[A-α-PW$_9$O$_{34}$]·7H$_2$O. Elemental analysis calcd (%) for **2**: C, 12.34; H, 4.35; and N, 3.00 (based on [Ni(DACH)$_2$][Ni$_6$(OH)$_3$(DACH)$_3$(HMIP)$_2$(H$_2$O)$_2$(PW$_9$O$_{34}$)]·56 H$_2$O). Found: C, 14.10; H, 3.37; N, 3.39. IR(KBr, cm^{-1}): 3458(s), 3338(w), 3264(w), 2929(w), 2859(w), 1562(w), 1361(w), 1033(w), 935(vs), 841(vs), 796(vs), and 715(vs). Elemental analysis calcd (%) for **3**: C, 10.53; H, 2.36; and N, 3.27 (based on [Ni(DACH)$_2$][Ni$_6$(OH)$_3$(DACH)$_2$(AP)(H$_2$O)$_5$(PW$_9$O$_{34}$)]·2H$_2$O). Found: C, 10.44; H, 2.63; and N, 3.34. IR(KBr, cm^{-1}): 3423(s), 3240(w), 2923(w), 2860(w), 1591(w), 1546(w), 1413(w), 1037(s), 943(vs), 848(vs), 796(vs), and 710(vs).

2.3. X-ray Crystallography

The single-crystal diffraction data of **1–3** were collected on a Gemini A Ultra CCD diffractometer with graphite monochromated Mo Kα (Λ = 0.71073 Å) radiation at 296(2) K. The structures were solved by direct methods and refined by the full-matrix least-squares fitting on F^2 method with the SHELX-2008 program package [26]. Anisotropic displacement parameters were refined for all atomic sites except for some disordered atoms. The contribution of the disordered solvent molecules in **1** and **2** was treated with the SQUEEZE method in PLATON (Utrecht University, Utrecht, The Netherlands). In the refinements, 0, 1, and 2 lattice water molecules were found for **1–3** from the Fourier maps, respectively. Based on the potential solvent-accessible voids and electron counts from the SQUEEZE reports, there were 31 and 55 lattice water molecules removed for **1** and **2**, respectively. According to the elemental analyses and TGA, there are 27 and 34 lattice water molecules lost from efflorescence in **1** and **2**, respectively. In **3**, 4 absorbed water molecules were found. Basic crystallographic data and structural refinement data are listed in Table 1. Detailed crystallographic data have been deposited on the Cambridge Crystallographic Data Centre: CCDC 2171028 (for **1**), 2170965 (for **2**), and 2170966 (for **3**). These data can be obtained free of charge via http://www.ccdc.cam.ac.uk/conts/retrieving.html accessed on 29 June 2022 or from the Cambridge Crystallographic Data Centre, 12 Union Road, Cambridge CB2 1EZ, UK; Fax: +44-1223-336-033; or email: deposit@ccdc.cam.ac.uk.

Table 1. Crystallographic data and structural refinements for **1–3**.

	1	2	3
Formula	$Ni_6PW_9O_{74}C_{18}H_{119}N_6$	$Ni_7PW_9O_{103}C_{48}H_{203}N_{10}$	$Ni_7PW_9O_{48}C_{30}H_{81}N_8$
Molecular weight	3642.06	4665.78	3418.61
Crystal system	Trigonal	Monoclinic	Orthorhombic
Space group	$P\text{-}3c1$	$P2_1/c$	$P2_12_12_1$
$a/\text{Å}$	18.1775	24.9850	36.6105
$b/\text{Å}$	18.1775	14.9500	14.1605
$c/\text{Å}$	21.3261	26.6410	13.8492
$\alpha/°$	90	90	90
$\beta/°$	90	98.697	90
$\gamma/°$	120	90	90
$V/\text{Å}^3$	6102.5	9837	7179.7
Z	4	4	4
$Dc/\text{g cm}^{-3}$	3.964	3.151	3.163
μ/mm^{-1}	18.880	11.956	16.263
$F(000)$	6840	9048	6312
Goodness-of-fit on F^2	1.075	1.099	1.131
R indices $[I > 2\sigma(I)]$ [1]	0.0355 (0.0946)	0.0633 (0.1624)	0.0601 (0.1350)
R indices (all data)	0.0504 (0.1043)	0.0996 (0.1777)	0.0704 (0.1413)

[1] $R_1 = \Sigma||F_0| - |F_c||/\Sigma|F_0|$. $wR_2 = \{\Sigma w[(F_0)^2 - (F_c)^2]^2/\Sigma w[(F_0)^2]^2\}^{1/2}$.

3. Result and Discussion

3.1. Structure of 1 and Designed Syntheses for 2

X-ray diffraction analyses reveal that **1** crystallizes in the trigonal space group $P\text{-}3c1$, consisting of the neutral $[Ni_6(\mu_3\text{-OH})_3(DACH)_3(H_2O)_6(PW_9O_{34})]$ (**1a**, Figure 1a) cluster. **1a** can be seen as the classical trilacunary Keggin $[B\text{-}\alpha\text{-}PW_9O_{34}]^{9-}$ fragment being capped by a triangular $[Ni_6(\mu_3\text{-OH})_3]^{9+}$ cluster. Due to the trigonal C_3 symmetry of **1**, there are only two independent Ni^{2+} in the Ni_6 cluster (Figure S1, Supplementary Materials). Each Ni1 and Ni2 interconnect with each other by edge-sharing, producing three edge-sharing $\{Ni_3O_4\}$ truncated cubanes. Three $Ni1O_6$ octahedra locate on the three lacunary sites of the $\{PW_9\}$ unit, while three $Ni2O_4N_2$ octahedra are on the three vertexes of the triangular Ni_6 cluster, further decorated by three DACH ligands, respectively (Figure 1b). According to BVS calculations [27], the bond valance of μ_3-O4 is 1.12, indicating its protonation. **1a** exhibits two opposite orientations, which are alternately arranged with a shoulder-to-shoulder arrangement along the a-axis and [110] direction (Figure 1c,d). Such arrangements construct the snowflake-like supramolecular channels with S_6 symmetry and are the hydrophobic voids as well (Figure 1e).

The presence of six terminal water molecules on the Ni_6 cluster provides abundant substituted sites for organic ligands. We started to incorporate organic ligands into the reaction system of **1**, attempting to construct POMCOFs with proper organic linkers. In our previous work, we have successfully made two 1D POMCOCs $\{[Ni_6(OH)_3(H_2O)_2\text{-}(enMe)_3(PW_9O_{34})](1,3\text{-bdc})\}[Ni(enMe)_2]\cdot 4H_2O$ (**4**, enMe = 1,2-diaminopropane, 1,3-bdc = 1,3-benzenedicarboxylate acid) and $\{[Ni_6(OH)_3(H_2O)(en)_4(PW_9O_{34})](Htda)\}\cdot H_3O\cdot 4H_2O$ (**5**, en = ethylenediamine, tda = thiodiglycolic acid) based on $\{Ni_6PW_9\}$ SBUs and V-type rigid dicarboxylate ligands (1,3-bdc and tda) [7]. To analyze these structures carefully, we found that in **4** and **5**, $\{Ni_6PW_9\}$ SBUs are arranged in shoulder-to-shoulder and face-to-face modes, respectively, which are further bridged by the V-type dicarboxylate ligands to 1D POMCOCs. In **1**, though the opposite-orientated $\{Ni_6PW_9\}$ units exhibit shoulder-to-shoulder arrangements along the a-axis, the interunit distances are too close to accommodate the organic ligands. Hence, we choose the similar V-type ligand MIP to see if the methyl group can further spread out the opposite orientated POM units and if the carboxyl groups can bridge adjacent same orientated units to 1D chains at the same time. By adding MIPA into the reaction of **1**, **2** was obtained. The observation of **2** confirms part of our speculations; though HMIP still acts as a decoration group, it changes the orientations of adjacent POM units such that two different orientated units both arrange in shoulder-to-shoulder modes separately with moderate interunit distances.

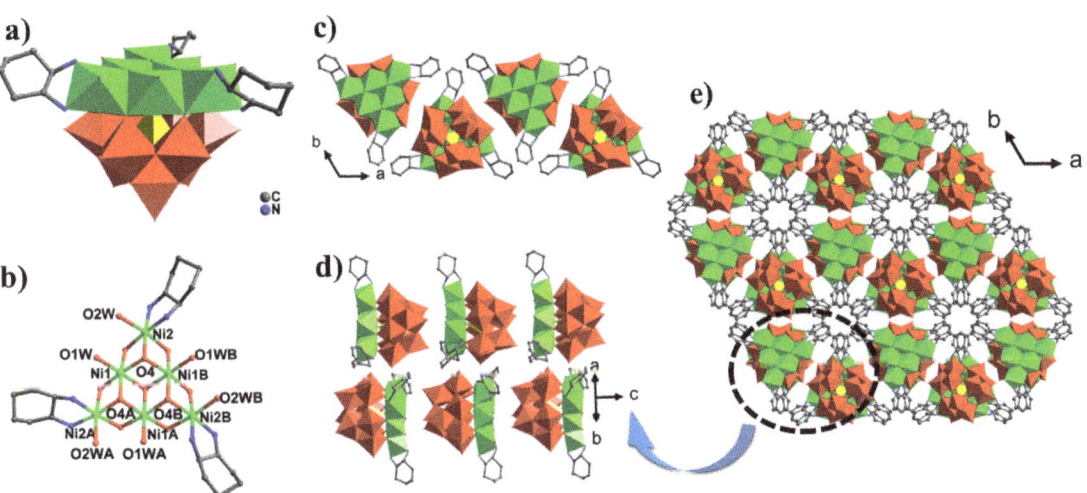

Figure 1. (**a**) A polyhedral view of polyoxoanion **1a**.; (**b**) View of the Ni$_6$ cluster in **1a**; (**c**,**d**) Shoulder-to-shoulder arrangements of **1a** with opposite orientations along the *a*-axis and [110] direction; and (**e**) The 3D supramolecular framework of **1**. Color code of polyhedral: WO$_6$: red; NiO$_6$/NiO$_4$N$_2$: green; and PO$_4$: yellow. Hydrogen atoms of the ligands are not shown for better clarity.

3.2. Structure of 2 and Designed Syntheses for 3

X-ray diffraction analyses reveal that **2** crystallizes in its monoclinic space group $P2_1/c$. Its polyoxoanionic cluster [Ni$_6$(OH)$_3$(DACH)$_3$(HMIP)$_2$(H$_2$O)$_2$(PW$_9$O$_{34}$)] (**2a**) is similar to that of **1a**, except for four water molecules in **1a** being replaced by two HMIP ligands in **2a** (Figure 2a,b and Figure S1). This difference makes **2a** an anionic cluster, accompanied by the charge-balancing [Ni(DACH)$_2$]$^{2+}$ complex, in which Ni^{2+} exhibit the planar square coordination geometry (Figure S2, Supplementary Materials).

Due to the large steric hindrance of HMIP, adjacent opposite-orientated POM clusters are spread out and adopt face-to-face arrangements with each other, while the same orientated units still maintain shoulder-to-shoulder arrangements (Figure 2c,d), which are ideal arrangements for making POMCOFs based on our previous research [7,20]. Using another organic linker with a longer length may help to achieve our aims, but the longer length corresponds to the larger steric hindrances, which may affect the orientations of POM units or increase the interunit distances. Rigid aromatic carboxylic ligands seem unlikely to satisfy our design. Hence, we transfer our focus to chainlike aliphatic dicarboxylic acids. Their higher flexibilities may facilitate their bridging functions on POM SBUs with more flexible orientations and interunit distances and may further result in some intriguing interpenetrating or helical structures that cannot be obtained with rigid aromatic carboxyl ligands. We found that the bilateral DACH molecules on each Ni$_6$ cluster prevent the bridging of adjacent same-orientated SBUs with organic linkers (Figure 2d,e). Additionally, the distance between two terminal –COOH groups from adjacent opposite-orientated POM SBUs is 6.20 Å (Figure 2e), which is nearly matchable with that of AP in the reported polymers (6.30 Å, Figure 2f) [28]. Using AP to replace HMIP ligand in **2** may achieve our goals. Based on the above considerations, AP was used as a linker in the synthesis of **3**. Under similar synthetic conditions with **1** and **2**, **3** was obtained. AP ligand successfully bridges adjacent opposite orientated POM cluster units to the unpreceded 1D helical chains.

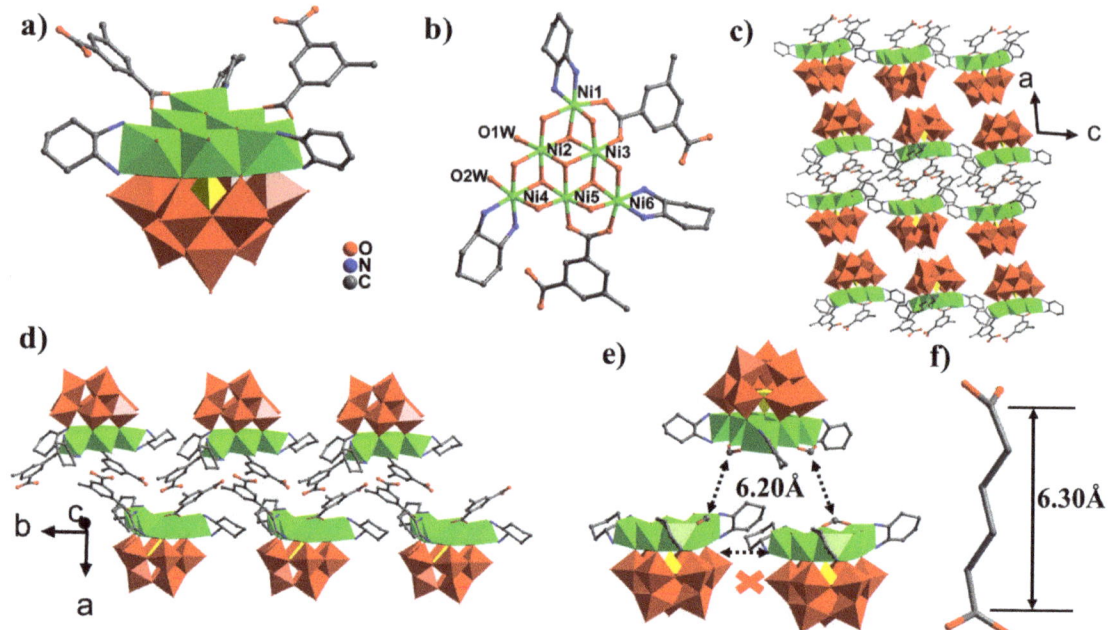

Figure 2. (**a**) A polyhedral view of polyoxoanion **2a**.; (**b**) View of the Ni$_6$ cluster in **2a**; (**c**) Spatial arrangements of **2a** along with the *a*-and *c*- axes; (**d**) Spatial arrangements of **2a** along the *b*-axis; (**e**) Interunit distances of **2a**; and (**f**) Matchable distance of AP ligand. Color code of polyhedral: WO$_6$: red; NiO$_6$/NiO$_4$N$_2$: green; and PO$_4$: yellow. Hydrogen atoms of the ligands are not shown for better clarity.

3.3. Structure of 3

3 crystallizes in the orthorhombic space group *P*2$_1$2$_1$2$_1$. Its asymmetric unit contains a [Ni$_6$(OH)$_3$(DACH)$_2$(AP)(H$_2$O)$_5$(PW$_9$O$_{34}$)] (**3a**) cluster (Figure 3a), a [Ni(DACH)$_2$]$^{2+}$ complex, and two lattice water molecules (Figure S1, Supplementary Materials). Compared with **1a** and **2a**, only four-terminal water molecules are substituted by two bidentate DACH ligands on the Ni$_6$ cluster of **3a** (Figure 3b). Each Ni$_6$ cluster links with two AP, of which, one terminal carboxyl group of the AP replaces two terminal water molecules on Ni5 and Ni6, while another carboxyl group replaces only one water molecule on Ni1 (Figure 3b). Each AP ligand bridges two Ni$_6$ clusters (Figure 3c). Such substitution and linkage successfully construct the 1D helical chain with left-hand helices around a 2$_1$-screw axis (Figure 3d,e). Adjacent 1D chains stack in -ABAB- and -AAA- sequences along the *a*-and *c*-axis, respectively (Figure 3f,g). It is worth noting that the orientation of each POM SBU and interunit distance have been continually adjusted to the face-to-shoulder arrangements with shorter interunit distance to match the linkage of AP ligand, which are different from those in rigid dicarboxylate ligand-bridged POMCOFs. Such special arrangements of POM SBUs, and the good flexibility of AP, synergistically contribute to the 1D helical chains with left-hand helices. Similar to that in **2**, [Ni(DACH)$_2$]$^{2+}$ complexes with planar square configuration locate interchain to compensate for the negative charges of the chains (Figure S2, Supplementary Materials).

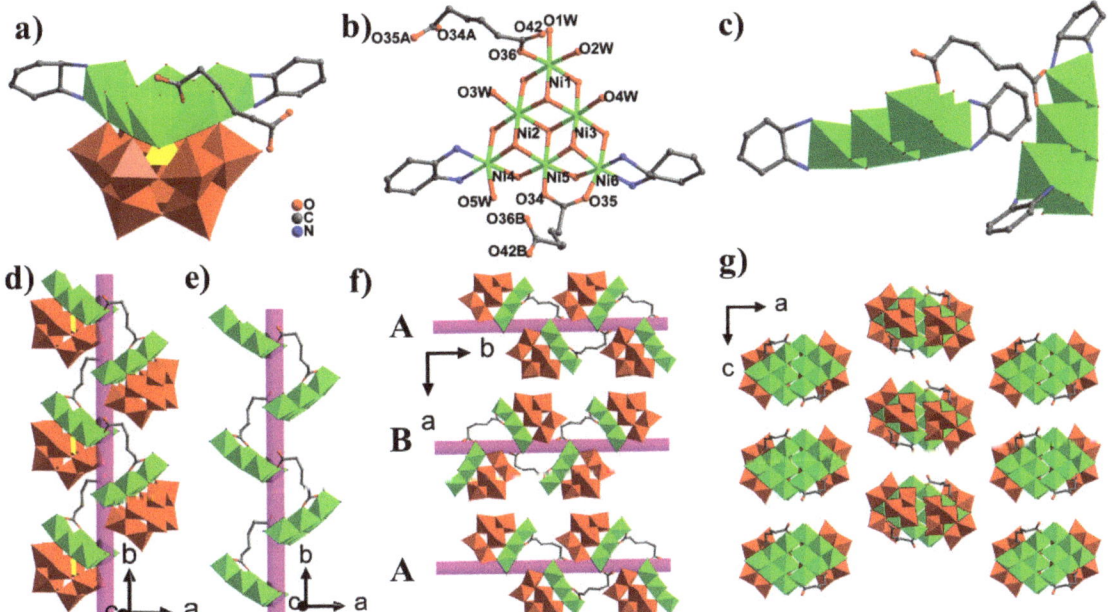

Figure 3. (a) A polyhedral view of polyoxoanion **3a**.; (b) View of the Ni$_6$ cluster and its' linkage with AP; (c) The linkage of AP with adjacent Ni$_6$ cluster; (d) View of the 1D helical chains with left-hand helices along the *b*-axis; (e) The simplified sketch of the Ni$_6$-AP 1D chain; and (**f**,**g**) The -ABAB- and -AAA- stacking modes along *a*-and and *c*-axis, respectively. Color code of polyhedral: WO$_6$: red; NiO$_6$/NiO$_4$N$_2$: green; and PO$_4$: yellow. Hydrogen atoms of the ligands are not shown for better clarity.

3.4. Structural Comparisons

In TMSPs' abundant structural chemistry, POM clusters have various linkages with each other to generate different 1D/2D/3D structures:

First, the interconnections of POM clusters (including different structural types) and rigid aromatic organic ligands. This linkage produces most of the 3D POMCOFs, while 1D chains and 2D layers are relatively rare through this connection, except for these three examples: the 1D chains built from the {Ni$_6$PW$_9$} unit and 1,3-bdc, tda ligand (Figure 4a,b) [7], respectively, and the layer made by another ethylenediamine-func- tionalized {Ni$_6$PW$_9$} unit and 1,3-bdc ligand (Figure 4c) [20].

Second, the interconnections of TMSP cluster units through TM-O=W bonds. This linkage generates a series of 1D chains and 2D layers [29,30]. The 3D open frameworks constructed by the pure TM–O=W linkage are only observed in Cu^{II}-substituted TMSPs, including [{Cu$_6$(μ$_3$-OH)$_3$(en)$_3$(H$_2$O)$_3$}(B-α-PW$_9$O$_{34}$)]·7H$_2$O and [Cu$_6$(μ$_3$-OH)$_3$(en)$_3$(H$_2$O)$_3$(B-α-PW$_9$O$_{34}$)]·4H$_2$O (Figure 4d), which are caused by the unique Jahn–Teller effect of CuO$_4$N$_2$ octahedra with the axial elongation [31,32].

Third, the TMSP frameworks with TM complex-bridges. TM complex-bridges are common in TMSPs' frameworks. They can extend the POM units to 1D/2D/3D frameworks through TM–O=W, TM–O–TM, and TM–N···N-TM linkages [33–36].

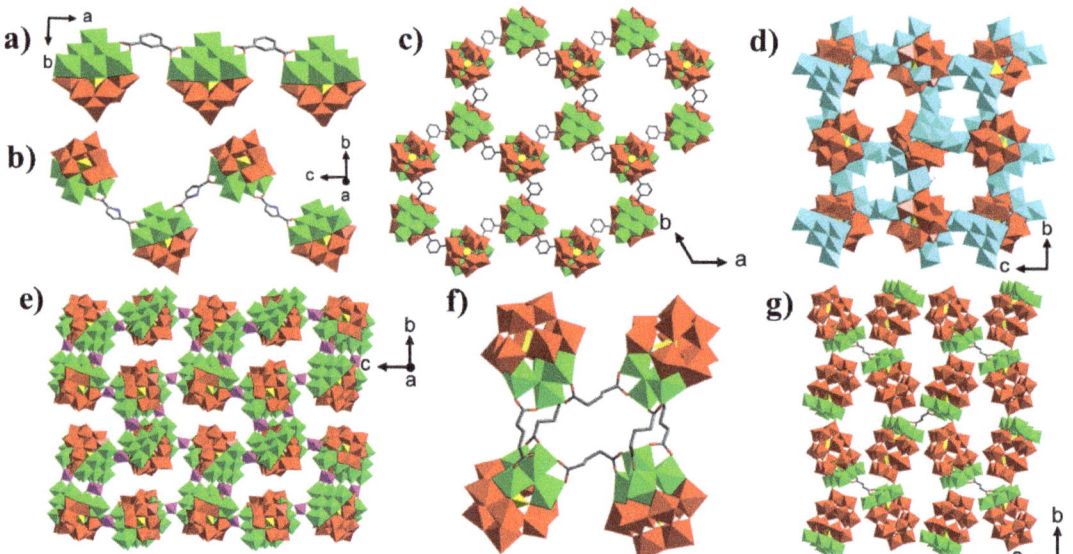

Figure 4. (**a**) One-dimensional chain built from {Ni$_6$PW$_9$} unit and 1,3-bdc ligand; (**b**) One-dimensional chain built from {Ni$_6$PW$_9$} unit and tda ligand; (**c**) Two-dimensional layer built from {Ni$_6$PW$_9$} unit and 1,3-bdc ligand; (**d**) Three-dimensional framework built from the interconnection of {Cu$_6$PW$_9$} unit through Cu–O=W linkage; (**e**) Three-dimensional framework built from {Ni$_6$PW$_9$} unit and WO$_4$ tetrahedron; (**f**) Tetramer built from {Ni$_4$SiW$_9$} and glutaric acid linker; and (**g**) Two-dimensional layer built from Dawson-type {Ni$_6$P$_2$W$_{15}$} unit and succinic acid linker. Color code of polyhedral: WO$_6$: red; NiO$_6$/NiO$_4$N$_2$: green; PO$_4$: yellow; CuO$_6$: light blue; and WO$_4$: purple. Hydrogen atoms of the ligands are not shown for better clarity.

Fourth, is the TMSP framework with WO$_4$ bridges. However, to the best of our knowledge, it was only found in the first chiral 3D framework of [Ni(enMe)$_2$]$_3$[WO$_4$]$_3$-[Ni$_6$(enMe)$_3$(OH)$_3$PW$_9$O$_{34}$]$_2$·9H$_2$O (Figure 4e) [37].

Compared with these TMSP-based frameworks with four different linkages, the 1D helical chains in **3** represent a new structural type of POMCOCs. Aliphatic dicarboxylic acid ligands are rare not only in POMCOFs but also in POMs. Limited evidence includes the glutaric acid-bridged tetramer [{(SiW$_9$O$_{34}$)Ni$_4$(OH)$_3$}$_4$(OOC(CH$_2$)$_3$COO)$_6$] (Figure 4f) and the succinic acid-bridge hexa-substituted Dawson-type layer [Ni$_6$(µ$_3$-OH)$_3$-(dap)$_2$(en)(H$_2$O){OOC(CH$_2$)$_2$COO}$_{0.5}$(CH$_3$COO)(P$_2$W$_{15}$O$_{56}$)] (Figure 4g) [23,24]. The AP in **3** is the longest aliphatic dicarboxylic acid being incorporated in POMs family. Moreover, it differs from those 1D chains with a TM–O=W linkage and 1D POMCOCs featuring strict chains [7,29]; the 1D helical chains in **3** are the first 1D POMCOC with helical features.

Since the hexa-Ni^{II} cluster of **1–3** is similar to those in the reported hexa-Ni-substituted TMSPs, we compared their bond lengths and bond angles to speculate the magnitude properties of the title compounds. As shown in Table S1 (Supplementary Materials), the Ni–O bond lengths and Ni–O–Ni bond angles of **1–3** are in the ranges of 1.915–2.295 Å and 90.9–114.2°, respectively. According to the previous research [4,7,31,38,39], when the Ni–O–Ni bond angles are in the range of 90–104°, ferromagnetic exchange interactions are dominant. When Ni–O–Ni bond angles are larger than 104°, anti-ferromagnetic exchange interactions may exist. When ferromagnetic and antiferromagnetic behaviors coexist, the overall magnetic behaviors are determined by which one is dominant. Normally, most of the Ni–O–Ni bond angles in the hexa-Ni^{II} cluster are in the ferromagnetic dominant ranges when ferromagnetic and antiferromagnetic coupling coexistences appear. Ferromagnetic exchange behaviors are expected for hexa-Ni^{II} clusters, which have been proved by the measurements in our previous research [4,7,31,38,39]. In **1**, since all the Ni–O–Ni bond angles are in the range of 92.5–102.1°, ferromagnetic exchange behaviors are expected. In **2** and **3**, the Ni–O–Ni bond angles are in the range of 90.9–106.8° and 91.6–114.2°, respectively, indicating the coexistences of ferromagnetic and antiferromagnetic couplings. There are only 1 and 2 Ni–O–Ni bond angles larger than 104°, indicating that the ferromagnetic exchange behaviors are dominant in **2** and **3**, similar to those reported in hexa-Ni^{II}-substituted TMSPs.

3.5. Powder XRD Patterns

As shown in Figure S3 (Supplementary Materials), the experimental PXRD patterns of **1–3** were all consistent with the simulated patterns obtained from single-crystal data, which confirm the purities of the samples. The differences in the intensities were attributable to the preferred orientations.

3.6. IR Spectra

As shown in Figure S4 (Supplementary Materials), the IR spectra of **1–3** show a series of similar absorption bands ranging from 4000–400 cm^{-1}. The wide absorption bands from 3528 to 3134 cm^{-1} are assigned to the stretching vibrations of the –OH, –CH$_2$, and –NH$_2$ groups. The sharp absorption peaks from 3005 to 2817 cm^{-1} are the stretching vibrations of the –CH$_2$ and –NH$_2$ groups. The peaks ranging from 1652–1363 cm^{-1} are the characteristic peaks of the bending vibrations of –NH$_2$ and –CH$_2$ groups in **1–3** and the carboxylate groups of the carboxylate ligands in **2–3**. Four intense characteristic peaks of v(P–O), v(W–O$_t$), v(W–O$_b$), and v(W–O$_c$) of the Keggin-fragments are observed at 1039, 941, 842, 796, and 716 cm^{-1} for **1**, 1033, 935, 841, 796, and 715 cm^{-1} for **2**, and 1037, 943, 848, 796, and 710 cm^{-1} for **3**, respectively.

3.7. UV-Vis Absorption Spectra

In order to investigate the optical properties of the title compounds, UV-Vis absorption and optical diffuse reflectance spectra of **1–3** were obtained in the wavelength range of 190–800 nm. As shown in Figure S5 (Supplementary Materials), the optical band gaps of **1–3** are 2.60, 2.59, and 2.56 eV, respectively, which are comparable to other Ni$_6$-substitute POMs, including [Ni$_6$(μ_3-OH)$_3$(en)$_2$(dien)(H$_2$O)$_5$(B-α-PW$_9$O$_{34}$)]·3H$_2$O (2.42 eV), [Ni$_6$(μ_3-OH)$_3$(dap)$_2$(py)$_6$- (H$_2$O)(B-α-PW$_9$O$_{34}$)]·H$_2$O (2.37 eV), and [Ni(en)$_2$][Ni$_6$(μ_3-OH)$_3$(en)$_3$(1,3-bdc)(H$_2$O)$_2$(B-α- PW$_9$O$_{34}$)]·9H$_2$O (2.53 eV) [20,29]. It was found that the band gaps of **1–3** are in the order of **3** < **2** < **1**, which conforms to the band gaps of the compounds decreasing with the increasing dimensionality or complexity of the structures, as proposed by Kanatzidis and Papavassiliou [40].

4. Conclusions

In summary, three new TMSPs containing {Ni$_6$PW$_9$} units were designed and synthesized from monomers to 1D POMCOC under hydrothermal conditions. **1** is a monomer with DACH molecules decorating the Ni$_6$ cluster. In order to construct POMCOF on the basis of **1**, the rigid aromatic MIP ligand was first incorporated and the anionic monomeric POM **2** was obtained. HMIP still acts as a decorating group on the Ni$_6$ cluster but fails to bridge adjacent {Ni$_6$PW$_9$} units. By analyzing the orientations and steric hindrance between adjacent {Ni$_6$PW$_9$} units of **2**, the aliphatic AP ligand was purposely chosen to replace HMIP on the base of **2**, which resulted in the formation of **3**, a new 1D POMCOC with novel helical chain. Owing to the good flexibility of the AP linker, **3** represents the first 1D POMCOC with a helical chain. This work is an example of our continued work of constructing POMCOFs with hexa-Ni^{II} substituted TMSP SBUs. The successfully designed syntheses from **1** to **3** provide us with a new strategy of using chainlike dicarboxylate acid as a linker to make POMCOFs, which may lead to some intriguing structures that cannot be found with rigid aromatic linkers. Further works with this strategy are in progress.

Supplementary Materials: The following supporting information can be downloaded at: https://www.mdpi.com/article/10.3390/molecules27134295/s1, Table S1: Comparisons of the bond lengths and bond angles in Ni_6-substi-tuted TMSPs; Figure S1: Asymmetric units of **1**–**3**; Figure S2: $[Ni(DACH)_2]^{2+}$ complex in **2** and **3**; Figure S3: PXRD of **1**–**3**. Figure S4: IR spectra of **1**–**3**; Figure S5: UV-Vis spectra of **1**–**3**; Figure S6: TG curves of **1**–**3**. [41–43] are cited in the Supplementary Materials.

Author Contributions: Conceptualization, C.-A.C.; methodology, C.-A.C. and Y.L.; formal analysis, C.-A.C. and Y.L.; data curation, Y.L.; writing—original draft, preparation, C.-A.C.; writing—reviewing and editing, supervision, and funding acquisition: G.-Y.Y. All authors have read and agreed to the published version of the manuscript.

Funding: This research was funded by the National Natural Science Foundation of China, grant numbers 21831001, 21571016, 91122028 and 20725101.

Institutional Review Board Statement: Not applicable.

Informed Consent Statement: Not applicable.

Data Availability Statement: Not applicable.

Conflicts of Interest: The authors declare no conflict of interest.

Sample Availability: Samples of the compounds are available from the authors.

References

1. Zheng, S.-T.; Yang, G.-Y. Recent advances in paramagnetic-TM-substituted polyoxometalates (TM = Mn, Fe, Co, Ni, Cu). *Chem. Soc. Rev.* **2012**, *41*, 7623–7646. [CrossRef] [PubMed]
2. Wang, S.-S.; Yang, G.-Y. Recent advances in polyoxometalate-catalyzed reactions. *Chem. Rev.* **2015**, *115*, 4893–4962. [CrossRef] [PubMed]
3. Liu, J.X.; Zhang, X.B.; Li, Y.L.; Huang, S.L.; Yang, G.Y. Polyoxometalate functionalized architectures. *Coord. Chem. Rev.* **2020**, *414*, 213260–213275. [CrossRef]
4. Zheng, S.-T.; Yuan, D.-Q.; Jia, H.-P.; Zhang, J.; Yang, G.-Y. Combination between lacunary polyoxometalates and high-nuclear transition metal clusters under hydrothermal conditions: I. from isolated cluster to 1-D chain. *Chem. Commun.* **2007**, *18*, 1858–1860. [CrossRef]
5. Pichon, C.; Mialane, P.; Dolbecq, A.; Marrot, J.; Rivière, E.; Bassil, B.S.; Kortz, U.; Keita, B.; Nadjo, L.; Sécheresse, F. Octa- and Nonanuclear nickel(II) polyoxometalate clusters: Synthesis and electrochemical and magnetic characterizations. *Inorg. Chem.* **2008**, *47*, 11120–11128. [CrossRef]
6. Ibrahim, M.; Xiang, Y.; Bassil, B.S.; Lan, Y.H.; Powell, A.K.; Oliveira, P.D.; Keita, B.; Kortz, U. Synthesis, magnetism, and electro-chemistry of the Ni_{14}- and Ni_5-Containing heteropolytungstates $[Ni_{14}(OH)_6(H_2O)_{10}(HPO_4)_4(P_2W_{15}O_{56})_4]^{34-}$ and $[Ni_5(OH)_4(H_2O)_4$-$(\beta$-$GeW_9O_{34})(\beta$-$GeW_8O_{30}(OH))]^{13-}$. *Inorg. Chem.* **2013**, *52*, 8399–8408. [CrossRef]
7. Zheng, S.T.; Zhang, J.; Yang, G.-Y. Designed synthesis of POM–organic frameworks from {Ni_6PW_9} building blocks under hydrothermal conditions. *Angew. Chem. Int. Ed.* **2008**, *47*, 3909–3913. [CrossRef]
8. Li, X.-X.; Wang, Y.-X.; Wang, R.-H.; Cui, C.-Y.; Tian, C.-B.; Yang, G.-Y. Designed assembly of heterometallic cluster organic frameworks based on Anderson-type polyoxometalate clusters. *Angew. Chem. Int. Ed.* **2016**, *55*, 6462–6466. [CrossRef]
9. Li, X.-X.; Zhao, D.; Zheng, S.-T. Recent advances in POM-organic frameworks and POM-organic polyhedra. *Coord. Chem. Rev.* **2019**, *397*, 220–240. [CrossRef]
10. Schubert, U. Cluster-based inorganic–organic hybrid materials. *Chem. Soc. Rev.* **2011**, *40*, 575–582. [CrossRef]
11. Fang, W.-H.; Yang, G.-Y. Induced aggregation and synergistic coordination strategy in cluster organic architectures. *Acc. Chem. Res.* **2018**, *51*, 2888–2896. [CrossRef] [PubMed]
12. Lin, L.-D.; Zhao, D.; Li, X.-X.; Zheng, S.-T. Recent advances in zeolite-like cluster organic frameworks. *Chem. Eur. J.* **2019**, *25*, 442–453. [CrossRef] [PubMed]
13. Dolbecq, A.; Mellot-Draznieks, C.; Mialane, P.; Marrot, J.; Férey, G.; Sécheresse, F. Hybrid 2D and 3D frameworks based on ε-Keggin polyoxometallates: Experiment and simulation. *Eur. J. Inorg. Chem.* **2005**, *2005*, 3009–3018. [CrossRef]
14. Wang, Y.-R.; Huang, Q.; He, C.-T.; Chen, Y.-F.; Liu, J.; Shen, F.-C.; Lan, Y.-Q. Oriented electron transmission in polyoxometalate-metalloporphyrin organic framework for highly selective electroreduction of CO_2. *Nat. Commun.* **2018**, *9*, 4466–4474. [CrossRef] [PubMed]
15. Li, X.-X.; Zhang, L.-J.; Cui, C.-Y.; Wang, R.-H.; Yang, G.-Y. Designed construction of cluster organic frameworks from Lindqvist-type polyoxovanadate cluster. *Inorg. Chem.* **2018**, *57*, 10323–10330. [CrossRef]
16. Han, J.-W.; Hill, C.-L. A Coordination network that catalyzes O_2-based oxidations. *J. Am. Chem. Soc.* **2007**, *129*, 15094–15095. [CrossRef]

17. Marleny Rodriguez-Albelo, L.; Ruiz-Salvador, A.R.; Sampieri, A.; Lewis, D.W.; Gómez, A.; Nohra, B.; Mialane, P.; Marrot, J.; Sécheresse, F.; Mellot-Draznieks, C.; et al. Zeolitic polyoxometalate-based metal–organic frameworks (Z-POMOFs): Computational evaluation of hypothetical polymorphs and the successful targeted synthesis of the redox-active Z-POMOF. *J. Am. Chem. Soc.* **2009**, *131*, 16078–16087. [CrossRef]
18. Wang, Y.-Y.; Zhang, M.; Li, S.-L.; Zhang, S.-R.; Xie, W.; Qin, J.-S.; Su, Z.-M.; Lan, Y.-Q. Diamondoid-structured polymolybdate-based metal–organic frameworks as high-capacity anodes for lithium-ion batteries. *Chem. Commun.* **2017**, *53*, 5204–5207. [CrossRef]
19. Cheng, W.; Shen, F.-C.; Xue, Y.-S.; Luo, X.; Fang, M.; Lan, Y.-Q.; Xu, Y. A pair of rare three-dimensional chiral polyoxometalate-based metal–organic framework enantiomers featuring superior performance as the anode of lithium-ion battery. *ACS Appl. Energy Mater.* **2018**, *1*, 4931–4938. [CrossRef]
20. Zhang, Z.; Wang, Y.-L.; Li, H.-L.; Sun, K.-N.; Yang, G.-Y. Syntheses, structures and properties of three organic–inorganic hybrid polyoxotungstates constructed from {Ni$_6$PW$_9$} building blocks: From isolated clusters to 2-D layers. *CrystEngComm* **2019**, *21*, 2641–2647. [CrossRef]
21. Li, X.-X.; Shen, F.-C.; Liu, J.; Li, S.-L.; Dong, L.-Z.; Fu, Q.; Su, Z.-M.; Lan, Y.-Q. A highly stable polyoxometalate-based metal–organic framework with an ABW zeolite-like structure. *Chem. Commun.* **2017**, *53*, 10054–10057. [CrossRef] [PubMed]
22. Wang, X.-L.; Liu, X.-J.; Tian, A.-X.; Ying, J.; Lin, H.-Y.; Liu, G.-C.; Gao, Q. A novel 2D→3D {Co$_6$PW$_9$}-based framework extended by semi-rigid bis(triazole) ligand. *Dalton Trans.* **2012**, *41*, 9587–9589. [CrossRef] [PubMed]
23. Rousseau, G.; Oms, O.; Dolbecq, A.; Marrot, J.; Mialane, P. Route for the elaboration of functionalized hybrid 3d-substituted trivacant Keggin anions. *Inorg. Chem.* **2011**, *50*, 7376–7378. [CrossRef] [PubMed]
24. Wang, X.-Q.; Liu, S.-X.; Liu, Y.-W.; He, D.-F.; Li, N.; Miao, J.; Ji, Y.-J.; Yang, G.-Y. Planar {Ni$_6$} cluster-containing polyoxometalate-based inorganic–organic hybrid compound and its extended structure. *Inorg. Chem.* **2014**, *53*, 13130–13135. [CrossRef]
25. Ginsberg, A.-P. *Inorganic Syntheses*; John Wiley & Sons: New York, NY, USA, 1990; p. 108.
26. Sheldrick, G.M. A short history of SHELX. *Acta Crystallogr. Sect. A Found. Crystallogr.* **2008**, *64*, 112–122. [CrossRef]
27. Brown, I.D.; Altermatt, D. Bond-valence parameters obtained from a systematic analysis of the inorganic crystal structure database. *Acta Crystallogr. Sect. B* **1985**, *41*, 244–247. [CrossRef]
28. Kariem, M.; Kumar, M.; Yawer, M.; Sheikh, H.M. Solvothermal synthesis and structure of coordination polymers of Nd(III) and Dy(III) with rigid isophthalic acid derivatives and flexible adipic acid. *J. Mol. Struct.* **2017**, *1150*, 438–446. [CrossRef]
29. Sun, J.J.; Wang, Y.L.; Yang, G.-Y. Two new hexa-Ni-substituted polyoxometalates in the form of an isolated cluster and 1-D chain: Syntheses, structures, and properties. *CrystEngComm* **2020**, *22*, 8387–8393. [CrossRef]
30. Sun, J.-J.; Wang, W.-D.; Li, X.-Y.; Yang, B.-F.; Yang, G.-Y. {Cu$_8$} cluster-sandwiched polyoxotungstates and their polymers: Syntheses, structures, and properties. *Inorg. Chem.* **2021**, *60*, 10459–10467. [CrossRef]
31. Zhao, J.-W.; Jia, H.-P.; Zhang, J.; Zheng, S.-T.; Yang, G.-Y. A combination of lacunary polyoxometalates and high-nuclear transition-metal clusters under hydrothermal conditions. Part II: From double cluster, dimer, and tetramer to three-dimensional frameworks. *Chem. Eur. J.* **2007**, *13*, 10030–10045. [CrossRef]
32. Li, B.; Zhao, J.-W.; Zheng, S.-T.; Yang, G.-Y. Combination chemistry of hexa-copper-substituted polyoxometalates driven by the Cu^{II}-polyhedra distortion: From tetramer, 1D chain to 3D framework. *Inorg. Chem.* **2009**, *48*, 8294–8303. [CrossRef] [PubMed]
33. Zhao, J.-W.; Wang, C.-M.; Zhang, J.; Zheng, S.-T.; Yang, G.-Y. Combination of lacunary polyoxometalates and high-nuclear transition metal clusters under hydrothermal conditions: IX. A series of novel polyoxotungstates sandwiched by octa-copper clusters. *Chem. Eur. J.* **2008**, *14*, 9223–9239. [CrossRef] [PubMed]
34. Zhao, J.-W.; Li, B.; Zheng, S.-T.; Yang, G.-Y. Two-dimensional extended (4,4)-topological network constructed from tetra-Ni^{II}-substituted sandwich-type Keggin polyoxometalate building blocks and Ni^{II}-organic cation bridges. *Cryst. Growth Des.* **2007**, *7*, 2658–2664. [CrossRef]
35. Li, Y.-X.; Zhang, Z.; Yang, B.-F.; Li, X.-X.; Zhang, Q.; Yang, G.-Y. A two-dimensional (4,4)-network built by tetra-Ni- substituted sandwich-type Keggin polyoxoanions linked by different Ni-organoamine complexes. *Inorg. Chem. Commun.* **2017**, *75*, 12–15. [CrossRef]
36. Zheng, S.-T.; Zhang, J.; Clemente-Juan, J.M.; Yuan, D.-Q.; Yang, G.-Y. Poly(polyoxotungstate)s with 20 nickel centers: From nanoclusters to one-dimensional chains. *Angew. Chem. Int. Ed.* **2009**, *48*, 7176–7179. [CrossRef]
37. Li, X.-X.; Fang, W.-H.; Zhao, J.-W.; Yang, G.-Y. The first 3-connected SrSi$_2$-type 3D chiral framework constructed from {Ni$_6$PW$_9$} building units. *Chem. Eur. J.* **2015**, *21*, 2315–2318. [CrossRef]
38. Zheng, S.-T.; Zhang, J.; Li, X.-X.; Fang, W.-H.; Yang, G.-Y. Cubic polyoxometalate—Organic molecular cage. *J. Am. Chem. Soc.* **2010**, *132*, 15102–15103. [CrossRef]
39. Huang, L.; Zhang, J.; Cheng, L.; Yang, G.-Y. Poly(polyoxometalate)s assembled by {Ni$_6$PW$_9$} units: From ring-shaped Ni$_{24}$-tetramers to rod-shaped Ni$_{40}$-octamers. *Chem. Commun.* **2012**, *48*, 9658–9660. [CrossRef]
40. Axtell, E.A.; Park, Y.; Chondroudis, K.; Kanatzidis, M.G. Incorporation of A$_2$Q into HgQ and dimensional reduction to A$_2$Hg$_3$Q$_4$ and A$_2$Hg$_6$Q$_7$ (A = K, Rb, Cs; Q = S, Se). Access of Li Ions in A$_2$Hg$_6$Q$_7$ through topotactic ion-exchange. *J. Am. Chem. Soc.* **1998**, *120*, 124–136. [CrossRef]

41. Müller, A.; Beckmann, E.; Bögge, H.; Schmidtmann, M.; Dress, A. Inorganic Chemistry Goes Protein Size: A Mo368 Nano-Hedgehog Initiating Nanochemistry by Symmetry Breaking. *Angew. Chem. Int. Ed.* **2002**, *41*, 1162–1167. [CrossRef]
42. Mialane, P.; Dolbecq, A.; Marrot, J.; Rivieère, E.; Seécheresse, F. A Supramolecular Tetradecanuclear Copper(II) Polyoxotungstate. *Angew. Chem. Int. Ed.* **2003**, *42*, 3523–3526. [CrossRef] [PubMed]
43. Huang, L.; Wang, S.-S.; Zhao, J.-W.; Cheng, L.; Yang, G.-Y. Synergistic Combination of Multi-Zr^{IV} Cations and Lacunary Keggin Germanotungstates Leading to a Gigantic Zr$_{24}$-Cluster-Substituted Polyoxometalate. *J. Am. Chem. Soc.* **2014**, *136*, 7637–7642. [CrossRef] [PubMed]

Article

Cations Modulated Assembly of Triol-Ligand Modified Cu-Centered Anderson-Evans Polyanions

Yiran Wang, Fengxue Duan, Xiaoting Liu and Bao Li *

State Key Laboratory of Supramolecular Structure and Materials, College of Chemistry, Jilin University, Changchun 130012, China; wangyr19@mails.jlu.edu.cn (Y.W.); duanfx17@mails.jlu.edu.cn (F.D.); liuxt16@mails.jlu.edu.cn (X.L.)
* Correspondence: libao@jlu.edu.cn

Abstract: Counter-cations are essential components of polyoxometalates (POMs), which have a distinct influence on the solubility, stabilization, self-assembly, and functionality of POMs. To investigate the roles of cations in the packing of POMs, as a systematic investigation, herein, a series of triol-ligand covalently modified Cu-centered Anderson-Evans POMs with different counter ions were prepared in an aqueous solution and characterized by various techniques including single-crystal X-ray diffraction. Using the strategy of controlling Mo sources, in the presence of triol ligand, NH_4^+, Cu^{2+} and Na^+ were introduced successfully into POMs. When $(NH_4)_6Mo_7O_{24}$ was selected, the counter cations of the produced POMs were ammonium ions, which resulted in the existence of clusters in the discrete state. Additionally, with the modulation of the pH of the solutions, the modified sites of triol ligands on the cluster can be controlled to form δ- or χ-isomers. By applying MoO_3 in the same reaction, Cu^{2+} ions served as linkers to connect triol-ligand modified polyanions into chains. When $Na_4Mo_8O_{26}$ was employed as the Mo source to react with triol ligands in the presence of $CuCl_2$, two 2-D networks were obtained with {$Na_4(H_2O)_{14}$} or {{$Na_2(H_2O)_4$} sub-clusters as linkers, where the building blocks were δ/δ- and χ/χ-isomers, respectively. The present investigation reveals that the charges, sizes and coordination manners of the counter cations have an obvious influence on the assembled structure of polyanions.

Keywords: polyoxometalate; Anderson-Evans; triol ligand; cation-modulation

1. Introduction

As an important and basic type of polyoxometalates (POMs), Anderson-Evans clusters have been synthesized and characterized for nearly one hundred years, with a general formula of $[X^{n+}M_6O_{24}H_m]^{(12-n-m)-}$, in which X expresses the heteroatom and M the addenda atoms (Mo or W in most cases) [1–3]. Compared with the terminal O atoms in Anderson-Evans polyanion, the bi- or tri-bridging O atoms (μ₂- or μ₃-O) have a higher reactive activity and can be replaced by some organic species with hydroxyl groups under specific circumstances to form various decoration types (Figure S1, Supplementary Materials) [4–6]. Through this method, different organic functional groups can be introduced into the inorganic skeleton, which not only enriches the structural figures of the final adducts but also integrates the properties of both parts, resulting in novel functionalities [7–11]. Because the organic functional groups have different hydrophilic and hydrophobic properties, they need to be introduced into polyanions under different solvent environments, such as water or organic solvents [12,13], while due to the use of $TBA_4Mo_8O_{26}$ (TBA = tetrabutylammonium cation) as the Mo source in most cases, triol-ligand modified Anderson-Evans polyanions are mostly prepared in organic solvents such as CH_3CN, CH_3OH and N,N-dimethylformamide, which limits their applications [14,15]. So, it is necessary to find new Mo sources, for example Na_2MoO_4, $(NH_4)_6Mo_7O_{24}$, and $Na_4Mo_8O_{26}$, suitable for reactions in aqueous solution, which may also bring new architectures. On the other hand, compared with the wide range of investigations on organic components covalently

modified Anderson-Evans polyanions with trivalent heteroatoms, examples with divalent center atoms are less focused on and reported [16–18]. Aside from the initial instances with Zn^{2+} and Ni^{2+} used as heteroatoms [19], investigations on the triol-ligand covalently modified Cu^{II}-, Co^{II}- and Ni^{II}-centered Anderson-Evans polyanions are conducted, including finding new modification types, their transformation between different modification architectures and the co-modification of triol ligand and methanol or acetic acid [20–24].

Except developing new modification manners for discrete polyanions, it is also important to construct extended structures by employing the triol-ligand modified Anderson-Evans polyanions as building blocks in appropriate ways. As a comparison, one dimensional (1D) to three dimensional (3D) structures based on undecorated Anderson-Evans polyanions have been reported, in which terminal O atoms of the clusters are used to coordinate with transition metal ions for the formation of the extended organic-inorganic hybrids [25–28]. For the triol-ligand decorated Anderson-Evans polyanions, only a few cases have been reported. For example, through the introduction of pyridine groups, the discrete polyanions can be assembled into an extended structure through M–N bonds, where N atoms are sourced from the pyridine groups [29–32]. Another important strategy is linking amino functionalized Anderson-Evans clusters with 4-connected building units though imine condensation to form metal–organic frameworks [33,34]. As a more common method, rare earth ions (Ln^{3+}) can also be used as nodes to link this type of cluster for the formation of coordination polymers through Ln–O bonds [35]. However, there are still no systematical investigations available on the triol-ligand modified X^{II}-centered Anderson-Evans polyanions in aqueous solution, especially with regard to the influence of cations on the assembled structures of polyanions [36].

Considering the investigations in this field, in the present contribution, we investigate the synthesis of triol-ligand modified Cu^{II}-centered Anderson-Evans polyanions in detail in an aqueous solution. By selecting $(NH_4)_6Mo_7O_{24}$, MoO_3 and $Na_4Mo_8O_{26}$ as our Mo sources, which have not been widely applied before, we synthesized a series of divalent metal ion-centered clusters, and also realized the controlled modulation of an assembly of polyanions with NH_4^+, Cu^{2+} and Na^+ as cations (Scheme 1). In addition, the influences of the size and connecting mode of linkers was discussed, as well as a Hirshfeld analysis of the building blocks with various decoration types. The presented results not only provide an efficient synthetic route for triol-ligand modified Anderson-Evans polyanions with divalent heteroatoms, but also express the important role of cations in the assembly of polyanions.

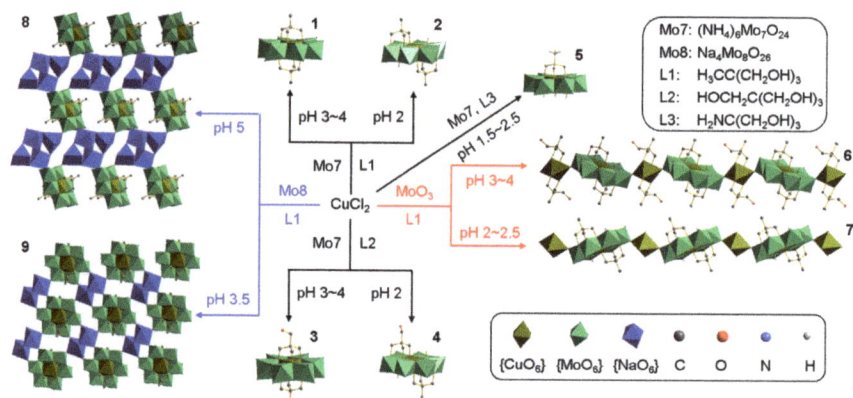

Scheme 1. Schematic illustration for the synthetic parameters in different decoration fashions and extended structures in the preparation of Cu-centered Anderson–Evans POMs **1–9**. All H atoms except those attaching to N and μ3-O atoms are omitted for clarity. The bold numbers are added to represent the corresponding compounds.

2. Results and Discussion

The prepared compounds **1–9** were characterized by single crystal X-ray diffraction analysis, elemental analysis, as well as IR spectra, which can be found in the Supplementary Materials as Figure S2. The thermal stability and the crystalline purity of compounds **1–9** were also evaluated, as shown in Figures S3–S11 and Figures S12–S14.

2.1. Preparation of Triol-Ligand Decorated Anderson-Evans POM Building Blocks

In order to investigate the assembly behaviors of triol-ligand covalently modified Anderson-Evans polyanions in the presence of different cations, we first investigated the synthesis of their building blocks in aqueous solutions in detail. Two important factors are considered in the selection of Mo sources. The first factor is that the used Mo sources can dissolve easily in water at the initial state or in the reaction process, which ensures the occurrence of the reaction and the acceptable yields. The second controls counter cations in the final adducts through the introduction of different Mo sources. For example, $(NH_4)_6Mo_7O_{24}$ and $Na_4Mo_8O_{26}$ result in NH_4^+ and Na^+ as counter cations, while MoO_3 does not lead to the formation of new metal ions, which brings about Cu^{2+} which serves as counter cation. Based on the above analysis, herein, to synthesize building blocks, $(NH_4)_6Mo_7O_{24}$ was used as an Mo source to react with $CuCl_2$ and triol ligand in the aqueous solution at 80 °C. When a triol ligand with a methyl as end group was used and the pH of the solution was adjusted to 3~4, compound **1** is obtained, in which triol ligands functionalized on the Anderson-Evans polyanion in a double-sided style to form δ/δ isomer (Figure 1a). In this case, two triol ligands distributed on both sides of the Anderson-Evans polyanion, and replaced all six hydroxyls around the central heteroatom. When the pH of the solution was lowered to 2, in the case of a stronger acidic environment, partial μ_2-O atoms were activated, and therefore each triol ligand replaced two μ_3-O atoms and one adjacent μ_2-O atom, resulting in a malposition modified structure compound **2** in χ/χ isomer (Figure 1b). The above results show that, at a higher pH value, triol ligands tend to replace all μ_3-O atoms to form a δ modification style; while in a lower pH environment, triol ligands are prone to substitute partial μ_3-O atoms, forming in an χ modification manner, where such phenomena are consistent with the literature [20]. In order to verify the universality of this method and the reliability of the conclusion, we used a triol ligand with a terminal hydroxyl group. Experiments show that, in accordance with the case of the triol ligand with the methyl group at the end, when the pH of the solution is 3~4, the hydroxyl-containing triol ligands replace all the μ_3-O atoms to obtain a double-sided modified Anderson-Evans POM in δ/δ isomer, compound **3** (Figure 1c). When the pH value of the solution is 2, the triol ligands replace partial μ_3-O atoms to obtain a bilaterally modified structure, compound **4**, in χ/χ isomer (Figure 1d). The above experiments show that the modified positions of the triol ligands on the Anderson-Evans polyanion can be modulated by adjusting the pH of the solution. Interestingly, it is different from those triol ligands with hydroxyl and methyl groups, when the end group is amino, we get a single-sided triol ligand decorated δ isomer, compound **5**, in the solution with a relatively lower pH of 1.5~2.5, in which adduct the amino group is in a protonated state (Figure 1e). Even the excessive triol ligands were added in this reaction, and the obtained products were still in a single-sided decoration state with the other side left free. A similar situation also exists in Al-centered Anderson-Evans POMs when modified by triol ligands in the aqueous solution [12]. When the pH of the solution rises with the amino group in the non-protonated state, the lone pair of electrons of the N atom combines easily with the d orbital of the transition metal ion Cu^{2+}, which results in the crosslinking between the generated adducts and makes it difficult to obtain single crystals. An effective way to introduce a non-protonated amino-containing triol ligand into Cu-centered Anderson-Evans cluster is through a two-step synthesis procedure. That is, an undecorated Cu-centered cluster can be synthesized firstly and then used to react with the triol ligand, through which method the coordination sites of Cu^{2+} are fully occupied by O atoms, losing the combining ability with other atoms or functional groups such as amino groups. With this synthetic route, the amino-containing triol ligand can be anchored on the Cu-centered Anderson-Evans cluster in a mono-decoration type through micro-assisted synthesis [37] and double-decoration type through a regular beaker reaction in aqueous solution [38].

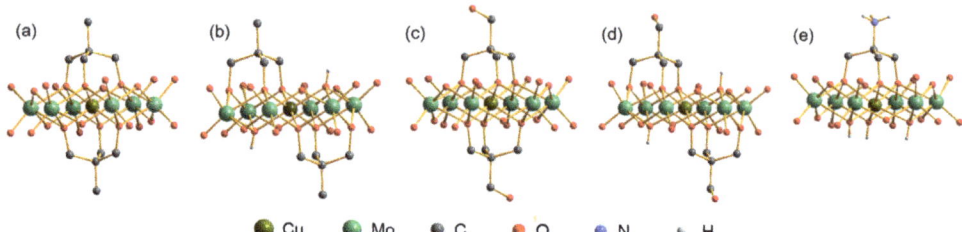

Figure 1. Polyanionic structures of compounds (a) **1**, (b) **2**, (c) **3**, (d) **4**, and (e) **5** in ball-and-stick representation. All H atoms except those attaching to N and μ_3-O atoms are omitted for clarity.

In all the five compounds, due to the compact and symmetric coordination environments of the Cu^{2+} ion, which is located at the center of the Anderson-Evans cluster, there is no obvious Jahn-Teller effect. Taking compound **5** as an example, the six Cu–O bond lengths were 1.958(2), 1.980(2), 1.992(2), 2.026(2), 2.195(2), and 2.213(3) Å, respectively, showing an averaged result without an extra-long Cu–O bond of over 2.4 Å. It is interesting that though half of the coordination sites of the Cu^{2+} ion were occupied by the triol ligand, the Cu–O bond lengths expressed no differences to those formed by hydroxyls. The similar coordination conditions also existed in compounds **1–4**.

It is worth noting that although the prepared organic–inorganic hybrids were all based on the same polyanion Cu-centered Anderson-Evans cluster, due to the different modification positions of the triol ligands on the polyanion, the charges of the obtained clusters were unequal. When the triol ligands are modified on the polyanion in δ isomer with all the hydroxyls around central heteroatom being replaced, the charge of the anion remains unchanged before and after the substitution. When the triol ligands modify in malposition on the polyanion, two unreacted hydroxyls are retained, accompanying with the replacement of two unprotonated μ_2-O (in −2 valence) by the O atom (in −1 valence) from the hydroxyl, and resulting in a decrease in the entire anion charge from −4 to −2. When the terminal group of a triol ligand is amino, although it substitutes all protonated μ_3-O atoms, the amino group is in a protonated state with an additional positive charge, so the charge of the entire anion is −3. That is, under different environmental conditions, we can make triol-ligand modified Cu-centered Anderson-Evans polyanions with 2~4 negative charges. This charge tunability is useful for the further development and utilization of polyanions, especially in terms of providing convenience to the controllable assembly based on the charge number.

2.2. Construction of 1D Structures Based on Triol-Ligand Decorated Anderson-Evans POMs

After obtaining the modification law of the triol ligands on the Anderson-Evans polyanion in aqueous solution, we attempted to obtain extended structures. When ammonium is applied as counter ion, it mainly combines with anion through electrostatic interactions. The lack of directionality and selectivity of the electrostatic interactions makes it unsuitable for the ordered assembly of polyanions. Therefore, we selected MoO_3 instead of $(NH_4)_6Mo_7O_{24}$ as the Mo source, thereby eliminating the possibility of ammonium as counter ions in the adduct. At low pH environments, two 1D chain structures with Cu^{2+} serving as linkers were obtained (Figure 2). The single-crystal X-ray diffraction results show that the Anderson-Evans polyanions were modified by triol ligands to form an χ/χ isomer, to which adjacent polyanions are further linked by Cu^{2+} through the terminal O atoms. The Cu^{2+} shows an octahedron coordination environment with an obvious Jahn-Teller effect, where two Cu–O bonds (bond lengths 2.323(2)–2.396(2) Å) connected to the polyanion are significantly elongated compared with the other four Cu–O bonds (bond length 1.924(2)–1.976(2) Å). The main difference between the two 1D compounds is that of the coordination environments of the linker Cu^{2+} ions. In compound **6**, except two terminal O atoms of clusters, four O atoms from two triol ligands complete the coordination environment of the Cu^{2+} ion; while in compound **7**, the four positions are occupied by coordinated water molecules. In compound **6**, only two hydroxyls of each triol ligand coordinate with Cu^{2+}, and the other one remains in a free state. Not only coordinated hydroxyls but also free hydroxyls are in the protonated state. In the two compounds, the linker Cu^{2+} ions have different coordination environments, and the reason is that at the environment of pH 3~4, although the hydroxyls are in a protonated state, they still have a certain coordination ability with the Cu^{2+} ion, and thus occupy the four coordination sites to obtain compound **6**. However, as the pH value decreases to 2~2.5, the interaction between the hydroxyls and the Cu^{2+} ion weakens to diminish the coordination ability, so that the water molecules occupy the corresponding coordination sites, resulting in compound **7**. This statement can be verified in experiments in which the aqueous solution of compound **6** was acidified and recrystallized to obtain compound **7**. The differences in the coordination modes of the Cu^{2+} ions in the two compounds also have a certain effect on its extending direction. Compound **6** stretches along the (111) direction, while the direction of the 1D chain in compound **7** is (100).

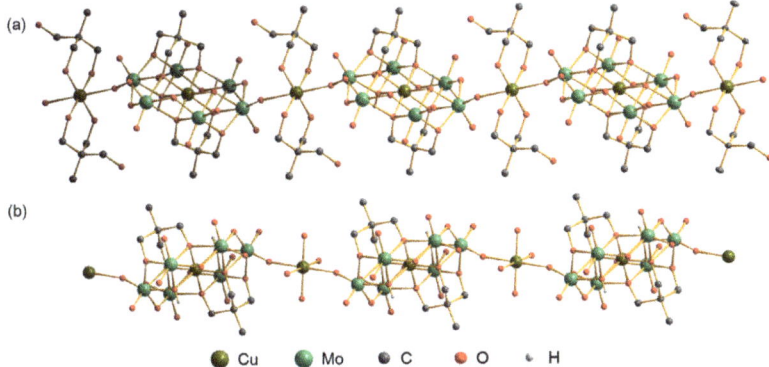

Figure 2. Ball-and-stick representations of compounds (**a**) **6** and (**b**) **7**, showing their 1D chain structures. All H atoms except those attaching to tri-bridging O atoms in polyanions are omitted for clarity.

In the two compounds, the charges of the polyanion and copper ion were −2 and +2, respectively. According to the theory of electrical neutrality, the two components are more easily combined in a 1:1 manner to form the 1D chain structure or the two-dimensional (2D) planar structure in crystallography, as shown in Figure S15, which has a relative low energy and is more stable. In the present case, because the coordination radius of the Cu^{2+} ion was not large enough, and four polyanions cannot be uniformly arranged around one Cu^{2+} ion due to the large steric hindrance, only a 1D chain structure was formed. On another hand, due to the low pH value of the solution, in both 1D structures, the triol ligands were modified on the polyanion in the χ/χ isomer. In order to obtain 1D structures based on the δ/δ isomer, we attempted to reduce the acidity of the solution. However, when the pH increased, due to the higher concentration of Cu^{2+} in the solution, it became easier to obtain a precipitate and the expected structure could not be obtained.

2.3. Construction of 2D Structures Based on Triol-Ligand Decorated Anderson-Evans POMs

As mentioned above, when Cu^{2+} is used as a linker, its small coordination range represents a disadvantage for the formation of a 2D structure, and its synthetic environment with a low pH value is also not conducive to obtaining the triol ligand modified Anderson-Evans clusters in the δ/δ isomer. Considering these points, we selected $Na_4Mo_8O_{26}$ as the Mo source, and used Na^+ with a high solubility in a wide pH range in an aqueous solution and a large range of connection to obtain 2D structures based on triol-ligand modified Anderson-Evans clusters in δ/δ and χ/χ isomer. As expected, when the pH of the solution was 5, we obtained compound **8**, in which the triol ligand double-sided decorated Anderson-Evans polyanion was an δ/δ isomer. In this case, four Na^+ ions aggregated to form a $\{Na_4(H_2O)_{14}\}$ cluster, which linked the adjacent four polyanions through the terminal O atoms into a 2D planar network along the (01$\bar{1}$) direction (Figure 3a). When the pH of the solution was lowered to 3.5, the triol ligand modified Anderson-Evans polyanions in χ/χ isomer was obtained as in compound **9**. In this adduct, two Na^+ ions form a dimer $\{Na_2(H_2O)_4\}$ and connect the adjacent four polyanions to form a 2D planar structure extending along the (100) direction (Figure 3b). It can be seen from the above two examples that the charges of the anions and cations have obvious matching characteristics. For the δ/δ isomer, because the charge of the polyanion was −4, four one-charged Na^+ ions combined to form a tetramer, thereby matching the charge with the anion. While for the χ/χ isomer, because the charge of polyanion was −2, two Na^+ ions combined to form a dimer with two positive charges, and were neutralized with the anion in a 1:1 mode. In addition, compared with Cu^{2+}, the tetramer and dimer of Na^+ are larger and have wider connecting ranges, so that four polyanions can be uniformly arranged around them to form 2D planar structures, which can also be seen as the adjacent 1D chains connecting to each other to form 2D structures.

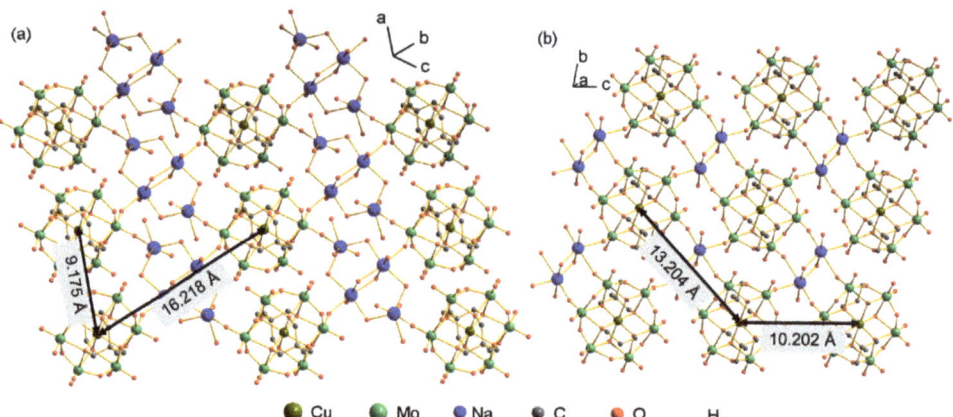

Figure 3. Ball-and-stick representations of compounds (**a**) **8** and (**b**) **9**, showing their 2D planar structures. Numbers in the diagram represent the distances of adjacent clusters in two dimensions. All H atoms except those attaching to tri-bridging O atoms in polyanions are omitted for clarity.

The size of the linking group and the bridging manner have an important effect on the distances between the order of the arranged anions. When the linking group was a Cu^{2+} ion, as in compounds **6** and **7**, the distances between two adjacent polyanions in the 1D chains were 14.090(2) Å and 13.312(2) Å, respectively (Figure S16). The slight difference between the two values is due to the steric hindrance caused by the triol ligand coordinated with linking Cu^{2+} (Figure 2). For inter-chains, and the distances between the two adjacent polyanions in compounds **6** and **7** are 11.128(2) Å and 8.872(2) Å, respectively (Figure S16). For compound **7**, the smaller distance between the adjacent polyanions is mainly due to the different orientations of the two polyanions in adjacent chains, which reduces the steric hindrance based on the rotation of one cluster, so that the distance between the two center heteroatoms decreases. When the linking group was a $\{Na_4(H_2O)_{14}\}$ tetramer, as in Compound **8**, the closest distances of the two adjacent polyanions in the 2D structure were 16.218(2) Å and 9.175(2) Å, respectively (Figure 3). In compound **9**, when the linking group was a {Na$_2$(H$_2$O)$_4$} dimer, the distances between the two adjacent polyanions were 13.204(2) Å and 10.202(2) Å, respectively. As shown in Figure 4, from the chemical environments of linkers in compounds **6–9**, we can see that when the bridging group is a single metal ion, the mode of bridging polyanions is simple, and the distance between polyanions mainly depends on the radius of the bridging metal ions. When the linker is a cluster formed by multiple metal ions, the bridging range becomes significantly larger, and it can interact with the polyanion through various modes, so that the polyanions have richer assembled structures.

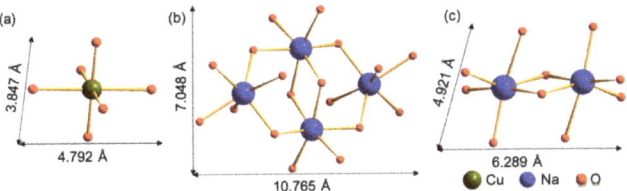

Figure 4. Ball-and-stick representations of coordination environments of linkers for compounds (**a**) **6** and **7**, (**b**) **8**, and (**c**) **9**, showing their different sizes and bridging manners.

2.4. Hirshfled Surface Analysis

As demonstrated by the analysis above, the triol-ligand modified Anderson-Evans POMs can serve as building blocks for the construction of 1D or 2D assemblies based on different metal ions and their various combinations. In fact, with similar building blocks, 3D assemblies have also been prepared, through which ionic frameworks form and exhibit selective adsorption capacity to CO_2 over N_2, H_2 and CH_4 [39,40]. All these 1D to 3D assemblies are constructed based on a strong ionic bond or coordination bond and provide firm connections between each other, while in the absence of metal ions, there are still relatively weak contacts between building blocks, which also have an important influence on their packing styles. Herein, the Hirshfeld surface analysis was applied to illustrate the supramolecular interactions between triol-ligand modified Anderson-Evans polyanions. To exclude the effects of metal ions, only compounds **1–5** were analyzed in which an ammonium ion serves as the counter cation and cannot provide obvious directional interactions to the assembly behavior of building blocks such as that of metal ions. As the important supramolecular interactions, the hydrogen bonds in compounds **1–5** are summarized in Tables S1–S5 in the Supplementary Materials.

Hirshfeld surfaces mapped with the d_{norm} of compounds **1–5** were firstly investigated, in which d_{norm} was the normalized sum of d_i and d_e, and is defined as follows [41]:

$$d_{norm} = (d_i - r_I^{vdw})/r_I^{vdw} + (d_e - r_E^{vdw})/r_E^{vdw}$$

d_i is the distance from Hirshfeld surface to the nearest atom I internal to the surface, d_e is the distance from Hirshfeld surface to the nearest atom E external to the surface, r_I^{vdw} is the van der Waals radius of the nearest atom I closest to and inside the Hirshfeld surface, and r_E^{vdw} is the van der Waals radius of the nearest atom E closest to and outside the Hirshfeld surface. As shown in Figure 5, when a methyl-containing triol ligand was used in compounds **1** and **2**, the main interaction sites (marked with red cones) were concentrated at the lateral edge of the disk-shaped cluster, where the terminal O atoms can serve as hydrogen bonding acceptors, while for compounds with hydroxyl or protonated amino groups such as compounds **3–5**, their ability to serve as hydrogen bonding donors resulted in a relatively uniform distribution of strong contact sites surrounding the cluster.

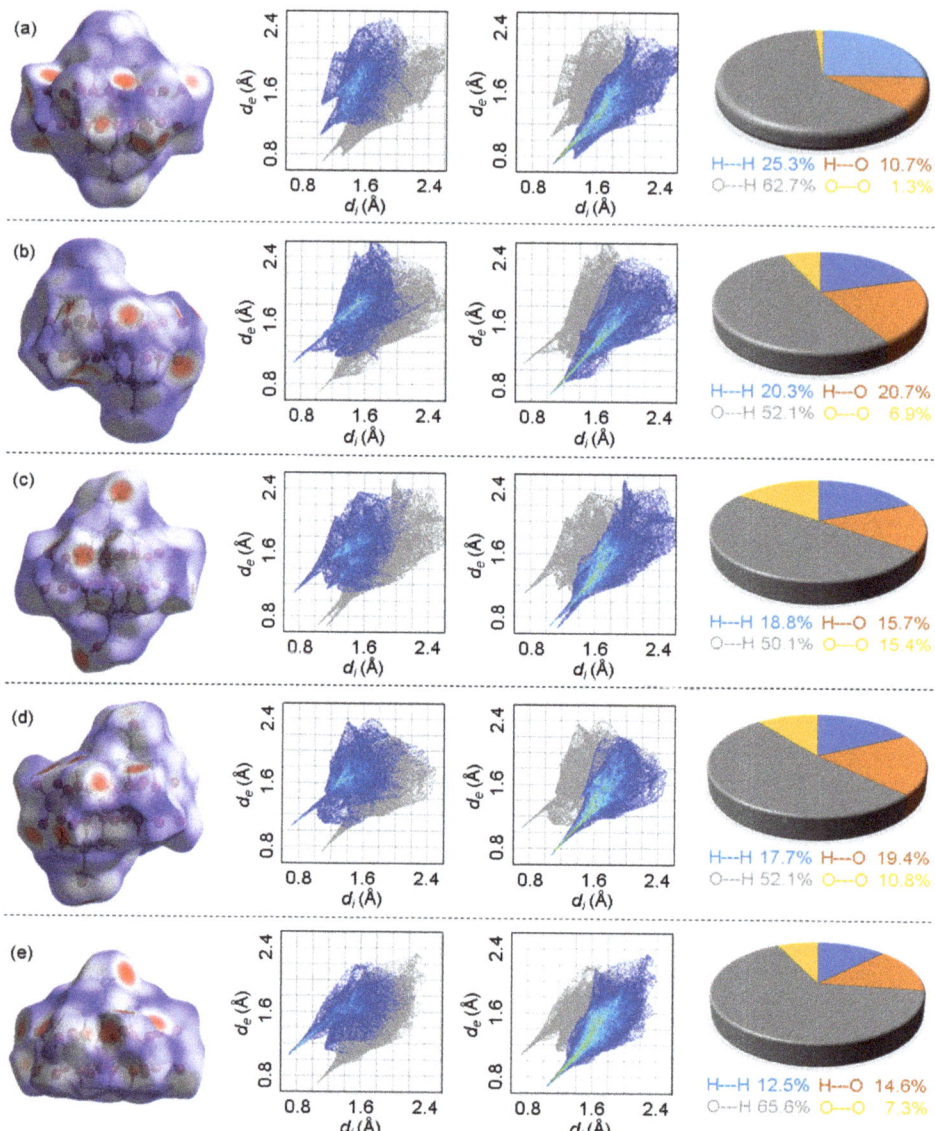

Figure 5. Hirshfeld analysis results for compounds (**a**) **1**, (**b**) **2**, (**c**) **3**, (**d**) **4**, and (**e**) **5**, respectively. For each column, from left to right, the four images are as follows: Hirshfeld surface mapped with d_{norm} in a transparent mode, colored 2D fingerprint plot showing contacts between H (internal to the surface) and H and O (external to the surface), colored 2D fingerprint plot showing contacts between O (internal to the surface) and H and O (external to the surface), and percentage distribution of various short contacts (the detailed percentages of contacts are labeled with the corresponding colors).

Hirshfeld surface images provide qualitative descriptions of the supramolecular interactions of clusters, and 2D fingerprint plots in the range of 0.6–2.6 Å for d_i and d_e were further applied to analyze the detailed contacts in quantitative accuracy (Figure 5). All five compounds have some common characteristics, such as for interactions of H atoms internal to the surface with H or O atoms external to the surface with generally small d_i values (0.6–2.0 Å) and large d_e values (1.0–2.4 Å), while O atoms internal to the surface with H or O atoms external to the surface have an opposite trend with large d_i values (1.0–2.6 Å) and small d_e values (0.6–2.4 Å). These features indicate that the triol-ligand modified clusters are preferred as hydrogen bonding acceptors than donors, which is also in accordance with the traits of O-rich surfaces. It should be noted that the modification sites of the triol ligand also have an obvious influence on the intermolecular interactions. For example, compounds **1** and **2** have the same triol ligand on the cluster and different positions to the δ/δ and χ/χ isomers, respectively. In compound **1**, trio ligands were located at the center position of the cluster, which enlarged the distances of methene and methyl H to other species and results in relatively large d_i and d_e values (both more than 1.0 Å) in 2D fingerprint plot. As a comparison, in compound **2**, triol ligands were located at the edge of the disk-like cluster, which enhanced their abilities as hydrogen-bonding donors and generated relatively stronger contacts between H atoms internal to the surface with H and O atoms external to the surface. In addition, the χ/χ isomer in compound **2** also provides two extra protonated μ_3-O atoms as hydrogen bonding donors compared with compound **1**, which are important components of short contacts. On the contrary, the positions of triol ligands have little influence on the hydrogen-bonding-acceptor ability of the cluster, and the 2D fingerprint plots showing contacts between O (internal to the surface) and H and O (external to the surface) of compounds **1** and **2** are very similar. Lastly, the discoid distributions of various contacts indicate that all five triol-ligand modified clusters are mainly used as hydrogen-bonding acceptors with the percentages of O—H contacts ranging from 50.1% to 65.6%.

3. Materials and Methods

3.1. General Methods and Materials

All chemicals were purchased from Aladdin and used without further purification. Double distilled water was used in all the reactions. Fourier transform infrared spectra were obtained with a Bruker Vertex 80v spectrometer equipped with a DTGS detector with a resolution of 4 cm^{-1} in KBr pellets. Elemental analysis of C, H, and N was conducted using a vario MICRO cube from Elementar Company of Germany. Elemental analysis for Cu, Mo and Na was carried out on a PLASMA-SPEC (I) inductively coupled plasma atomic emission spectrometer. Thermogravimetric analysis curves were obtained with a Q500 Thermal Analyzer (New Castle TA Instruments, New Castle, DE, USA) in a flowing N$_2$ under a heating rate of 10 °C·min^{-1}. Powder X-ray diffraction data were recorded on a Rigaku SmartLab X-ray diffractometer using Cu K$_\alpha$ radiation at a wavelength of 1.54 Å. Single-crystal X-ray diffraction data were collected on a Bruker D8 VENTURE diffractometer with graphite-monochromated Mo K$_\alpha$ (λ = 0.71073 Å) at 293 K. All crystals were solved by *SHELXT* and refined by full-matrix-least-squares fitting for F^2 using the Olex2 software [42,43]. All non-H atoms were refined with anisotropic thermal parameters. A summary of the crystallographic data and structural refinements for compounds **1–9** is listed in Table 1. The detailed CCDC 2162383–2162391 contains the supplementary crystallographic data for this paper. These data can be obtained free of charge from www.ccdc.cam.ac.uk/data_request/cif, accessed on 27 March 2022, or by emailing data_request@ccdc.cam.ac.uk, or by contacting The Cambridge Crystallographic Data Centre, 12 Union Road, Cambridge CB2 1EZ, UK.

Table 1. The summary of crystal data and structure refinements for compounds **1–9**.

	1	2	3	4	5
Formula	$[NH_4]_4[CuMo_6O_{18}(CH_3C(CH_2O)_3]_2]\cdot 5.5H_2O$	$[NH_4]_2[CuMo_6O_{16}(OH)_2(CH_3C(CH_2O)_3]_2]\cdot 2H_2O$	$[NH_4]_4[CuMo_6O_{18}(HOCH_2C(CH_2O)_3]_2]\cdot 5.5H_2O$	$[NH_4]_2[CuMo_6O_{16}(OH)_2(HOCH_2C(CH_2O)_3]_2]\cdot 4H_2O$	$(NH_4)_3[CuMo_6O_{18}(OH)_3(NH_3C(CH_2O)_3]_2]\cdot 6H_2O$
F.W.	1332.68	1235.55	1364.68	1303.59	1259.55
S.G.	$P4_2/n$	$P2_1/n$	$P4_2/n$	$P\bar{1}$	$P\bar{1}$
a (Å)	16.6461(3)	8.8076(5)	16.6651(3)	8.7546(2)	10.9061(8)
b (Å)	16.6461(3)	17.1452(9)	16.6651(3)	9.8069(3)	12.1946(9)
c (Å)	12.9739(4)	11.1324(6)	13.0675(5)	10.3904(3)	12.9648(9)
α (deg)	90	90	90	95.823(1)	72.139(3)
β (deg)	90	110.302(2)	90	98.085(1)	62.020(2)
γ (deg)	90	90	90	112.510(1)	82.620(3)
V (Å³)	3595.0(2)	1544.0(2)	3629.2(2)	804.1(1)	1510.8(2)
Z	4	2	4	1	2
D_c (g cm⁻³)	2.462	2.658	2.498	2.692	2.769
F(000)	2600	1190	2664	631	1218
Reflections coll./unique	43,727/4424	6373/1635	27,878/3123	25,824/3694	46,661/7481
R_{int}	0.0172	0.0255	0.0291	0.0224	0.0275
GOF on F^2	1.146	1.251	1.310	1.079	1.045
R_1 [a] [$I > 2\sigma(I)$]	0.0176	0.0483	0.0470	0.0197	0.0261
wR_2 [b] (all data)	0.0438	0.1082	0.1063	0.0517	0.0779
CCDC no.	2,162,383	2,162,384	2,162,385	2,162,386	2,162,387
	6	7	8	9	

	6	7	8	9
Formula	$Cu[CH_3C(CH_2OH)_3]_2[CuMo_6O_{16}(OH)_2(CH_3C(CH_2O)_3]_2]\cdot 6H_2O$	$Cu(H_2O)_4[CuMo_6O_{16}(OH)_2(CH_3C(CH_2O)_3]_2]\cdot 10H_2O$	$Na_2(H_2O)_{14}[CuMo_6O_{18}(CH_3C(CH_2O)_3]_2]\cdot 2H_2O$	$Na_2(H_2O)_4[CuMo_6O_{16}(OH)_2(CH_3C(CH_2O)_3]_2]\cdot 6H_2O$
F.W.	1575.36	1479.20	1541.64	1389.58
S.G.	$P\bar{1}$	$P2_1/c$	$P\bar{1}$	$P\bar{1}$
a (Å)	10.4512(9)	13.3123(9)	9.1747(2)	9.6583(3)
b (Å)	10.7305(9)	13.4631(8)	10.4014(2)	10.2021(4)
c (Å)	11.1280(9)	11.5585(8)	11.6954(2)	10.5633(4)
α (deg)	110.643(3)	90	96.479(1)	78.845(1)
β (deg)	91.194(4)	99.210(3)	97.716(1)	72.565(1)
γ (deg)	105.688(3)	90	92.964(1)	67.093(1)
V (Å³)	1115.0(2)	2044.9(2)	1096.4(1)	911.4(1)
Z	1	2	1	1
D_c (g cm⁻³)	2.672	2.402	2.335	2.532
F(000)	774	1444	755	675
Reflections coll./unique	16,695/5112	32,538/5059	14,364/5416	114,40/4184
R_{int}	0.0219	0.0296	0.0205	0.0212
GOF on F^2	1.032	1.032	1.039	1.065
R_1 [a] [$I > 2\sigma(I)$]	0.0205	0.0190	0.0219	0.0222
wR_2 [b] (all data)	0.0528	0.0481	0.0544	0.0552
CCDC no.	216,238	2,162,389	2,162,390	2,162,391

[a] $R_1 = \Sigma||F_o|-|F_c||/\Sigma|F_o|$. [b] $wR_2 = \{\Sigma[w(F_o^2-F_c^2)^2]/\Sigma[w(F_o^2)^2]\}^{1/2}$.

3.2. Synthesis of Compounds

3.2.1. Synthesis of $[NH_4]_4[CuMo_6O_{18}(CH_3C(CH_2O)_3)_2]\cdot 5.5H_2O$ (**1**)

A mixture of $CuCl_2\cdot 2H_2O$ (0.34 g, 2.00 mmol), $(NH_4)_6Mo_7O_{24}\cdot 4H_2O$ (2.47 g, 2.00 mmol) and 0.48 g of $CH_3C(CH_2OH)_3$ (0.48 g, 4.00 mmol) was dissolved in 20 mL of deionized water under stirring. The pH of the mixture was adjusted to 3~4 with the concentrated HCl, and the resulting solution was heated to 80 °C, and maintained for 2 h. After a hot filtration process, a blue solution was obtained, which provided blue crystals suitable for X-ray single-crystal diffraction analysis after about one day at room temperature. Elemental analysis calcd. (%) for $[NH_4]_4[CuMo_6O_{18}(CH_3C(CH_2O)_3)_2]\cdot 5.5H_2O$ (Mw = 1332.68 g mol^{-1}): C 9.01%, H 3.40%, N 4.20%, Cu 4.77%, and Mo 43.19%; found C 9.20%, H 3.40%, N 4.25%, Cu 4.83%, and Mo 43.86%.

3.2.2. Synthesis of $[NH_4]_2\{CuMo_6O_{16}(OH)_2[CH_3C(CH_2O)_3]_2\}\cdot 2H_2O$ (**2**)

The synthetic procedure of compound **2** was similar to that of compound **1** except that the pH of the solution was adjusted to 2. Elemental analysis calcd. (%) for $[NH_4]_2\{CuMo_6O_{16}(OH)_2[CH_3C(CH_2O)_3]_2\}\cdot 2H_2O$ (Mw = 1235.58 g mol^{-1}): C 9.72%, H 2.61%, N 2.27%, Cu 5.14%, and Mo 46.59%; found C 9.31%, H 2.80%, N 2.31%, Cu 5.21%, and Mo 46.76%.

3.2.3. Synthesis of $[NH_4]_4[CuMo_6O_{18}(HOCH_2C(CH_2O)_3)_2]\cdot 5.5H_2O$ (**3**)

The synthetic procedure of compound **3** was similar with that of compound **1** except that $HOCH_2C(CH_2OH)_3$ (0.54 g, 4.00 mmol) was used for substituting $CH_3C(CH_2OH)_3$. Elemental analysis calcd. (%) for $[NH_4]_4[CuMo_6O_{18}(HOCH_2C(CH_2O)_3)_2]\cdot 5.5H_2O$ (Mw = 1364.68 g mol^{-1}): C 8.80%, H 3.32%, N 4.11%, Cu 4.66%, and Mo 42.18%; found C 8.91%, H 3.36%, N 4.18%, Cu 4.70%, and Mo 42.24%.

3.2.4. Synthesis of $[NH_4]_2\{CuMo_6O_{16}(OH)_2[HOCH_2C(CH_2O)_3]_2\}\cdot 4H_2O$ (**4**)

The synthetic procedure of compound **4** was similar with that of compound **3** except that the pH of the solution was adjusted to 2. Elemental analysis calcd. (%) for $[NH_4]_2\{CuMo_6O_{16}(OH)_2[HOCH_2C(CH_2O)_3]_2\}\cdot 4H_2O$ (Mw = 1303.59 g mol^{-1}): C 9.21%, H 2.78%, N 2.15%, Cu 4.87%, and Mo 44.16%; found C 9.28%, H 2.87%, N 2.22%, Cu 4.90%, and Mo 44.21%.

3.2.5. Synthesis of $(NH_4)_3\{CuMo_6O_{18}(OH)_3[NH_3C(CH_2O)_3]\}\cdot 6H_2O$ (**5**)

$CuCl_2\cdot 2H_2O$ (0.34 g, 2.00 mmol), $(NH_4)_6Mo_7O_{24}\cdot 4H_2O$ (2.47 g, 2.00 mmol) and $NH_2C(CH_2OH)_3$ (0.48 g, 4.00 mmol) were dissolved in 20 mL of deionized water under stirring. The pH of the mixture was adjusted to about 1.5~2.5 by the concentrated HCl, and the resulting solution was heated to 80 °C gradually for 2 h. A blue solution was obtained with a hot filtration process, and provided blue crystals in one day. Elemental analysis calcd. (%) for $[NH_4]_3[CuMo_6O_{18}(OH)_3(NH_3C(CH_2O)_3)_2]\cdot 6H_2O$ (Mw = 1259.55 g mol^{-1}): C 3.81%, H 2.88%, N 4.45%, Cu 5.05%, and Mo 45.70%; found C 3.88%, H 2.95%, N 4.55%, Cu 5.10%, and Mo 45.77%.

3.2.6. Synthesis of $Cu[CH_3C(CH_2OH)_3]_2\{CuMo_6O_{16}(OH)_2[CH_3C(CH_2O)_3]_2\}\cdot 6H_2O$ (**6**)

To a stirred solution of $CuCl_2\cdot 2H_2O$ (0.34 g, 2.00 mmol) and $CH_3C(CH_2OH)_3$ (0.48 g, 4.00 mmol) in water (20 mL), MoO_3 (0.86 g, 12.00 mmol) was added. The pH of the resulting solution was adjusted to 3~4 by the concentrated HCl. After heating the mixture to 80 °C for 2 h, a blue solution was obtained using hot filtration, which generates blue crystals in several days. Elemental analysis calcd. (%) for $Cu[CH_3C(CH_2OH)_3]_2\{CuMo_6O_{16}(OH)_2[CH_3C(CH_2O)_3]_2\}\cdot 6H_2O$ (Mw = 1575.36 g mol^{-1}): C 15.25%, H 3.58%, Cu 8.07%, and Mo 36.54%; found C 15.58%, H 3.65%, Cu 8.14%, and Mo 36.55%.

3.2.7. Synthesis of Cu(H$_2$O)$_4$ {CuMo$_6$O$_{16}$(OH)$_2$[CH$_3$C(CH$_2$O)$_3$]$_2$}·10H$_2$O (**7**)

The synthetic procedure of compound **7** was similar with that of compound **6** except that the pH of the solution was adjusted to 2~2.5. Elemental analysis calcd. (%) for Cu(H$_2$O)$_4$ {CuMo$_6$O$_{16}$(OH)$_2$[CH$_3$C(CH$_2$O)$_3$]$_2$}·10H$_2$O (Mw = 1479.20 g mol^{-1}): C 8.12%, H 3.27%, Cu 8.59%, and Mo 38.92%; found C 8.21%, H 3.35%, Cu 8.64%, and Mo 39.01%.

3.2.8. Synthesis of Na$_4$(H$_2$O)$_{14}$ {CuMo$_6$O$_{18}$[CH$_3$C(CH$_2$O)$_3$]$_2$}·2H$_2$O (**8**)

CuCl$_2$·2H$_2$O (0.34 g, 2.00 mmol), Na$_4$Mo$_8$O$_{26}$ (1.92 g, 1.50 mmol) and CH$_3$C(CH$_2$OH)$_3$ (0.48 g, 4.00 mmol) were dissolved in 20 mL of deionized water under stirring. Concentrated HCl was used to adjust the pH of the mixture to 5, and then the resulting solution was heated to 80 °C for 2 h. A blue solution was obtained using a hot filtration process, and generated blue crystals in several days. Elemental analysis calcd. (%) for Na$_4$(H$_2$O)$_{14}$ {CuMo$_6$O$_{18}$[CH$_3$C(CH$_2$O)$_3$]$_2$}·2H$_2$O (Mw = 1541.64 g mol^{-1}): C 7.79%, H 3.27%, Na 5.97%, Cu 4.12%, and Mo 37.34%; found C 7.84%, H 3.35%, Na 6.05%, Cu 4.06%, and Mo 37.18%.

3.2.9. Synthesis of Na$_2$(H$_2$O)$_4$ {CuMo$_6$O$_{16}$(OH)$_2$[CH$_3$C(CH$_2$O)$_3$]$_2$}·6H$_2$O (**9**)

The synthetic procedure of compound **9** was similar with that of compound **8** except that the pH of the solution was adjusted to 3.5. Elemental analysis calcd. (%) for Na$_2$(H$_2$O)$_4$ {CuMo$_6$O$_{16}$(OH)$_2$[CH$_3$C(CH$_2$O)$_3$]$_2$}·6H$_2$O (Mw = 1389.58 g mol^{-1}): C 8.64%, H 2.90%, Na 3.31%, Cu 4.57%, and Mo 41.43%; found C 8.70%, H 2.94%, Na 3.38%, Cu 4.66%, and Mo 41.52%.

4. Conclusions

By synthesizing a series of Cu^{II}-centered Anderson-Evans POMs modified with triol ligands in the presence of different cations in aqueous solution, ordered assemblies of polyanions from 0D to 2D under cation regulation were successfully realized. The results show that NH$_4^+$ directs the formation of 0D building blocks, and Cu^{2+} leads the 1D chain structures, while Na$^+$ dominates 2D networks. During the assembled process, the charge ratio between cations and anions has an important effect on the assembled structures. When the size of a cation is large, its wider connection range can link the four adjacent polyanions together to form a 2D structure, while when the size of cation is small, the polyanions can only be arranged in a 1D chain. The research work in this section not only provides an important method and idea for the construction of an ordered assembly of polyanions, but also lays the foundation for the development of specific functional systems.

Supplementary Materials: The following supporting information is available online at https://www.mdpi.com/article/10.3390/molecules27092933/s1. Figure S1: The schematic drawing of possible decoration types when triol ligands bind to an Anderson-Evans POM cluster, where the blue octahedron represents {MO$_6$} (M = Mo or W) and yellow octahedron denotes heteroatom-oxygen {XO$_6$}; Figure S2: FT-IR spectra of compounds **1–9**; Figures S3–S11: TGA curves of compounds **1–9**; Figures S12–S14: XRD patterns of as-synthesized compounds **1–9** and their simulated patterns from the corresponding single-crystal X-ray diffraction data; Figure S15: Two possible arrangements of linkers and building blocks with 1:1 charge ratio. Spheres and disks represent linkers and building blocks, respectively. The solid and dotted lines show the interactions between the adjacent components; Figure S16: Ball-and-stick representations of compounds (a) **6** and (b) **7**, showing their neighboring chains. Numbers in the diagram represent the distances of adjacent clusters. All H atoms except those attached to tri-bridging O atoms in polyanions are omitted for clarity. Tables S1–S5: Hydrogen bonds for compounds **1–5**.

Author Contributions: Conceptualization, B.L.; formal analysis, Y.W.; investigation, F.D. and X.L.; writing—original draft preparation, Y.W.; writing—review and editing, B.L.; supervision, B.L.; funding acquisition, B.L. All authors have read and agreed to the published version of the manuscript.

Funding: This research was funded by Natural Science Foundation of China, grant number 22172060.

Institutional Review Board Statement: Not applicable.

Informed Consent Statement: Not applicable.

Data Availability Statement: Crystallographic data can be obtained free of charge from the Cambridge Crystallographic Data Centre via www.ccdc.cam.ac.uk/data_request/cif, accessed on 27 March 2022.

Conflicts of Interest: The authors declare no conflict of interest.

Sample Availability: Samples of the compounds are available from the authors.

References

1. Anderson, J.S. Constitution of the poly-acids. *Nature* **1937**, *140*, 850. [CrossRef]
2. Evans, H.T. The crystal structures of ammonium and potassium molybdotellurates. *J. Am. Chem. Soc.* **1948**, *70*, 1291–1292. [CrossRef]
3. Blazevic, A.; Rompel, A. The Anderson–Evans polyoxometalate: From inorganic building blocks via hybrid organic–inorganic structures to tomorrows "bio-POM". *Coord. Chem. Rev.* **2016**, *307*, 42–64. [CrossRef]
4. Zhang, J.W.; Huang, Y.C.; Li, G.; Wei, Y.G. Recent advances in alkoxylation chemistry of polyoxometalates: From synthetic strategies, structural overviews to functional applications. *Coord. Chem. Rev.* **2019**, *378*, 395–414. [CrossRef]
5. Wu, P.F.; Wang, Y.; Huang, B.; Xiao, Z.C. Anderson-type polyoxometalates: From structures to functions. *Nanoscale* **2021**, *13*, 7119–7133. [CrossRef]
6. Lin, C.G.; Hutin, M.; Busche, C.; Bell, N.L.; Long, D.L.; Cronin, L. Elucidating the paramagnetic interactions of an inorganic-organic hybrid radical-functionalized Mn-Anderson cluster. *Dalton Trans.* **2021**, *50*, 2350–2353. [CrossRef]
7. Gao, B.; Li, B.; Wu, L.X. Layered supramolecular network of cyclodextrin triplets with azobenzene-grafting polyoxometalate for dye degradation and partner-enhancement. *Chem. Commun.* **2021**, *57*, 10512–10515. [CrossRef]
8. Liu, D.J.K.; Zhang, G.H.; Gao, B.; Li, B.; Wu, L.X. From achiral to helical bilayer self-assemblies of a 1,3,5-triazine-2,4,6-triphenol-grafted polyanionic cluster: Countercation and solvent modulation. *Dalton Trans.* **2019**, *48*, 11623–11627. [CrossRef]
9. Lin, C.-G.; Fura, G.D.; Long, Y.; Xuan, W.; Song, Y.-F. Polyoxometalate-based supramolecular hydrogels constructed through host–guest interactions. *Inorg. Chem. Front.* **2017**, *4*, 789–794. [CrossRef]
10. Marcano, D.E.S.; Lentink, S.; Moussawi, M.A.; Parac-Vogt, T.N. Solution dynamics of hybrid Anderson-Evans polyoxometalates. *Inorg. Chem.* **2021**, *60*, 10215–10226. [CrossRef]
11. Winter, A.; Endres, P.; Schroter, E.; Jager, M.; Gorls, H.; Neumann, C.; Turchanin, A.; Schubert, U.S. Towards covalent photosensitizer-polyoxometalate dyads-bipyridyl-functionalized polyoxometalates and their transition metal complexes. *Molecules* **2019**, *24*, 4446. [CrossRef] [PubMed]
12. Ai, H.; Wang, Y.; Li, B.; Wu, L.X. Synthesis and characterization of single-side organically grafted Anderson-type polyoxometalates. *Eur. J. Inorg. Chem.* **2014**, *2014*, 2766–2772. [CrossRef]
13. Xu, Q.H.; Yuan, S.S.; Zhu, L.; Hao, J.; Wei, Y.G. Synthesis of novel bis(Triol)-functionalized Anderson clusters serving as potential synthons for forming organic–inorganic hybrid chains. *Chem. Commun.* **2017**, *53*, 5283–5286. [CrossRef]
14. Boulmier, A.; Vacher, A.; Zang, D.; Yang, S.; Saad, A.; Marrot, J.; Oms, O.; Mialane, P.; Ledoux, I.; Ruhlmann, L.; et al. Anderson-type polyoxometalates functionalized by tetrathiafulvalene groups: Synthesis, electrochemical studies, and NLO properties. *Inorg. Chem.* **2018**, *57*, 3742–3752. [CrossRef] [PubMed]
15. Guan, W.M.; Li, B.; Wu, L.X. Chiral hexamers of organically modified polyoxometalates via ionic complexation. *Dalton Trans.* **2022**, *51*, 4541–4548. [CrossRef] [PubMed]
16. Lin, C.G.; Chen, W.; Long, D.L.; Cronin, L.; Song, Y.F. Step-by-step covalent modification of Cr-templated Anderson-type polyoxometalates. *Dalton Trans.* **2014**, *43*, 8587–8590. [CrossRef] [PubMed]
17. Zhou, Y.; Zhang, G.H.; Li, B.; Wu, L.X. Two-dimensional supramolecular ionic frameworks for precise membrane separation of small nanoparticles. *ACS Appl. Mater. Interfaces* **2020**, *12*, 30761–30769. [CrossRef] [PubMed]
18. Li, Q.; Wei, Y.G. Unprecedented monofunctionalized beta-Anderson clusters: [$R_1R_2C(CH_2O)_2Mn^{IV}W_6O_{22}$]$^{6-}$, a class of potential candidates for new inorganic linkers. *Chem. Commun.* **2021**, *57*, 3865–3868. [CrossRef]
19. Hasenknopf, B.; Delmont, R.; Herson, P.; Gouzerh, P. Anderson-type heteropolymolybdates containing tris(alkoxo) ligands: Synthesis and structural characterization. *Eur. J. Inorg. Chem.* **2002**, *2002*, 1081–1087. [CrossRef]
20. Wang, Y.; Liu, X.T.; Xu, W.; Yue, Y.; Li, B.; Wu, L.X. Triol-ligand modification and structural transformation of Anderson-Evans oxomolybdates via modulating oxidation state of Co-heteroatom. *Inorg. Chem.* **2017**, *56*, 7019–7028. [CrossRef]
21. Wang, Y.; Li, B.; Qian, H.J.; Wu, L.X. Controlled triol-derivative bonding and decoration transformation on Cu-centered Anderson-Evans polyoxometalates. *Inorg. Chem.* **2016**, *55*, 4271–4277. [CrossRef] [PubMed]
22. Gumerova, N.I.; Roller, A.; Rompel, A. Synthesis and characterization of the first Nickel(II)-centered single-side tris-functionalized Anderson-type polyoxomolybdate. *Eur. J. Inorg. Chem.* **2016**, *2016*, 5507–5511. [CrossRef]
23. Gumerova, N.I.; Roller, A.; Rompel, A. [$Ni(OH)_3W_6O_{18}(OCH_2)_3CCH_2OH$]$^{4-}$: The first tris-functionalized Anderson-type heteropolytungstate. *Chem. Commun.* **2016**, *52*, 9263–9266. [CrossRef] [PubMed]
24. Ali Khan, M.; Shakoor, Z.; Akhtar, T.; Sajid, M.; Muhammad Asif, H. Exploration on χ-Anderson type polyoxometalates based hybrids towards photovoltaic response in solar cell. *Inorg. Chem. Commun.* **2021**, *133*, 108875. [CrossRef]

25. Zhao, J.-W.; Zhang, J.-L.; Li, Y.-Z.; Cao, J.; Chen, L.-J. Novel one-dimensional organic–inorganic polyoxometalate hybrids constructed from heteropolymolybdate units and copper–aminoacid complexes. *Cryst. Growth Des.* **2014**, *14*, 1467–1475. [CrossRef]
26. An, H.Y.; Li, Y.G.; Wang, E.B.; Xiao, D.R.; Sun, C.Y.; Xu, L. Self-assembly of a series of extended architectures based on polyoxometalate clusters and silver coordination complexes. *Inorg. Chem.* **2005**, *44*, 6062–6070. [CrossRef]
27. Cao, R.G.; Liu, S.X.; Xie, L.H.; Pan, Y.B.; Cao, J.F.; Ren, Y.H.; Xu, L. Organic-inorganic hybrids constructed of Anderson-type polyoxoanions and oxalato-bridged dinuclear copper complexes. *Inorg. Chem.* **2007**, *46*, 3541–3547. [CrossRef]
28. Zhang, J.-Y.; Chang, Z.-H.; Wang, X.-L.; Wang, X.; Lin, H.-Y. Different Anderson-type polyoxometalate-based metal–organic complexes exhibiting –OH group-directed structures and electrochemical sensing performance. *New J. Chem.* **2021**, *45*, 3328–3334. [CrossRef]
29. Li, X.X.; Ma, X.; Zheng, W.X.; Qi, Y.J.; Zheng, S.T.; Yang, G.Y. Composite hybrid cluster built from the integration of polyoxometalate and a metal halide cluster: Synthetic strategy, structure, and properties. *Inorg. Chem.* **2016**, *55*, 8257–8259. [CrossRef]
30. Li, X.X.; Wang, Y.X.; Wang, R.H.; Cui, C.Y.; Tian, C.B.; Yang, G.Y. Designed assembly of heterometallic cluster organic frameworks based on Anderson-type polyoxometalate clusters. *Angew. Chem. Int. Ed.* **2016**, *55*, 6462–6466. [CrossRef]
31. Zhu, M.-C.; Huang, Y.-Y.; Ma, J.-P.; Hu, S.-M.; Wang, Y.; Guo, J.; Zhao, Y.-X.; Wang, L.-S. Coordination polymers based on organic–inorganic hybrid rigid rod comprising a backbone of Anderson-Evans POMs. *Cryst. Growth Des.* **2018**, *19*, 925–931. [CrossRef]
32. Yazigi, F.-J.; Wilson, C.; Long, D.-L.; Forgan, R.S. Synthetic considerations in the self-assembly of coordination polymers of pyridine-functionalized hybrid Mn-Anderson polyoxometalates. *Cryst. Growth Des.* **2017**, *17*, 4739–4748. [CrossRef]
33. Xu, W.; Pei, X.; Diercks, C.S.; Lyu, H.; Ji, Z.; Yaghi, O.M. A metal-organic framework of organic vertices and polyoxometalate linkers as a solid-state electrolyte. *J. Am. Chem. Soc.* **2019**, *141*, 17522–17526. [CrossRef] [PubMed]
34. Ma, R.; Liu, N.; Lin, T.-T.; Zhao, T.; Huang, S.-L.; Yang, G.-Y. Anderson polyoxometalate built-in covalent organic frameworks for enhancing catalytic performances. *J. Mater. Chem. A* **2020**, *8*, 8548–8553. [CrossRef]
35. Merkel, M.P.; Anson, C.E.; Kostakis, G.E.; Powell, A.K. Taking the third route for construction of POMOFs: The first use of carboxylate-functionalized Mn^{III} Anderson–Evans POM-hybrid linkers and lanthanide nodes. *Cryst. Growth Des.* **2021**, *21*, 3179–3190. [CrossRef]
36. Misra, A.; Kozma, K.; Streb, C.; Nyman, M. Beyond charge balance: Counter-cations in polyoxometalate chemistry. *Angew. Chem. Int. Ed.* **2020**, *59*, 596–612. [CrossRef]
37. Yu, W.D.; Zhang, Y.; Han, Y.Y.; Li, B.; Shao, S.; Zhang, L.P.; Xie, H.K.; Yan, J. Microwave-assisted synthesis of tris-Anderson polyoxometalates for facile CO_2 cycloaddition. *Inorg. Chem.* **2021**, *60*, 3980–3987. [CrossRef]
38. Fang, Q.; Fu, J.; Wang, F.; Qin, Z.; Ma, W.; Zhang, J.; Li, G. Tris functionalized Cu-centered cyclohexamolybdate molecular armor as a bimetallic catalyst for rapid p-nitrophenol hydrogenation. *New J. Chem.* **2019**, *43*, 28–36. [CrossRef]
39. Duan, F.X.; Liu, X.T.; Qu, D.; Li, B.; Wu, L.X. Polyoxometalate-based ionic frameworks for highly selective CO_2 capture and separation. *CCS Chem.* **2020**, *2*, 2676–2687. [CrossRef]
40. Cheng, M.; Xiao, Z.; Yu, L.; Lin, X.; Wang, Y.; Wu, P. Direct syntheses of nanocages and frameworks based on Anderson-type polyoxometalates via one-pot reactions. *Inorg. Chem.* **2019**, *58*, 11988–11992. [CrossRef]
41. Spackman, P.R.; Turner, M.J.; McKinnon, J.J.; Wolff, S.K.; Grimwood, D.J.; Jayatilaka, D.; Spackman, M.A. CrystalExplorer: A program for Hirshfeld surface analysis, visualization and quantitative analysis of molecular crystals. *J. Appl. Crystallogr.* **2021**, *54*, 1006–1011. [CrossRef] [PubMed]
42. Dolomanov, O.V.; Bourhis, L.J.; Gildea, R.J.; Howard, J.A.K.; Puschmann, H. OLEX2: A complete structure solution, refinement and analysis program. *J. Appl. Crystallogr.* **2009**, *42*, 339–341. [CrossRef]
43. Sheldrick, G.M. SHELXT-integrated space-group and crystal-structure determination. *Acta Cryst. A* **2015**, *71*, 3–8. [CrossRef] [PubMed]

Article

Organic/Inorganic Species Synergistically Supported Unprecedented Vanadomolybdates

Tian Chang, Di Qu, Bao Li * and Lixin Wu

State Key Laboratory of Supramolecular Structure and Materials, College of Chemistry, Jilin University, Changchun 130012, China
* Correspondence: libao@jlu.edu.cn

Abstract: Vanadomolybdates (VMos), comprised of Mo and V in high valences with O bridges, are one of the most important types of polyoxometalates (POMs), which have high activity due to their strong capabilities of gaining/losing electrons. Compared with other POMs, the preparation of VMos is difficult due to their relatively low structural stability, especially those with unclassical architectures. To overcome this shortcoming, in this study, triol ligands were applied to synthesize VMos through a beaker reaction in the presence of V_2O_5, Na_2MoO_4, and organic species in the aqueous solution. The single-crystal X-ray diffraction results indicate that two VMo clusters, $Na_4\{V_5Mo_2O_{19}[CH_3C(CH_2O)_3]\}\cdot 13H_2O$ and $Na_4\{V_5Mo_2O_{19}[CH_3CH_2C(CH_2O)_3]\}\cdot 13H_2O$, with a similar architecture, were synthesized, which were both stabilized by triol ligand and $\{MoO_6\}$ polyhedron. Both clusters are composed of five V ions and one Mo ion in a classical Lindqvist arrangement with an additional Mo ion, showing an unprecedented hepta-nuclear VMo structure. The counter Na^+ cations assemble into one-dimensional channels, which facilitates the transport of protons and was further confirmed by proton conductivity experiments. The present results provide a new strategy to prepare and stabilize VMos, which is applicable for developing other compounds, especially those with untraditional architectures.

Keywords: polyoxometalate; vanadomolybdate; triol ligand; covalent modification

1. Introduction

Organic species covalent functionalization on polyoxometalates (POMs) is an efficient method of constructing POMs that possess novel architectures and properties, and results in abundant achievements in the fields of organic and inorganic hybrids as well as materials science [1,2]. To date, several strategies have been developed in accordance with the structural features of POMs, and the classical structures are shown in Figure 1. By utilizing the strong Mo≡N bond, some N-containing organic molecules are capable of being anchored on a Lindqvist-type cluster $[Mo_6O_{19}]^{2-}$ (Figure 1a) [3,4]. In some lacunary POMs, P, Si, Sn, or Ge occupy the vacant position and introduce organically functional groups through P–C, Si–C, Sn–C, or Ge–C bonds (Figure 1b) [5]. As a comparison, modification through M–O–C (M = Mo, W, V) is a more convenient and more common method which provides volatile connection modes between organic and inorganic parts and is applicable to different types of POMs. The simplest modification through an M–O–C bond is that methanol or ethanol molecules replace bi- or tri-bridging O atoms of clusters, resulting in partially substituted POMs, which is applicable to nearly all types of POMs (Figure 1c) [6]. Beyond a single M–O–C bond, carboxyl is also a good candidate to covalently bond on the cluster through two M–O–C bonds sourced from one molecule, which provides a stronger combination (Figure 1d) [7,8]. In addition, triol ligands have the ability to supply three M–O–C bonds linking to the same C center through three methene units, which largely improves the stability of the formed complexes, and therefore has obtained rapid progress in the last two decades [9–11]. To date, Lindqvist-(Figure 1e), Anderson-(Figure 1f), Keggin-(Figure 1g), and Dawson-type clusters (Figure 1h) have all been involved in the successful modification by triol ligands with various architectures, which express excellent properties and various applications [12–14]. The unique coordination behavior of the triol ligand endows its ability to stabilize structures and makes it a good candidate for building POMs with novel or metastable clusters.

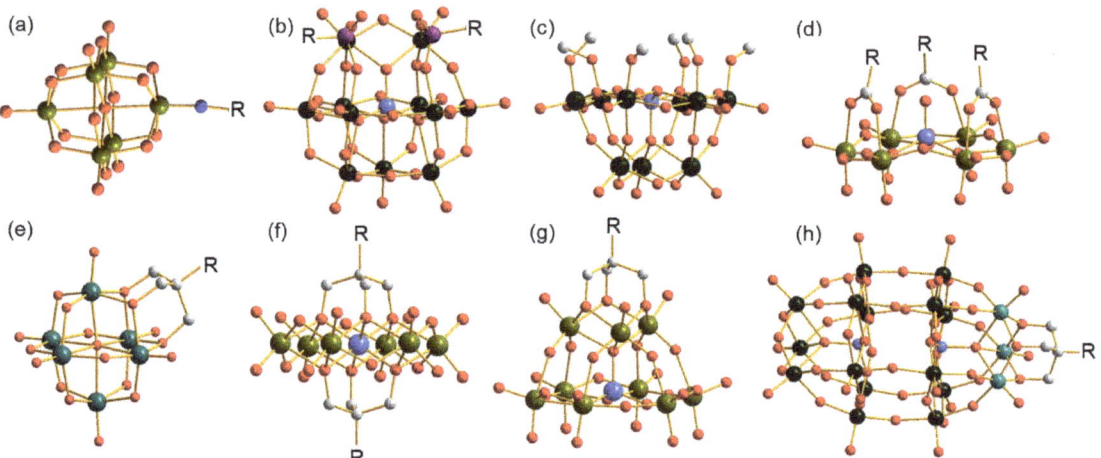

Figure 1. Ball-and-stick representation of several classical organo-functionalized clusters. (**a**) Imido-functionalized Lindqvist-type polyanion, (**b**) organophosphorus-functionalized lacunary Keggin-type polyanion, (**c**) methanol-functionalized lacunary Keggin-type polyanion, (**d**) carboxyl-functionalized Anderson-like-type polyanion; triol-ligand-functionalized (**e**) Lindqvist-type, (**f**) Anderson-type, (**g**) lacunary Keggin-type, and (**h**) Dawson-type polyanions. All H atoms are omitted for clarity. R represents the functional group. Dark cyan ball: V, olive ball: Mo, dark green ball: W, purple ball: P, light grey ball: C, red ball: O, blue ball: N, light blue ball: heteroatom.

Vanadomolybdates (VMos), as one of the important branches of POM chemistry, have attracted extensive attention due to their diverse structures and multi-functions in the fields of electrochemistry, magnetism, and catalysis [15–17]. Compared with other POMs, VMos have a stronger ability to gain or lose electrons, and therefore possess a higher activity, resulting in their excellent properties [18,19]. On the other hand, the structures of VMos have relatively low stability, especially those of lacunary species, which need additional components to support their architectures. For instance, with the aid of a transition metal ion dimer {Mn$_2$}, two unstable trivacant Keggin-type VMo clusters were linked to form a sandwich structure, which shows an unusual ferromagnetic coupling [20]. Another strategy for preparing VMos is utilizing the organic species, which are covalently anchored on the clusters and stabilize the obtained hybrids. A classical case was reported by Wei et al., where an unprecedented {V$_4$Mo$_3$} was synthesized in the presence of anilines as stabilizers, which replaced six terminal or bi-bridging O atoms of the {Mo$_3$} cluster [21]. With a similar strategy, an Anderson-type VMo cluster was also isolated, in which two triol ligands replaced all hydroxyls surrounding the heteroatom V, resulting in doubly decorated hybrids [22]. In another example, triethanolamine was applied to stabilize a {VMo$_6$O$_{25}$} cluster, forming an organic component single-functionalized product [23]. We have also made a contribution in this field showing that a triol ligand is used to anchor on the top position of [VMo$_9$O$_{34}$]$^{9-}$ cluster with the maintaining of the vacant sites, which are considered to be more active in the reaction [24]. These results indicate that organically covalent modification is an efficient method to prepare VMos, especially those without stable structures in the isolated state. More importantly, with the help of the coordination ability of organic species, VMos with novel structures are expectable.

During our investigation on triol-ligand-modified VMos, we discovered one type of novel Lindqvist-type derivative with V_2O_5, Na_2MoO_4, and triol ligand as reactants. In this work, the synthesis and structural characterization of $Na_4\{V_5Mo_2O_{19}[CH_3C(CH_2O)_3]\}\cdot 13H_2O$ (**1**) and $Na_4\{V_5Mo_2O_{19}[CH_3CH_2C(CH_2O)_3]\}\cdot 13H_2O$ (**2**) are presented. The polyanions of compounds **1** and **2** are isostructural except the terminal group of the triol ligand, which can both be seen as a $\{MoO_6\}$ polyhedron attached to a Lindqvist cluster by sharing three O atoms. As far as we know, this type of hepta-nuclear VMo has never been reported in the POM family. The results presented here would provide a new route for preparing triol-ligand-modified VMos with unusual architectures.

2. Results and Discussion
2.1. Synthesis of Compounds **1** and **2**

In the presence of the triol ligand, V_2O_5 and $Na_2MoO_4\cdot 2H_2O$ were added to water, which resulted in the formation of a light brown solution under heating conditions. Crystals of compounds **1** and **2** were obtained by standing the solutions for a week. Several factors are supposed to play important roles in the successful synthesis of compounds. Firstly, the molar ratio of V/Mo in the reactants is one such factor. Only when the molar ratio of V/Mo is in the range of 5:2 and 5:3 can the targeted products be obtained. If a higher percentage of V exists in the reactants, a triol-ligand-modified hexavanadate cluster is obtained, which has the same architecture as that reported in the literature [25]. If reactants with lower percentages of V are used, some VMo clusters are obtained and most of them cannot provide high-quality crystals for further accurate determination of their structures. Secondly, the control of pH is another essential factor. When the pH of the reaction solution is controlled between 4–5, the quality of the crystals is good for giving high-quality X-ray diffraction data for analysis. If the pH decreases to 3–4, the quality of the crystals becomes poor and is not very stable in air, which quickly degenerates into powder after the crystals leave the mother liquid due to weathering. In a more acidic environment with a smaller pH value, crystals cannot be obtained. On the other hand, when pH is higher than 5, the solubility of the reactants in the water becomes poor, and nearly no reaction occurs. In addition, the amount of triol ligand seems to have so little influence on the product that excessive organic species cannot change its number on the cluster, and only one triol ligand remains attached to the polyanion. On the other hand, the presence of the triol ligand is the essential factor for the formation of the present architecture. We have also conducted experiments under similar conditions without a triol ligand, and as a result, only some VMo species with usual structures were obtained. All these factors indicate that the compounds presented here are only successfully synthesized under relatively harsh conditions, which need good modulation to the reactions.

2.2. Structures of Compounds **1** and **2**

Single crystal X-ray diffraction analysis reveals that compounds **1** and **2** possess very similar architectures except the difference of the terminal groups of triol ligand, methyl for compound **1** and ethyl for compound **2**, as shown in Figure 2. Here, compound **1** is used as a representative for illustrating their structural characteristics.

2.2.1. Inorganic Architecture

The inorganic architecture of the polyanion of compound **1** is composed of five $\{VO_6\}$ and two $\{MoO_6\}$ polyhedra in the edge-sharing style. A Lindqvist-type cluster including five $\{VO_6\}$ and one $\{MoO_6\}$ polyhedra can be distinguished with a μ_6-O as the center, as shown in Figure S1a. The other $\{MoO_6\}$ polyhedron attaches to this cluster through three μ_2-O atoms in a face-sharing style, in which one O atom is from the V–O–V unit and the other two O atoms are sourced from V–O–Mo units. This connection makes the two $\{MoO_6\}$ polyhedra neighbors, and therefore the polyanion of compound **1** can also be seen as the combination of a mono-lacunary Lindqvist $\{V_5\}$ cluster with a metal dimer $\{Mo_2\}$ in the presence of five shared O atoms, as shown in Figure S1b. In addition, all bond lengths and angles in the cluster are in the normal ranges, which are listed as Tables S1–S4 in the Supporting Information.

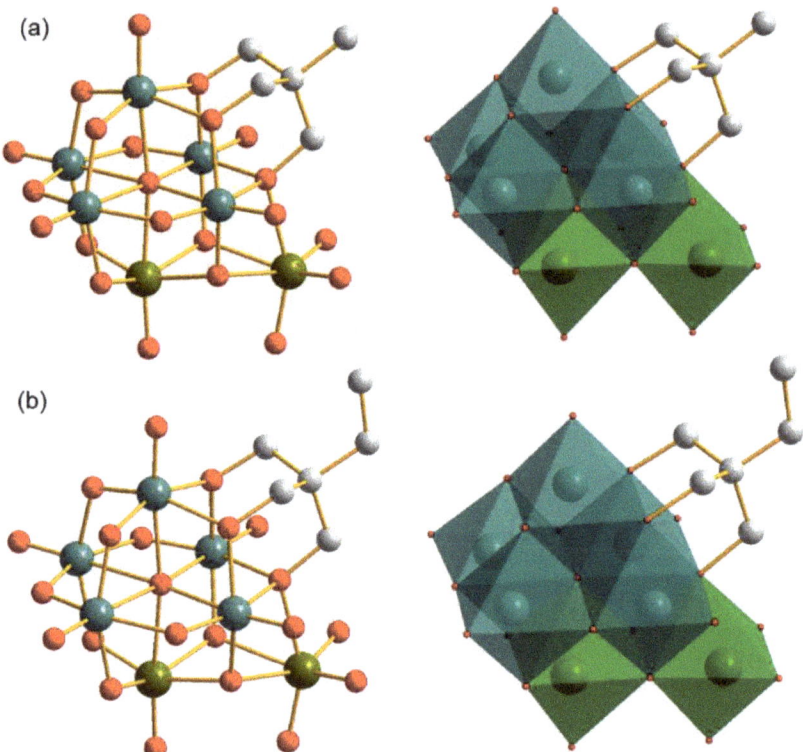

Figure 2. Polyanionic structures of compounds (**a**) **1** and (**b**) **2** in ball-and-stick representation (left) and combined ball-and-stick and polyhedron representation (right). All H atoms are omitted for clarity. Dark cyan ball: V, olive ball: Mo, light grey ball: C, red ball: O, dark cyan polyhedron: {VO$_6$}, olive polyhedron: {MoO$_6$}.

The unique combination of compound **1** results in an unreported atomic V/Mo ratio of 5:2. As an important sub-class of POMs, many VMos have been synthesized and reported, which possess various atomic V/Mo ratios. The change of atomic V/Mo ratio has an obvious influence on the architectures of the clusters, as well as their properties, which generally originates from the following aspects: (1) The V ion is used as a heteroatom in VMos, such as 1:6 for an Anderson-type POM, 1:12 for a Keggin-type POM, and 2:18 for a Dawson-type POM. It can also be applied in the lacunary cluster, such as 1:9 for a trivacant Keggin-type POM. (2) The V ion is used to replace one or more addenda atoms of clusters. For isopolyoxometalates, the Lindqvist-type [Mo$_6$O$_{19}$]$^{2-}$ can be substituted by one or two V ions to form a cluster with 1:5 or 2:4 atomic V/Mo ratios, while for octamolybdates, the substitution of one or two V ions results in products having 1:7 or 2:6 atomic V/Mo ratios, which derivatives have been widely investigated. For the heteropolyoxometalates, there are more possibilities due to the increase of the number of addenda atoms, which generate a series of VMos derivatives. (3) A V ion is used to build VMo clusters with nonclassical architectures. With this strategy, {VO$_4$}/{VO$_6$} can combine with {MoO$_6$} in various manners, and with the aid of other organic or inorganic species, more clusters with novel structures are obtained. Through the above-mentioned strategies, VMos with different atomic V/Mo ratios have been achieved, which are briefly summarized in Table 1.

It should be noted that partial V^V ions can be reduced to V^{IV} ions, which enriches the diversity of the structures. In the present case, by utilizing the strong coordination ability of the triol ligand, an extra {MoO_6} octahedron is introduced into the $[V_5MoO_{19}]^{7-}$ polyanion, resulting in the formation of architecture with an unreported atomic V/Mo ratio of 5:2. This result also proves that the atomic V/Mo ratio has an important influence on the structure of the cluster, which can be tuned and applied for finding new VMo species.

Table 1. Summary of atomic V/Mo ratios in some classical VMos.

Atomic V/Mo Ratio ^{a}	Formula of Polyanion ^{b}	Valence of V	Valence of Mo	Ref.
1:4	$[VMo_4O_{17}]^{5-}$	+5	+6	[26]
1:5	$[VMo_5O_{19}]^{3-}$	+5	+6	[27]
1:6	$[VMo_6O_{24}]^{6-}$	+5	+6	[16]
1:7	$[VMo_7O_{26}]^{5-}$	+5	+6	[15]
1:9	$\{VMo_9O_{31}[CH_3C(CH_2O)_3]\}^{6-}$	+5	+6	[24]
1:11	$[P(VMo_{11})O_{40}]^{4-}$	+5	+6	[28]
1:12	$[VMo_{12}O_{40}]^{3-}$	+5	+6	[29]
2:2	$[V_2O_2(\mu\text{-MeO})_2(\mu\text{-MoO}_4)_2(4,4'\text{-tBubpy}_2)]$	+5	+6	[30]
2:4	$[V_2Mo_4O_{19}]^{4-}$	+5	+6	[31]
2:6	$[V_2Mo_6O_{26}]^{6-}$	+5	+6	[32]
2:8	$[HV_2Mo_8O_{32}]^{5-}$	+5	+6	[26]
2:10	$[HV_2Mo_{10}O_{38}]^{5-}$	+5	+6	[26]
2:16	$[V_2Mo_{16}O_{58}]^{10-}$	+5	+6	[33]
2:18	$[V_2Mo_{18}O_{62}]^{6-}$	+5	+6	[34]
2:22	$[Fe_5CoMo_{22}V_2O_{87}(H_2O)]\}^{12-}$	+5	+6	[35]
3:3	$[V_3Mo_3O_{16}(C_5H_9O_3)]^{2-}$	+5	+6	[36]
3:9	$[V_3Mo_9O_{38}]^{7-}$	+5	+6	[26]
3:10	$[V(V^{IV}V^V Mo_{10}O_{40})]^{6-}$	+4, +5	+6	[37]
3:17	$[H_2V^{IV}Mo_{17}O_{54}(V^VO_4)_2]^{6-}$	+4, +5	+6	[38]
4:3	$[V_4Mo_3O_{14}(NAr)_3(\mu_2\text{-NAr})_3]^{9-}$	+5	+6	[21]
4:8	$[As_2V_4Mo_8AsO_{40}]^{5-}$	+5	+6	[39]
5:2	$\{V_5Mo_2O_{19}[CH_3C(CH_2O)_3]\}^{4-}$	+5	+6	This work
5:4	$[V_5Mo_4O_{27}]^{5-}$	+5	+6	[40]
5:8	$[V_5Mo_8O_{40}]^{7-}$	+5	+6	[41]
6:57	$[V_6Mo_{57}O_{183}(NO)_6(H_2O)_{18}]^{6-}$	+5	+6	[42]
7:8	$[(V^VMo_8V_4^{IV}O_{40})(V^{IV}O)_2]^{7-}$	+4, +5	+6	[43]
7:11	$[V^V_5V^{IV}_2Mo_{11}O_{52}(SeO_3)]^{7-}$	+4, +5	+6	[44]
8:2	$[V_8Mo_2O_{28}]^{4-}$	+5	+6	[26]
8:4	$[V_8Mo_4O_{36}]^{8-}$	+5	+6	[45]
9:1	$[V_9MoO_{28}]^{5-}$	+5	+6	[31]
9:8	$[(V^VMo^{VI}_4Mo^V_4V_4^{IV}O_{40})(V^{IV}O)_4]^{7-}$	+4, +5	+6	[43]
10:12	$[Mo_{12}V_{10}O_{58}(SeO_3)_8]^{10-}$	+5	+6	[44]
14:16	$[V_{12}^{IV}V^V_2Mo_{16}O_{84}]^{14-}$	+4, +5	+6	[46]

^{a} The atomic V/Mo ratios used here are the same with as those in formulas without an approximate process.
^{b} The apparent sequence of V and Mo in the formulas of polyanions are unified.

2.2.2. The Decoration of Triol Ligand on the Cluster

As mentioned above, compounds **1** and **2** possess a unique inorganic architecture, as well as an unreported atomic V/Mo ratio. Another feature of these two compounds is that a triol ligand attaches to the inorganic cluster, which plays a key role in the formation of polyanions. As is well known, bare Lindqvist-type polyoxovanadates (POVs) are not stable, and they can only be achieved in the presence of alcohols replacing one or more µ$_2$-O atoms to reduce the charge density of the surface. The used alcohols can be methanol and ethanol, as well as a triol ligand, which can provide a strong coordination environment and anchors on the cluster up to four times. As the derivatives of Lindqvist-type POVs, triol ligand is also the essential factor for the successful preparation of compounds **1** and **2**. It is the same with those in the Lindqvist-type POVs; the triol ligands in compounds **1** and **2** also replace three μ_2-O atoms of V–O–V units and cover on a {V$_3$} cluster. The decoration of triol ligand has an obvious influence on the V–(μ_2-O) bond lengths concerned, which possess an average value of 2.038 Å (in compound **1**). As a comparison, other V–(μ_2-O) bonds, which are not involved in the coordination with triol ligands, have a relatively shorter average bond length of 1.862 Å. A similar phenomenon is also found in triol-ligand-decorated trivacant Kegging-type cluster, showing the strong coordination ability of triol ligand [24].

A more interesting feature of the polyanions of compounds **1** and **2** is that the present cluster can be seen as the triol ligand and the {MoO$_6$} polyhedron co-supported $[V_5MoO_{19}]^{7-}$ polyanion. As shown in Figure S2a, both the triol ligand and {MoO$_6$} polyhedron share three O atoms with the $[V_5MoO_{19}]^{7-}$ cluster. This structure is different from the two-triol-ligand covalently decorated POVs [47]. As shown in Figure S2b,c, the triol ligands can anchor on the Lindqvist-type POVs in cis or trans configurations by replacing six different μ_2-O atoms, while in the present case, the triol ligand shares an O atom with the additional {MoO$_6$} polyhedron, resulting in their combination with the $[V_5MoO_{19}]^{7-}$ cluster through a total of five shared O atoms. In addition, it should be noted that all three terminal O atoms of {MoO$_6$} polyhedron are involved in the formation of Mo=O double bonds, with the bond lengths of 1.730, 1.730, and 1.744 Å, which are different from those in the triol-ligand-decorated heptavanadate cluster, with one water molecule serving as a terminal group with a much longer V–O bond length of 1.995 Å [48]. The triol ligand modification on the cluster also influences the packing model of the polyanions. As shown in Figure S3, a double layer comprised of polyanions forms, in which triol ligands face opposite directions towards the inside of the layer. This packing model can reduce the exposure possibility of methyl groups in an aqueous solution, which benefits the minimalization of interface energy and the crystallization of the compounds.

2.2.3. The Assembly Structure of Cations in Compounds **1** and **2**

Both compounds **1** and **2** crystallize with Na$^+$ as counter cations, and four crystally independent Na$^+$ ions exist in every asymmetric unit. Taking compound **1** as the example, Na1, Na2, and Na3 are located on the symmetric plane and therefore possess the site occupancy of 0.5. For Na4, though it is not located at any symmetry element, and has an apparent site occupancy of 1, it is treated as a disorder due to the following properties: (1) Na4 and its symmetric atom Na4′ have a distance of 2.128 Å, which is much shorter than that between two normal sodium cations with a distance over 3.0 Å, and is even shorter than the Na–O bond length with a general value over 2.3 Å. The abnormally short distance between Na4 and Na4′ indicates that the two Na$^+$ ions should exist alternately, showing a disorder in space. (2) Na4 has a thermal displacement parameter over 0.1, which is much higher than those of Na$_1$, Na$_2$, and Na$_3$ with an average value of about 0.04. After the disorder treatment, the thermal displacement parameter of Na4 decreases to about 0.04 and remains stable in the anisotropic state after several refinement cycles. (3) The whole polyanion has a total charge of −4, which needs two positive charges in the asymmetric unit for charge neutralization. Additionally, in the presence of three Na$^+$ ions with half site occupancies, the fourth one should also occupy the 0.5 site, which is also in accordance with the above analysis conditions. Finally, the elemental analysis result of Na also supports this treatment.

Four Na$^+$ ions express different coordination environments as well as coordination numbers (Figure 3a). Na1 is coordinated by six water molecules and one terminal O atom of polyanion, showing a seven-coordination type. Na2, Na3, and Na4 are all in a six-coordination type, in which all the coordination sites of Na2 and Na3 are occupied by water molecules. As a comparison, the coordination environment of Na4 is completed by five water molecules and one O atom from polyanion. One important feature of these Na$^+$ ions is that the coordinated water molecules are generally shared by two or three cations, resulting in their possessing wide edge- or face-connection with each other.

This feature also facilitates the formation of the extended structure. As shown in Figure 3b, a regular honeycomb structure formed by Na⁺ ions and its coordinating O atoms can be obtained, showing one-dimensional (1D) channels along the c axis. Here, to better exhibit the architecture of the packing of Na⁺ ions, polyanions are omitted for clarity except for two O atom links, with Na⁺ being kept. The passageways constructed here are helpful for the fast migration of ions, which can be used as a proton conductor.

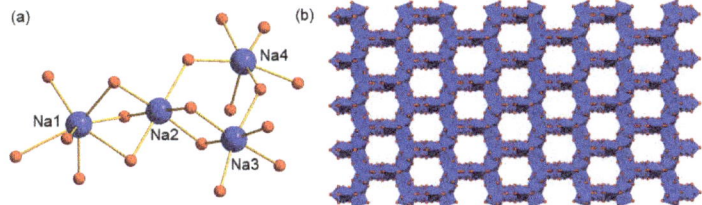

Figure 3. (a) The coordination environments of four independent Na⁺, and (b) the packing model of counter cation Na⁺ in compound **1**. The polyanions are omitted for clarity. Blue ball: Na, red ball O, blue polyhedron: {NaO₆} or {NaO₇}.

2.2.4. FT-IR, XPS and TGA Curves of Compounds 1 and 2

Except for single crystal X-ray diffraction analysis, the structures of compounds **1** and **2** are also confirmed by FT-IR spectra. As shown in Figure S4a, the FT-IR spectrum of compound **1** in the low wavenumber region expresses characteristic vibration of the inorganic cluster sourced from V–O, Mo–O, and Mo=O. The existence of wavenumbers over 1000 cm^{-1} can be ascribed to the vibrations of C–O, C–H, and H–O, showing the occurrence of organic components in the compound. Compound **2** has a similar FI-IR pattern as that of compound **1**, which is shown in Figure S4b.

XPS was used to identify the bond valence of Mo and V in the clusters. For compound **1**, as shown in Figure S5a, the binding energy peaks located at 235 and 232 eV are sourced from Mo^{VI}3d$_{3/2}$ and Mo^{VI}3d$_{5/2}$, indicating that the bond valence of Mo is +6. The bond valence of V is determined in Figure S5b, in which two binding energy peaks at 517 and 524 eV are V^{V}2p$_{1/2}$ and V^{V}2p$_{3/2}$, respectively, showing a +5 bond valence for V. As shown in Figure S6, the XPS spectra of compound **2** are similar to those of compound **1**, which also indicate the existence of Mo^{VI} and V^{V} in the cluster.

To verify the purity of the prepared compounds, the powder X-ray diffraction (PXRD) patterns of compounds **1** and **2** was checked. As shown in Figure S7, the as-synthesized and simulated PXRD patterns are similar for each compound, indicating that the powder products maintained the same architectures as those in the single crystal state.

The thermal stability of compounds **1** and **2** were evaluated through TGA analysis. As shown in Figure S8, both compounds **1** and **2** quickly lose their lattice water molecules in the range of room temperature to 100 °C. Additionally, the organic species leave the cluster, followed by the decomposition of polyanion, showing the supporting role of triol ligand in the cluster.

2.2.5. Stability of Compounds 1 and 2 in Aqueous Solution

The existing states of compounds **1** and **2** in an aqueous solution were examined using ¹H NMR spectra. As shown in Figure S9, compound **1** is not very stable in the aqueous solution and decomposition occurred in the process of dissolving. The signals belonging to the free triol ligand appeared at the beginning and increased over time. Calculated from the integration value of the peaks, the initial decomposition ratio is about 6.6%, and reaches 7.9% after 3 days. The different chemical environments of methene groups in compound **1** generate the splitting of the signal, showing double peaks at around 4.9 ppm. Compound **2** shows a similar behavior to that of compound **1** in an aqueous solution (Figure S10). The difference is that compound **2** seems to have a relatively higher stability, and only

2.1% decomposes at the beginning, reaching 6.2% after 3 days. It should be noted that the decomposition process can be accelerated by the addition of acid or base.

2.3. Proton Conductivity of Compounds 1 and 2

Proton-conducting materials play essential roles in fuel cells, which have seen rapid progress due to the demand for clean energy. For building high-proton-conductivity materials, an extensive hydrogen bonding network is needed, which accelerates the transport of the proton and induces energy loss. POMs are good candidates for proton-conducting materials due to their abundant O atoms on the surface of the cluster, which can serve as hydrogen bonding acceptors. Here, the proton conductivity of compounds **1** and **2** under different temperatures are investigated in the condition of a controlled 75% relative humidity (RH). As shown in Figure 4a, with the temperature changing from 25 to 60 °C, the proton conductivity of compound **1** increased from 2.25×10^{-5} to 7.74×10^{-5} S cm^{-1}. Based on the proton conductivities at various temperatures, the activation energy (E_a) of the proton conductivity of compound **1** was evaluated. As shown in Figure S11a, a linear fitness can be obtained with the ln(σT) as longitudinal ordinate and 1/T as horizontal ordinate, from which the activity energy E_a is calculated as 0.32 eV. The relatively low E_a value (smaller than 0.4 eV) indicates that the proton conduction process in compound **1** is ascribed to a Grotthuss mechanism. This result is also in accordance with the crystal structure described above, in which a 1D channel exists and facilitates the transfer of protons. In addition, the water molecules on the wall of the channel also promote proton transport. The proton conductivity behavior of compound **2** was also investigated to evaluate the influence of the terminal group of the triol ligand. As shown in Figure 4b, in the range of 25 to 60 °C, the proton conductivity increases from 1.97×10^{-5} to 7.48×10^{-5} S cm^{-1}. It can be seen that compounds **1** and **2** have similar proton conductivity, which sources their similar crystal structures as well as packing models. The slight decrease of proton conductivity of compound **2** compared to that of compound **1** is ascribed to originate from the slightly higher hydrophobicity of ethyl to methyl. In addition, the E_a of proton conduction for compound **2** was calculated as 0.36 eV (Figure S11b), which also indicates a Grotthuss mechanism.

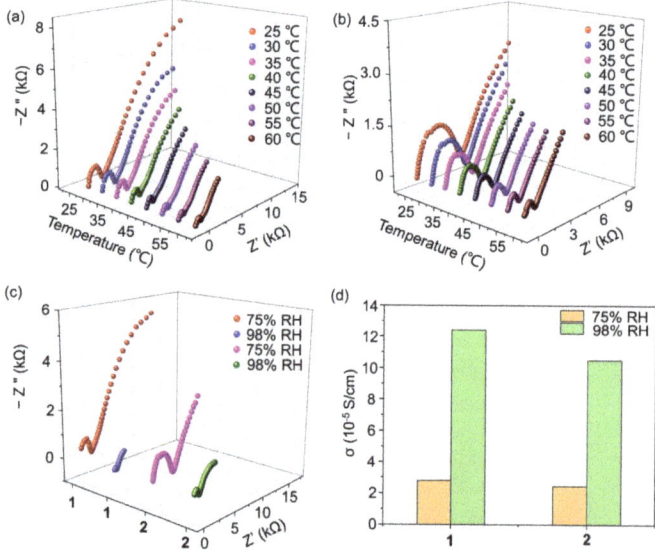

Figure 4. Nyquist plots of compounds (**a**) **1**, and (**b**) **2** at various temperatures and 75% RH; (**c**) Nyquist plots of compounds **1** and **2** under the conditions of 75% and 98% RH at 30 °C; (**d**) proton conductivity of compounds **1** and **2** under different conditions.

We further investigated the influence of RH on proton conductivity. As shown in Figure 4c, when the RH increases to 98%, the proton conductivity of compound **1** increases to 1.24×10^{-4} S cm^{-1} at 30 °C, showing a 4.4 times larger elevation compared with that under 75% RH. In a higher RH condition, the number of dissociated water molecules increases and promotes the transport of protons in the 1D channel through the stable hydrogen bonding network. Compound **2** also shows similar proton conductivity behaviors and has a value of 1.05×10^{-4} S cm^{-1} at 30 °C and 98% RH, which also shows a 4.3 times improvement. By collecting proton conductivity under different conditions, as shown in Figure 4d, it can be concluded that RH has a more obvious influence than that of temperature. In addition, as listed in Table S5, we have also collected the recent results of the proton conductivity of POMs, showing that compounds **1** and **2** possess a relatively high performance.

3. Materials and Methods

3.1. Instruments and Materials

All chemicals used here were purchased from Aladdin and used directly. All reactions were conducted with double-distilled water as solvent. C and H elemental analysis was made on a vario MICRO cube from the Elementar Company of Germany. Elemental analysis of V, Mo, and Na was conducted on a PLASMA-SPEC (I) inductively coupled plasma atomic emission spectrometer. FT-IR spectra were obtained from a Bruker Vertex 80v spectrometer in KBr pellets, in which the detector was DTGS and the resolution was 4 cm^{-1}. Thermal stability of compounds was evaluated by a Q500 Thermal Analyzer (New Castle TA Instruments, New Castle, DE, USA) with a flowing N$_2$ atmosphere and a heating rate of 10 °C·min^{-1}. XPS data were acquired on an ESCALAB-250 spectrometer with a monochromic X-ray source (Al Kα line, 1486.6 eV). The charging shift of XPS was corrected at a binding energy of C1s at 284.6 eV. Electrochemical impedance spectroscopy was recorded on a Solartron 1260 A impedance analyzer. Powder X-ray diffraction data were recorded on a Rigaku SmartLab X-ray diffractometer with a Cu Kα radiation at a wavelength of 1.54 Å. ^{1}H NMR spectra were obtained from a Bruker AVANCE 500 MHz spectrometer. The proton conductivities of samples were calculated according to the following equation:

$$\sigma = l/SR$$

where σ is the proton conductivity (S/cm), and l, and S, and R are the thickness, area, and resistance of the sample, respectively.

Single-crystal X-ray diffraction data of compounds **1** and **2** were collected on a Bruker D8 VENTURE diffractometer with graphite-monochromated Mo Kα (λ = 0.71073 Å) at 293 K. Both crystals were solved by *SHELXT* and refined by full-matrix-least-squares fitting for F^2 using the Olex^2 software. All non-H atoms were refined with anisotropic thermal parameters. A summary of the crystallographic data and structural refinements for compounds **1** and **2** is listed in Table 2. The files with CCDC numbers 2206198 and 2206199 contain the supplementary crystallographic data for this paper. These data are available free of charge from www.ccdc.cam.ac.uk/data_request/cif (accessed on 9 September 2022), or by emailing data_request@ccdc.cam.ac.uk, or by contacting The Cambridge Crystallographic Data Centre, 12 Union Road, Cambridge, CB2 1EZ, UK.

3.2. Synthesis of Compounds **1** and **2**

3.2.1. Synthesis of Na$_4${V$_5$Mo$_2$O$_{19}$[CH$_3$C(CH$_2$O)$_3$]}·13H$_2$O (1)

Na$_2$MoO$_4$·2H$_2$O (0.88 g, 2.00 mmol) and V$_2$O$_5$ (0.45 g, 2.50 mmol) were mixed in 10 mL of deionized water, and then CH$_3$C(CH$_2$OH)$_3$ (0.24 g, 2.00 mmol) was added under stirring. Concentrated HCl was used to adjust the pH of the mixture to 4~5, which was further heated to 60 °C and maintained for 4 h. After a hot filtration process, a light brown solution was obtained, which generated light brown crystals after about one week at room temperature with a yield of 54.1% based on Mo. Elemental analysis calcd. (%) for Na$_4${V$_5$Mo$_2$O$_{19}$[CH$_3$C(CH$_2$O)$_3$]}·13H$_2$O (Mw = 1193.87 g mol^{-1}): C 5.03%, H 2.96%,

Na 7.70%, V 21.33%, and Mo 16.07%; found C 5.06%, H 2.98%, Na 7.71%, V 21.50%, and Mo 16.01%.

Table 2. The summary of crystal data and structural refinements for compounds **1** and **2**.

Item	Compound 1	Compound 2
Formula	Na$_4${V$_5$Mo$_2$O$_{19}$[CH$_3$C-(CH$_2$O)$_3$]}·13H$_2$O	Na$_4${V$_5$Mo$_2$O$_{19}$[CH$_3$CH$_2$C-(CH$_2$O)$_3$]}·13H$_2$O
Formula weight	1193.87	1207.89
Crystal system	Monoclinic	Monoclinic
Space group	$C2/m$	$C2/m$
a (Å)	22.294 (5)	22.170 (1)
b (Å)	10.180 (2)	10.146 (1)
c (Å)	16.394 (4)	16.777 (1)
α (deg.)	90	90
β (deg.)	114.233 (8)	112.257 (2)
γ (deg.)	90	90
V (Å^3)	3392.6 (12)	3492.5 (3)
Z	4	4
D_c (g cm^{-3})	2.337	2.297
F (000)	2352	2384
Reflections coll./unique	24947/4045	21017/3261
R_{int}	0.0785	0.0315
GOOF on F^2	1.029	1.111
[a] R_1 [$I > 2\sigma(I)$]	0.0468	0.0410
[b] wR_2 (all data)	0.1154	0.1218
CCDC no.	2206198	2206199

[a] $R_1 = \Sigma||F_o| - |F_c||/\Sigma|F_o|$. [b] $wR_2 = {\Sigma[w(F_o^2 - F_c^2)^2]/\Sigma[w(F_o^2)^2]}^{1/2}$.

3.2.2. Synthesis of Na$_4${V$_5$Mo$_2$O$_{19}$[CH$_3$CH$_2$C(CH$_2$O)$_3$]}·13H$_2$O (**2**)

The synthetic procedure of compound **2** was similar to that of compound **1**, except that the triol ligand was replaced by CH$_3$CH$_2$C(CH$_2$OH)$_3$ (0.27 g, 2.00 mmol) and the yield was 55.2% based on Mo. Elemental analysis calcd. (%) for [Na$_4${V$_5$Mo$_2$O$_{19}$[CH$_3$CH$_2$C(CH$_2$O)$_3$]}·13H$_2$O (Mw = 1207.89 g mol^{-1}): C 5.97%, H 3.09%, Na 7.61%, V 21.09%, and Mo 15.89%; found C 5.89%, H 3.05%, Na 7.55%, V 21.03%, and Mo 15.94%.

4. Conclusions

In conclusion, triol-ligand and {MoO$_6$} polyhedron co-decorated [V$_5$MoO$_{19}$]$^{7-}$ polyanion is constructed with different terminal groups. The prepared compounds present a new type of derivative of Lindqvist-type clusters, which have never been reported previously. The organic and inorganic parts both attach to the main polyanion by replacing three O atoms in a *cis* conformation, in which one O atom is shared by three components and further improves the stability of the cluster. This manner of combination is different from those two triol-ligand-decorated Lindqvist-type clusters, in which no O atoms are shared by organic species. Another feature of the prepared compounds is that they express a new type of VMos with an unreported atomic ratio of 5:2. In addition, the Na$^+$ cations in compounds assemble into 1D channels, whose walls are full of water molecules. The further proton conductivity experiments show that this structure facilitates the transport of protons, which bears a Grotthuss mechanism for both compounds.

Supplementary Materials: The following supporting information can be downloaded at: https://www.mdpi.com/article/10.3390/molecules27217447/s1. Figure S1: Ball-and-stick representation of the cluster comprising of five {VO$_6$} and one {MoO$_6$} polyhedra, showing a Lindqvist-type structure, and polyhedron representation of inorganic architecture of compound **1** comprising of a mono-lacunary {V$_6$} cluster and a {Mo$_2$} dimer; Figure S2: Ball-and-stick representation of polyanion of compound **1**, and two triol ligands covalently modified Lindqvist {V$_6$} cluster in *cis* and *trans* conformations; Figure S3: The packing model of polyanion of compound **1** along *a* axis, showing a double-layer structure; Figure S4: FT-IR spectra of compounds **1** and **2**; Figures S5 and S6: XPS spectra for Mo and V of compounds **1** and **2**; Figure S7: PXRD patterns of the as-synthesized and simulated for (a) compound **1**, and (b) compound **2**; Figure S8: TGA curves of compounds **1** and **2**; Figure S9: ^{1}H NMR spectra of triol ligand TME and compound **1** after its dissolving in water for different amounts of time; Figure S10. ^{1}H NMR spectra of triol ligand TMP and compound **2** after its dissolving in water for different amounts of time; Figure S11: Linear fitness of ln(σT) versus 1/T of compounds **1** and **2**; Table S1: Important bond lengths in the cluster of compound **1**; Table S2: Important bond angles in the cluster of compound **1**; Table S3: Important bond lengths in the cluster of compound **2**; Table S4: Important bond angles in the cluster of compound **2**; Table S5: The proton conductivity performance of different POMs.

Author Contributions: Conceptualization, B.L.; formal analysis, T.C.; investigation, T.C. and D.Q.; writing—original draft preparation, T.C.; writing—review and editing, B.L. and L.W.; supervision, B.L.; funding acquisition, B.L. All authors have read and agreed to the published version of the manuscript.

Funding: This research is funded by the Natural Science Foundation of China, grant number 22172060.

Institutional Review Board Statement: Not applicable.

Informed Consent Statement: Not applicable.

Data Availability Statement: Crystallographic data is obtainable free of charge from the Cambridge Crystallographic Data Centre via www.ccdc.cam.ac.uk/data_request/cif (accessed on 9 September 2022).

Conflicts of Interest: The authors declare no conflict of interest.

Sample Availability: Samples of the compounds **1** and **2** are available from the authors.

References

1. Wei, Z.; Wang, J.; Yu, H.; Han, S.; Wei, Y. Recent Advances of Anderson-Type Polyoxometalates as Catalysts Largely for Oxidative Transformations of Organic Molecules. *Molecules* **2022**, *27*, 5212. [CrossRef] [PubMed]
2. Thorimbert, S.; Hasenknopf, B.; Lacôte, E. Cross-Linking Organic and Polyoxometalate Chemistries. *Isr. J. Chem.* **2011**, *51*, 275–280. [CrossRef]
3. Pardiwala, A.; Kumar, S.; Jangir, R. Insights into organic-inorganic hybrid molecular materials: Organoimido functionalized polyoxomolybdates. *Dalton Trans.* **2022**, *51*, 4945–4975. [CrossRef] [PubMed]
4. Zhang, J.; Xiao, F.; Hao, J.; Wei, Y. The chemistry of organoimido derivatives of polyoxometalates. *Dalton Trans.* **2012**, *41*, 3599–3615. [CrossRef]
5. Cameron, J.M.; Guillemot, G.; Galambos, T.; Amin, S.S.; Hampson, E.; Mall Haidaraly, K.; Newton, G.N.; Izzet, G. Supramolecular assemblies of organo-functionalised hybrid polyoxometalates: From functional building blocks to hierarchical nanomaterials. *Chem. Soc. Rev.* **2022**, *51*, 293–328. [CrossRef]
6. Wang, Y.; Liu, X.T.; Xu, W.; Yue, Y.; Li, B.; Wu, L.X. Triol-ligand modification and structural transformation of Anderson-Evans oxomolybdates via modulating oxidation state of Co-heteroatom. *Inorg. Chem.* **2017**, *56*, 7019–7028. [CrossRef]
7. Wang, Y.; Zhang, G.L.; Jiang, F.R.; Li, B.; Wu, L.X. A closed hollow capsule structure assembled by double acetate-decorated Anderson-like polyanions. *J. Coord. Chem.* **2017**, *70*, 25–35. [CrossRef]
8. Ma, P.T.; Hu, F.; Wang, J.P.; Niu, J.Y. Carboxylate covalently modified polyoxometalates: From synthesis, structural diversity to applications. *Coord. Chem. Rev.* **2019**, *378*, 281–309. [CrossRef]
9. Wang, Y.R.; Duan, F.X.; Liu, X.T.; Li, B. Cations Modulated Assembly of Triol-Ligand Modified Cu-Centered Anderson-Evans Polyanions. *Molecules* **2022**, *27*, 2933. [CrossRef]
10. Duan, F.X.; Liu, X.T.; Qu, D.; Li, B.; Wu, L.X. Polyoxometalate-based ionic frameworks for highly selective CO$_2$ capture and separation. *CCS Chem.* **2020**, *2*, 2676–2687. [CrossRef]
11. Zhang, J.W.; Huang, Y.C.; Li, G.; Wei, Y.G. Recent advances in alkoxylation chemistry of polyoxometalates: From synthetic strategies, structural overviews to functional applications. *Coord. Chem. Rev.* **2019**, *378*, 395–414. [CrossRef]

12. Jia, H.; Li, Q.; Bayaguud, A.; Huang, Y.; She, S.; Chen, K.; Wei, Y. Diversified polyoxovanadate derivatives obtained by copper(i)-catalysed azide-alkyne cycloaddition reaction: Their synthesis and structural characterization. *Dalton Trans.* **2018**, *47*, 577–584. [CrossRef] [PubMed]
13. Santoni, M.P.; Pal, A.K.; Hanan, G.S.; Tang, M.C.; Furtos, A.; Hasenknopf, B. A light-harvesting polyoxometalate-polypyridine hybrid induces electron transfer as its Re(I) complex. *Dalton Trans.* **2014**, *43*, 6990–6993. [CrossRef] [PubMed]
14. Gao, B.; Li, B.; Wu, L.X. Layered supramolecular network of cyclodextrin triplets with azobenzene-grafting polyoxometalate for dye degradation and partner-enhancement. *Chem. Commun.* **2021**, *57*, 10512–10515. [CrossRef] [PubMed]
15. Zhao, Y.; Gao, Q.; Tao, R.; Li, F.; Sun, Z.; Xu, L. Molybdovanadate [VMo$_7$O$_{26}$]$^{5-}$ cluster directed inorganic-organic hybrid: The highest coordination linkage and In Situ isoniazid dimerization. *Inorg. Chem. Commun.* **2018**, *89*, 94–98. [CrossRef]
16. Gao, Q.; Hu, D.H.; Duan, M.H.; Li, D.H. A novel organic-inorganic hybrid built upon both [VMo$_6$O$_{24}$]$^{6-}$ and [Mo$_7$O$_{24}$]$^{6-}$ units: Synthesis, crystal structure, surface photovoltage and electrocatalytic activities. *J. Mol. Struct.* **2019**, *1184*, 400–404. [CrossRef]
17. Krivosudsky, L.; Roller, A.; Rompel, A. Regioselective synthesis and characterization of monovanadium-substituted beta-octamolybdate [VMo$_7$O$_{26}$]$^{5-}$. *Acta Cryst. C* **2019**, *75*, 872–876. [CrossRef]
18. Amini, M.; Sheykhi, A.; Naslhajian, H.; Bayrami, A.; Bagherzadeh, M.; Hołyńska, M. A novel 12-molybdovanadate nanocluster: Synthesis, structure investigation and its application as an efficient heterogeneous sulfoxidation catalyst. *Inorg. Chem. Commun.* **2017**, *83*, 103–108. [CrossRef]
19. Odyakov, V.F.; Zhizhina, E.G.; Rodikova, Y.A.; Gogin, L.L. Mo-V-Phosphoric Heteropoly Acids and Their Salts: Aqueous Solution Preparation—Challenges and Perspectives. *Eur. J. Inorg. Chem.* **2015**, *2015*, 3618–3631. [CrossRef]
20. Yang, Y.; Xu, L.; Gao, G.; Li, F.; Liu, X.; Guo, W. An Unexpected Ferromagnetic Coupling in a Dinuclear Manganese(II) Linked Trivacant Heteropolymolybdate Derivative. *Eur. J. Inorg. Chem.* **2009**, *2009*, 1460–1463. [CrossRef]
21. Huang, Y.; Zhang, J.; Ge, J.; Sui, C.; Hao, J.; Wei, Y. [V$_4$Mo$_3$O$_{14}$(NAr)$_3$(μ_2-NAr)$_3$]$^{2-}$: The first polyarylimido-stabilized molybdovanadate cluster. *Chem. Commun.* **2017**, *53*, 2551–2554. [CrossRef]
22. Cheng, M.; Xiao, Z.; Yu, L.; Lin, X.; Wang, Y.; Wu, P. Direct syntheses of nanocages and frameworks based on Anderson-type polyoxometalates via one-pot reactions. *Inorg. Chem.* **2019**, *58*, 11988–11992. [CrossRef] [PubMed]
23. Gao, Q.; Li, F.; Wang, Y.; Xu, L.; Bai, J.; Wang, Y. Organic functionalization of polyoxometalate in aqueous solution: Self-assembly of a new building block of {VMo$_6$O$_{25}$} with triethanolamine. *Dalton Trans.* **2014**, *43*, 941–944. [CrossRef] [PubMed]
24. Qu, D.; Liu, X.T.; Duan, F.X.; Xue, R.; Li, B.; Wu, L.X. {VMo$_9$O$_{31}$[RC(CH$_2$O)$_3$]}$^{6-}$: The first class of triol ligand covalently-decorated Keggin-type polyoxomolybdates. *Dalton Trans.* **2020**, *49*, 12950–12954. [CrossRef]
25. Chen, Q.; Goshorn, D.P.; Scholes, C.P.; Tan, X.L.; Zubieta, J. Coordination compounds of polyoxovanadates with a hexametalate core. Chemical and structural characterization of [V$^{VI}_6$O$_{13}$[(OCH$_2$)$_3$CR]$_2$]$^{2-}$, [V$^{VI}_6$O$_{11}$(OH)$_2$[(OCH$_2$)$_3$CR]$_2$], [V$^{IV}_4$V$^V_2$O$_9$(OH)$_4$[(OCH$_2$)$_3$CR]$_2$]$^{2-}$, and [V$^{IV}_6$O$_7$(OH)$_6$](OCH$_2$)$_3$CR]$_2$]$_2$. *J. Am. Chem. Soc.* **1992**, *114*, 4667–4681. [CrossRef]
26. Howarth, O.W.; Pettersson, L.; Andersson, I. Aqueous molybdovanadates at high Mo: V ratio. *J. Chem. Soc. Dalton Trans.* **1991**, 1799–1812. [CrossRef]
27. Tucher, J.; Wu, Y.; Nye, L.C.; Ivanovic-Burmazovic, I.; Khusniyarov, M.M.; Streb, C. Metal substitution in a Lindqvist polyoxometalate leads to improved photocatalytic performance. *Dalton Trans.* **2012**, *41*, 9938–9943. [CrossRef] [PubMed]
28. Himeno, S.; Ishio, N. A voltammetric study on the formation of V(V)- and V(IV)-substituted molybdophosphate(V) complexes in aqueous solution. *J. Electroanal. Chem.* **1998**, *451*, 203–209. [CrossRef]
29. Zhang, S.W.; Huang, G.Q.; Wei, Y.G.; Shao, M.C.; Tang, Y.Q. Structure of Na$_3$[VMo$_{12}$O$_{40}$]·19H$_2$O. *Acta Cryst. C* **1993**, *49*, 1446–1448. [CrossRef]
30. Kodama, S.; Nomoto, A.; Yano, S.; Ueshima, M.; Ogawa, A. Novel Heterotetranuclear V$_2$Mo$_2$ or V$_2$W$_2$ Complexes with 4,4'-Di-tert-butyl-2,2'-bipyridine: Syntheses, Crystal Structures, and Catalytic Activities. *Inorg. Chem.* **2011**, *50*, 9942–9947. [CrossRef]
31. Maksimovskaya, R.I.; Chumachenko, N.N. ^{51}V and ^{17}O NMR studies of the mixed metal polyanions in aqueous V—Mo solutions. *Polyhedron* **1987**, *6*, 1813–1821. [CrossRef]
32. Björnberg, A. Multicomponent polyanions. 26. The crystal structure of Na$_6$Mo$_6$V$_2$O$_{26}$(H$_2$O)$_{16}$, a compound containing sodium-coordinated hexamolybdodivanadate anions. *Acta Cryst. B* **1979**, *35*, 1995–1999. [CrossRef]
33. Li, F.; Meng, F.; Ma, L.; Xu, L.; Sun, Z.; Gao, Q. 3D pure inorganic framework based on polymolybdovanadate possessing photoelectric properties. *Dalton Trans.* **2013**, *42*, 12079–12082. [CrossRef]
34. Liang, D.-J.; Liu, S.-X.; Ren, Y.-H.; Zhang, C.-D.; Xu, L. [Mo$_{18}$O$_{54}$(VO$_4$)$_2$]$^{6-}$: A conventional Dawson structure with unpredicted transition metal hetero-atoms based on V^{V}O$_4$ tetrahedra. *Inorg. Chem. Commun.* **2007**, *10*, 933–935. [CrossRef]
35. Gao, Q.; Li, F.; Xu, L. 1D pure inorganic helical chain framework built upon banana-shaped molybdovanadates: Synthesis, crystal structure, and magnetic properties. *Inorg. Chem. Commun.* **2015**, *59*, 50–52. [CrossRef]
36. Karoui, H.; Ritchie, C. Microwave-assisted synthesis of organically functionalized hexa-molybdovanadates. *New J. Chem.* **2018**, *42*, 25–28. [CrossRef]
37. Cindrić, M.; Kamenar, B.; Strukan, N.; Veksli, Z. Synthesis, structure and ESR spectrum of (Hmorph)$_6$[V^{IV}, V^V, Mo$_{10}$)VO$_{40}$]·3H$_2$O. *Polyhedron* **1995**, *14*, 1045–1049. [CrossRef]
38. Miras, H.N.; Long, D.L.; Kogerler, P.; Cronin, L. Bridging the gap between solution and solid state studies in polyoxometalate chemistry: Discovery of a family of [V$_1$M$_{17}$]-based cages encapsulating two {V^{V}O$_4$} moieties. *Dalton Trans.* **2008**, 214–221. [CrossRef]

39. Müller, A.; Krickemeyer, E.; Dillinger, S.; Bögge, H.; Plass, W.; Proust, A.; Dloczik, L.; Menke, C.; Meyer, J.; Rohlfing, R. New Perspectives in Polyoxometalate Chemistry by isolation of compounds containing very large moieties as transferable building blocks: $(NMe_4)_5[As_2Mo_8V_4AsO_{40}]\cdot 3H_2O$, $(NH_4)_{21}[H_3Mo_{57}V_6(NO)_6O_{183}(H_2O)_{18}]\cdot 65H_2O$, $(NH_2Me_2)_{18}(NH_4)_6[Mo_{57}V_6(NO)_6O_{183}(H_2O)_{18}]\cdot 14H_2O$, and $(NH_4)_{12}[Mo_{36}(NO)_4O_{108}(H_2O)_{16}]\cdot 33H_2O$. *Z. Anorg. Allg. Chem.* **1994**, *620*, 599–619. [CrossRef]
40. Kamenar, B.; Cindrić, M.; Strukan, N. Synthesis and structure of a molybdovanadate with the asymmetric $[Mo_4V_5O_{27}]^{5-}$ anion. *Polyhedron* **1994**, *13*, 2271–2275. [CrossRef]
41. Björnberg, A. Multicomponent polyanions. 28. The structure of $K_7Mo_8V_5O_{40}\cdot \sim 8H_2O$, a compound containing a structurally new potassium-coordinated octamolybdopentavanadate anion. *Acta Cryst. B* **1980**, *36*, 1530–1536. [CrossRef]
42. Zhang, S.-W.; Huang, G.-Q.; Shao, M.-C.; Tang, Y.-Q. Crystal structure of a novel mixed valence Mo^V–Mo^{VI} heteropolymolybdate cluster: $[H_3O^+]_6[Mo_{57}V_6O_{183}(NO)_6(H_2O)_{18}]^{6-}\cdot 89H_2O$. *J. Chem. Soc. Chem. Commun.* **1993**, 37–38. [CrossRef]
43. Yao, S.; Zhang, Z.; Li, Y.; Wang, E. Two dumbbell-like polyoxometalates constructed from capped molybdovanadate and transition metal complexes. *Inorg. Chim. Acta* **2010**, *363*, 2131–2136. [CrossRef]
44. Corella-Ochoa, M.N.; Miras, H.N.; Long, D.L.; Cronin, L. Controlling the self-assembly of a mixed-metal Mo/V-selenite family of polyoxometalates. *Chem. Eur. J.* **2012**, *18*, 13743–13754. [CrossRef] [PubMed]
45. Björnberg, A. Multicomponent polyanions. 22. The molecular and crystal structure of $K_8Mo_4V_8O_{36}\cdot 12H_2O$, a compound containing a structurally new heteropolyanion. *Acta Cryst. B* **1979**, *35*, 1989–1995. [CrossRef]
46. Zhang, Y.; Haushalter, R.C.; Clearfield, A. A heteropolyanion containing two linked mixed Mo/V pentadecaoxometalate clusters: Structure of $[Mo_{16}V_{14}O_{84}]^{14-}$. *J. Chem. Soc. Chem. Commun.* **1995**, 1149–1150. [CrossRef]
47. Salazar Marcano, D.E.; Moussawi, M.A.; Anyushin, A.V.; Lentink, S.; Van Meervelt, L.; Ivanović-Burmazović, I.; Parac-Vogt, T.N. Versatile post-functionalisation strategy for the formation of modular organic–inorganic polyoxometalate hybrids. *Chem. Sci.* **2022**, *13*, 2891–2899. [CrossRef]
48. Fernández-Navarro, L.; Nunes-Collado, A.; Artetxe, B.; Ruiz-Bilbao, E.; San Felices, L.; Reinoso, S.; San Jose Wery, A.; Gutierrez-Zorrilla, J.M. Isolation of the Elusive Heptavanadate Anion with Trisalkoxide Ligands. *Inorg. Chem.* **2021**, *60*, 5442–5445. [CrossRef] [PubMed]

Article

Vanadium-Substituted Dawson-Type Polyoxometalate–TiO$_2$ Nanowire Composite Film as Advanced Cathode Material for Bifunctional Electrochromic Energy-Storage Devices

Yu Fu, Yanyan Yang *, Dongxue Chu, Zefeng Liu, Lili Zhou, Xiaoyang Yu and Xiaoshu Qu *

College of Chemical and Pharmaceutical Engineering, Jilin Institute of Chemical Technology, Jilin 132022, China; fuyu1996211@163.com (Y.F.); chudongxue1997@163.com (D.C.); lzf122321@163.com (Z.L.); z13069189739@163.com (L.Z.); yangyangyu@jlict.edu.cn (X.Y.)
* Correspondence: yyy200409@163.com (Y.Y.); xiaoshuqu@jlict.edu.cn (X.Q.)

Abstract: Polyoxometalates (POMs) demonstrate potential for application in the development of integrated smart energy devices based on bifunctional electrochromic (EC) optical modulation and electrochemical energy storage. Herein, a nanocomposite thin film composed of a vanadium-substituted Dawson-type POM, i.e., $K_7[P_2W_{17}VO_{62}] \cdot 18H_2O$, and TiO$_2$ nanowires were constructed via the combination of hydrothermal and layer-by-layer self-assembly methods. Through scanning electron microscopy and energy-dispersive spectroscopy characterisations, it was found that the TiO$_2$ nanowire substrate acts as a skeleton to adsorb POM nanoparticles, thereby avoiding the aggregation or stacking of POM particles. The unique three-dimensional core−shell structures of these nanocomposites with high specific surface areas increases the number of active sites during the reaction process and shortens the ion diffusion pathway, thereby improving the electrochemical activities and electrical conductivities. Compared with pure POM thin films, the composite films showed improved EC properties with a significant optical contrast (38.32% at 580 nm), a short response time (1.65 and 1.64 s for colouring and bleaching, respectively), an excellent colouration efficiency (116.5 cm^2 C^{-1}), and satisfactory energy-storage properties (volumetric capacitance = 297.1 F cm^{-3} at 0.2 mA cm^{-2}). Finally, a solid-state electrochromic energy-storage (EES) device was fabricated using the composite film as the cathode. After charging, the constructed device was able to light up a single light-emitting diode for 20 s. These results highlight the promising features of POM-based EES devices and demonstrate their potential for use in a wide range of applications, such as smart windows, military camouflage, sensors, and intelligent systems.

Keywords: polyoxometalate; TiO$_2$ nanowire; composite film; bifunctional electrochromic energy storage

Citation: Fu, Y.; Yang, Y.; Chu, D.; Liu, Z.; Zhou, L.; Yu, X.; Qu, X. Vanadium-Substituted Dawson-Type Polyoxometalate–TiO$_2$ Nanowire Composite Film as Advanced Cathode Material for Bifunctional Electrochromic Energy-Storage Devices. *Molecules* **2022**, *27*, 4291. https://doi.org/10.3390/molecules27134291

Academic Editor: Xiaobing Cui

Received: 3 June 2022
Accepted: 1 July 2022
Published: 4 July 2022

Publisher's Note: MDPI stays neutral with regard to jurisdictional claims in published maps and institutional affiliations.

Copyright: © 2022 by the authors. Licensee MDPI, Basel, Switzerland. This article is an open access article distributed under the terms and conditions of the Creative Commons Attribution (CC BY) license (https:// creativecommons.org/licenses/by/ 4.0/).

1. Introduction

With the continuing development of sustainable resources, devices for energy storage and conversion, such as solar cells, supercapacitors, and electrochromic (EC) devices, have attracted increasing attention [1–3]. EC devices are known to change colour via charge insertion/extraction or reversible redox reactions driven by an external electric field [4,5]. Simultaneously, the ion intercalation/deintercalation steps taking place during the reversible redox reactions of the EC process can also generate a pseudocapacitive behaviour [6,7], thereby resulting in EC devices and supercapacitors having similar working mechanisms and device structures. Based on this principle, one can envisage that these two functions could be integrated into a single electrochromic energy-storage (EES) device using the same material. As such, several EES devices have been widely explored. For example, Feng et al. [8] utilised exfoliated graphene/V$_2$O$_5$ as the active material of a micro-supercapacitor to judge its charge-discharge state via the observed colour. In addition, Xue et al. [9] synthesised a smart EC supercapacitor device using a porous co-doped NiO film as the positive electrode. This device exhibited a high specific capacitance, high energy density, and good cycle stability. After charging, these two devices were able to light up light-emitting diodes (LEDs).

Among the various EES materials reported to date, polyoxometalates (POMs) demonstrate a multi-electron reaction specificity during the electrochemical redox process, which contributes to their chromatic transitions and high-efficiency energy-storage performances [10–12]. As an example, Ma et al. [13] synthesised a POMs-based supramolecular crystalline material, namely, $H_3PMo^{VI}_{12}O_{40} \cdot (BPE)_{2.5} \cdot 3H_2O$ (BPE = 1,2-Bis(4-pyridyl)ethylene), via a one-step hydrothermal method. The compound had a high specific capacitance (i.e., 137.5 F g^{-1} at 2 A g^{-1}) and good cycle stability (i.e., 92.0% after 1000 cycles) than parent $H_3PMo^{VI}_{12}O_{40}$. This work provided an alternative method for improving the performance of POMs-based capacitor electrode materials. In addition, Wang et al. [14] reported a high-performance PW_{12}-based EC device, wherein the optical contrast of the optimised device containing an I^-/I^{3-} redox couple in the electrolyte reached 59.4%. The PW_{12}-EC device also showed a fast response time for bleaching and colouration. However, POM materials tend to aggregate or stack to form dense structures, which can hinder ion diffusion and affect their electrochemical properties. To overcome this issue, the incorporation of POMs into nanostructures or composite materials has been investigated to increase their surface areas [15]. For example, L et al. [16] prepared graphene oxide/$W_{18}O_{49}$ nanorod (rGO-WNd) composites through the high-temperature thermal reduction of ammonium tungstate and graphene oxide (GO). Compared with the cycle stability, capacitance, and EC properties of the pure WNd film, the corresponding properties of the Rgo-WNd composite film were significantly enhanced. This could be attributed to a higher degree of ion diffusion and the acceleration of charge transfer after the addition of rGO. As a result, the response times of such materials are improved.

Titanium dioxide is recognised as a promising candidate for EC and energy-storage applications owing to its excellent electrochemical stability, optical modulation, reversibility, and mass transport properties, as well as the fact that it enhances contact with the electrolyte and improves the resulting reaction kinetics [17]. In recent years, various TiO_2 nanostructures, such as nanorods, nanotubes, and nanowires, have received attention as excellent composite materials because of their large specific surface areas and orderly structures. For example, Khanna et al. [18] fabricated a TiO_2@NiTi system for use as an electrode in energy-storage applications, and this material produced a specific capacitance of ~1 F g^{-1}. This result reveals that their system is a promising material for energy-storage applications. In addition, Ji et al. [19] designed and fabricated a novel bilayer composite with an excellent energy-storage performance by combining an aligned TiO_2 nanoarray (TNA) and random TiO_2 nanowires (TiO_2 NWs) with a poly(vinylidene fluoride) (PVDF) matrix. A superior discharge energy density of 16.13 J cm^{-3} was obtained for the 5 vol% TiO_2 NW/TNA-PVDF composite, which was 2.0 times higher than that of the pure PVDF matrix (8.23 J cm^{-3}). Furthermore, Lv et al. [20] synthesised TiO_2 nanotube membrane electrodes that exhibited excellent EC performances, combining a high colouration contrast with a transmittance of 65% in the visible spectrum, in addition to a good cycle stability (88.2% for initial optical modulation after 1000 cycles). Zhang et al. [21] reported a novel EC device based on polyaniline nanofibers wrapped with antimony-doped tin oxide/TiO_2 nanorods (ATO/TiO_2@PANI film) as an EC electrode material. Compared with the pure PANI film, the EC device based on ATO/TiO_2@PANI film shows better electrochromic performance.

Based on the above considerations, our group previously designed a series of POM-based EC thin film materials [22,23]. In 2020, we reported the first dual-function electrochromic-energy storage material based on POMs and TiO_2 nanowires [24]. However, the response time of the film is long, and its capacitive performance is relatively low. As we know, the structure and composition of POMs have a great influence on their electrochemical activity; therefore, the electrochromic-energy storage properties could be adjusted easily by changing the type of POMs. In general, the lacunary and substituted Dawson structures can show enhanced electrochromic performances [25]. Thus, in the current study, to improve the performances of these materials, we chose vanadium-substituted Dawson-type polyoxotungstate $K_7[P_2W_{17}VO_{62}]\cdot 18H_2O$ ($P_2W_{17}V$) and TiO_2 nanowires to fabricate a nanocomposite thin film via hydrothermal and layer-by-layer (LbL) self-assembly methods. The microstructure of TiO_2 is regulated by a hydrothermal treatment, allowing its nanowire array to be employed as the substrate for the composite film. The synergistic effects of the TiO_2 NWs and the POMs could improve the EC properties of the composite film. Scanning electron microscopy (SEM), transmission electron microscopy (TEM), atomic force microscopy (AFM), and X-ray photoelectron spectroscopy (XPS) are used to investigate the surface morphology, structure, and chemical properties of the obtained nanocomposite film. Finally, the EC and energy-storage properties of the composite film are compared with those of the pure $P_2W_{17}V$-modified fluorine-doped tin oxide (FTO) film.

2. Materials and Methods

2.1. Chemicals and Materials

All reagents were of analytical grade and used as received without further treatment. The FTO-coated glasses (<10 ohm sq^{-1}) were purchased from Pilkington (Toledo, Ohio, USA). (3-Aminopropyl)trimethoxysilane (APS), polyetherimide (PEI), and propylene carbonate (PC) were purchased from the Aladdin Chemical Co., Ltd. and were used without further treatment. $P_2W_{17}V$ was prepared according to a method reported in literature [26,27], and it was characterised by infrared (IR) spectroscopy (Figure S1), ultraviolet-visible (UV-vis) absorption spectroscopy (Figure S2), and cyclic voltammetry (CV) (Figure S3).

2.2. Preparation of the Composite Films

The TiO_2 nanowire arrays were prepared via a hydrothermal synthesis method according to our previous report [22], and this was followed by the preparation of the composite films. More specifically, the surface of the FTO substrate was cleaned in $NH_3/H_2O/H_2O/H_2O_2$ (volume ratio 1:1:1) at 80 °C for 20 min and then rinsed with deionised water. This step was repeated 3–5 times to remove any inorganic and organic impurities from the FTO substrate. The composite film was prepared via the LbL assembly method. Initially, the cleaned FTO substrate was modified with TiO_2 NWs. Subsequently, the pure FTO and the modified FTO were immersed in APS overnight. After this time, the samples were placed in HCl (pH 2.0) for 20 min, rinsed with deionised water, and dried under a stream of nitrogen to give the precursor. Finally, the composite film (NW–$P_2W_{17}V$) was constructed by depositing negatively charged $P_2W_{17}V$ (5×10^{-3} mol L^{-1} in 0.2 mol L^{-1} HOAc-NaAc at pH 3.99) and positively charged PEI (5×10^{-3} mol L^{-1} at pH = 4) onto the TiO_2 NWs, according to the LbL method. For comparison, an additional film was prepared on the pure FTO substrate using the same method, and this was designated as FTO–$P_2W_{17}V$. A schematic outline of the fabrication process is shown in Figure 1a.

2.3. Characterisation

SEM images were measured on FEI Verious 460 L scanning electron microscope (Hillsboro, OH, USA). AFM images were investigated by Icon Bruker microscope (Ettlingen, Germany). TEM images were measured on a FEI Tecnai G2F20 S-TWIN microscope equipped with an energy-dispersive spectrometer (EDS) (Hillsboro, OH, USA). XPS analysis were measured on a Thermo ESCALAB 250 spectrometer (Shanghai, China). The EC and capacitive properties of the films were determined by combining the in-situ TU-1901 PERSEE UV-vis spectrophotometer (Beijing, China) with an CHI660B Chenhua electrochemical workstation (Shanghai, China) in a three-electrode configuration, where the nanocomposites served as the working electrodes; a Pt plate/Pt wire acted as the counter electrode, and Ag/AgCl was used as the reference electrode.

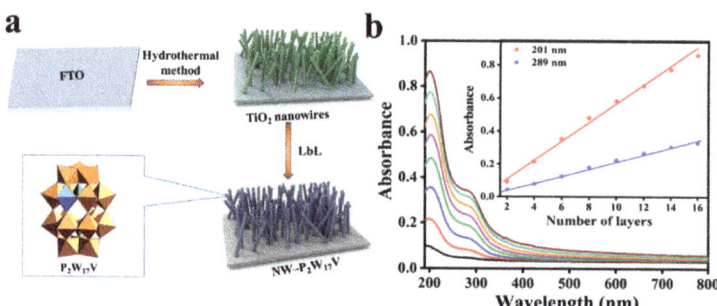

Figure 1. Characterization of material preparation process. (**a**) Schematic outline of the fabrication process of nanocomposite film; (**b**) UV-vis absorption spectra of composite film on quartz substrate (number of cycles: 2–16). Inset: plots of the absorbance values at 201 and 289 nm as a function of the layer number.

3. Results and Discussion

3.1. Characterisation of the NW−$P_2W_{17}V$ and FTO−$P_2W_{17}V$ Materials

The multilayer growth process of composite film on the precursor-coated quartz substrate (on both sides) was monitored by UV-Vis spectroscopy (Figure 1b). It exhibited strong absorption of $P_2W_{17}V$ with two characteristic absorption peaks at 201 and 289 nm. The peak at 201 nm originates from the terminal oxygen to tungsten charge-transfer transition ($O_d{\rightarrow}W$), whereas the peak at 289 nm corresponds with the charge-transfer transition from the bridging-oxygen to tungsten ($O_b/O_c{\rightarrow}W$). The inset of Figure 1b shows the plots of the absorbance values at 201 and 289 nm as a function of the layer number and suggests that growth is uniform during each cycle.

SEM-EDS and TEM were then performed to obtain the detailed information about the surface morphologies and homogeneities of the composite materials. The SEM images of the FTO−$P_2W_{17}V$ film are shown in Figure S4, wherein it can be visualised that the FTO substrate was covered by aggregated $P_2W_{17}V$ anions. In addition, the cross-sectional view of the FTO−$P_2W_{17}V$ film gave a thickness of ~150 nm. As shown in Figure S5, the FTO substrate was covered with densely grown TiO_2 NWs, and the cross-sectional image confirmed that the height of the nanowires was approximately 600 nm. After the LbL process, it was apparent that the interspaces of the NWs were filled, and the NWs became wider and more compact owing to the deposition of $P_2W_{17}V$ and PEI (Figure 2a). Moreover, the EDS mapping of P, W, Ti, and V confirmed the feasibility of the hydrothermal treatment and LbL process (Figure 2b), since the POMs and the TiO_2 NWs were evenly distributed on the surface of the FTO substrate.

Subsequently, AFM was employed to study the surface morphologies and roughness properties of the FTO−$P_2W_{17}V$ and NW−$P_2W_{17}V$ films (Figure 2c,d and Figure S6). Two-dimensional (2D) and three-dimensional (3D) images of the two films confirmed that their surface microstructures were quite different. More specifically, the AFM images of the FTO−$P_2W_{17}V$ film displayed some uniformly sized spherical particles, which resulted from the FTO substrate being covered with cross-linked POM anions with a thickness of 100 nm (Figure S6b). From Figure 2d, it was apparent that the surface of the NW−$P_2W_{17}V$ film shows a regular cylindrical microstructure, suggesting the presence of TiO_2 NWs substrate. The height of the NWs anchored with the POMs was ~500 nm, which corresponded well with the SEM observations. In addition, the root mean square (RMS) roughness for each film was calculated from an area of 5×5 μm² in the AFM image, wherein the surface roughness (i.e., RMS) values of the NW−$P_2W_{17}V$ and FTO−$P_2W_{17}V$ films were found to be 73.6 and 20.5 nm, respectively. A higher roughness could lead to a larger reactive surface area, thereby improving the electrochemical performance of the material.

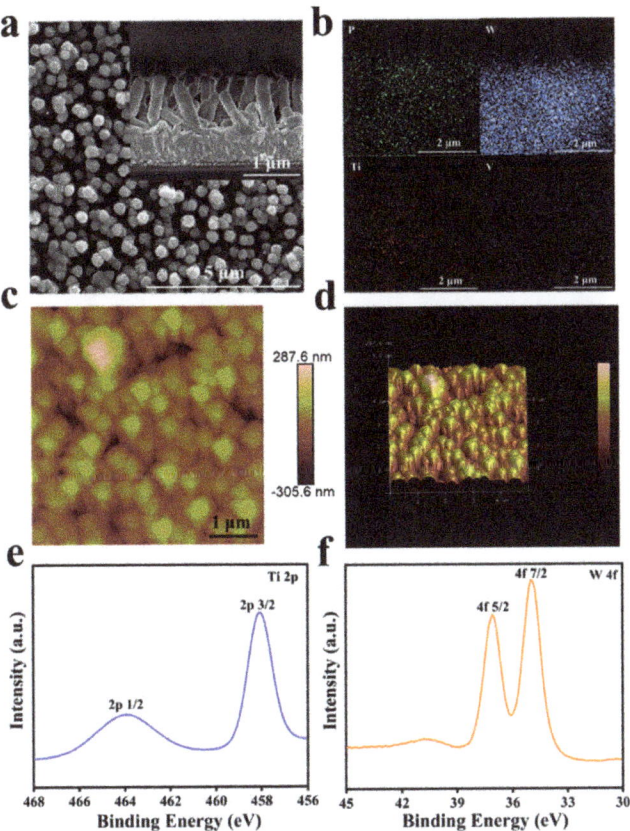

Figure 2. (**a**) SEM images of NW−P$_2$W$_{17}$V (inset: the cross-sectional images of prepared films); (**b**) EDS mapping of NW−P$_2$W$_{17}$V for P, W, Ti and V respectively; (**c**) 2D AFM images; and (**d**) 3D AFM images of NW−P$_2$W$_{17}$V films; High-resolution XPS spectra for Ti 2p (**e**) and W 4f (**f**).

The surface chemical compositions of the as-prepared films were further determined and quantified by XPS analysis. The high-resolution XPS spectra of the prepared composite film shown in Figure 2 indicates that the composite material mainly contains C, P, Ti, and W [28,29], wherein the Ti should originate from the TiO$_2$ NWs on the FTO substrate. This result further confirms that the POMs and the TiO$_2$ NWs are distributed on the surface of the FTO substrate. As shown in Figure 2e, the most intense doublet peaks are observed at 35.6 and 37.7 eV, which correspond to the binding energies of the electrons in the W4f$_{7/2}$ and W4f$_{5/2}$ levels of W in the W$^{(VI)}$ valence state. These results indicate that the majority of W atoms were in a highly oxidised state and could be reduced to W$^{(V)}$, which is the key reaction in the EC process of polyoxotungstate-based materials. With respect to the high-resolution Ti2p peaks, they could be split into peaks at 458.9 and 464.6 eV, which were both attributed to TiO$_2$ (Figure 2f), thereby indicating that the main matrix component was TiO$_2$. Furthermore, the prepared film exhibited a peak corresponding to the C1s level (284.8 eV) of the carbon present in the PEI polycation, whereas the P2p signal (at 133.0 eV) and the V2p signal (at 532.4 eV) [30] were ascribed to P$_2$W$_{17}$V (Figure S7). Thus, the XPS data suggest that PEI cations and P$_2$W$_{17}$V anions were incorporated into the TiO$_2$ NW substrate, which is consistent with the UV-vis results.

TEM is indispensable for the characterisation of nanostructured materials, particularly when the particle shape is important in determining its function, and so TEM was employed herein to evaluate the microstructure of the composite and the spatial relationship between TiO$_2$ NWs and P$_2$W$_{17}$V. Figure 3a–b show the typical TEM images of the TiO$_2$ NWs with a diameter of ~50 nm. The EDS elemental mapping patterns of the TiO$_2$ NW–P$_2$W$_{17}$V film were also recorded, as shown in Figure 3c–f and Figure S8. Combined with the TEM morphological observations, the distributions of W, P, Ti, and V suggest a uniform distribution of P$_2$W$_{17}$V on the TiO$_2$ NWs. As shown in the TEM image (Figure 3b), following the LbL assembly process, the P$_2$W$_{17}$V coating layer covered the surface of the NWs, forming a core-shell structure. As indicated by the arrows, the darker columnar area is a TiO$_2$ NW and the lighter part surrounding it are P$_2$W$_{17}$V particles. The selected area electron diffraction pattern showed the specific diffraction spots of TiO$_2$ nanowires, and it can be attributed to the rutile phase [23].

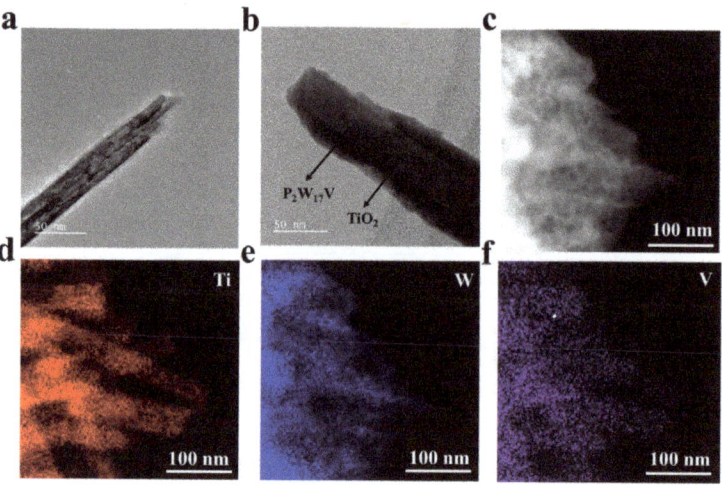

Figure 3. TEM images of TiO$_2$ nanowires (**a**) and NW–P$_2$W$_{17}$V (**b**); (**c–f**) EDS elemental mapping patterns of Ti, W, and V in the NW–P$_2$W$_{17}$V films.

3.2. EC Performance

To explore the potential of the prepared composite material for application as EC supercapacitor, its EC properties were investigated and compared with those of the FTO–P$_2$W$_{17}$V film. As demonstrated in Figure 4a–b, the transmittance was reduced along potentials ranging from 0 to −1.0 V. In addition, as shown in Figure 4c, the maximum transmittance modulation of the NW–P$_2$W$_{17}$V film (38.32%) was significantly higher than that of the FTO–P$_2$W$_{17}$V film (22.25%) at 580 nm, thereby indicating that the effective combination of two cathodic EC materials could indeed improve the overall performance. For the switching kinetics, the fast switching speed (i.e., the time required to achieve 90% of full modulation) for each of the two prepared films was determined, as shown in Figure 4d. Notably, FTO–P$_2$W$_{17}$V (t$_c$ = 1.49 s and t$_b$ = 1.65 s) and NW–P$_2$W$_{17}$V (t$_c$ = 1.65 s and t$_b$ = 1.64 s) films could undergo relatively rapid colouring and bleaching processes, which are important processes in the context of EC applications. Furthermore, as shown in the optical photograph presented in Figure 4e, the P$_2$W$_{17}$V-modified film turned blue, and became deeper in colour upon increasing the applied potential; this colour was attributed to the intervalence charge-transfer band (W^V–O–W^{VI} or W^{VI}–O–W^V). The transmittance showed a good linear relationship with the applied potential, indicating that the colouration state could be adjusted precisely, thereby rendering this system suitable for practical use in industry.

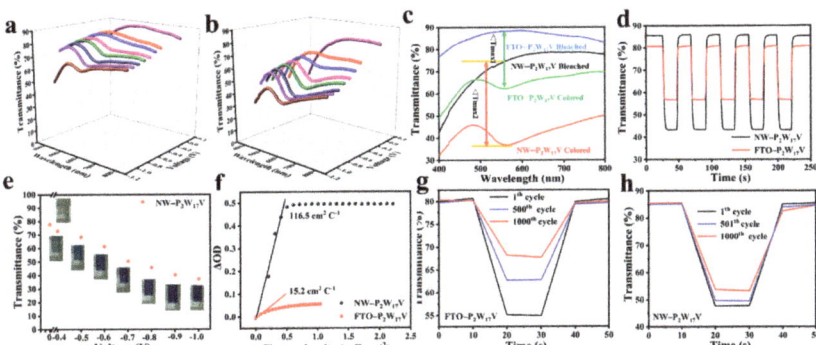

Figure 4. Visible transmittance spectrum of FTO−$P_2W_{17}V$ (**a**) and NW−$P_2W_{17}V$ (**b**) films at different potentials. (**c**) Visible spectra of prepared films at colored and bleached state; (**d**) Chronoamperometry measurements and corresponding in situ optical transmittance curves for FTO−$P_2W_{17}V$ and NW−$P_2W_{17}V$ films at 580 nm; (**e**) Plots of the transmittance value versus applied voltage for NW−$P_2W_{17}V$ and corresponding optical images; (**f**) Coloration efficiency at 580 nm of NW−$P_2W_{17}V$ and FTO−$P_2W_{17}V$ films during subsequent double-potential steps (−1 V and +1 V); Cycle stability of FTO−$P_2W_{17}V$ (**g**) and NW−$P_2W_{17}V$ films (**h**) at 580 nm under square wave potentials of −1 V and +1 V.

The CE is a crucial factor in evaluating the correlation between the change in colour and the number of injected charges. The CE can be calculated from Equations (1) and (2) [31–33]:

$$CE = \Delta OD/(Q/A) \qquad (1)$$

$$\Delta OD(\lambda) = \log T_b/T_c \qquad (2)$$

where Q is the charge density, A is the area of the composite film, and T_b and T_c are the transmittances of the film in the bleached and coloured states at a certain wavelength (λ), respectively. Figure 4f shows the variation in the optical density with respect to the extent of electric charge exchange from the electrolyte to the EC film. The CE can be obtained from the slope of the line that fits the linear region of the plot. Thus, the CE values of samples were calculated to be 116.5 $cm^2\ C^{-1}$ for NW−$P_2W_{17}V$ and 15.2 $cm^2\ C^{-1}$ for FTO−$P_2W_{17}V$, wherein the larger value obtained for the NW−$P_2W_{17}V$ system indicates that a large transmittance modulation can be realised through the introduction of a small amount of charge.

The electrochemical stability of a film is vital for determining its EC performance. Thus, the cycling stabilities of the FTO−$P_2W_{17}V$ and NW−$P_2W_{17}V$ films were tested by chronoamperometry at 580 nm over 1000 cycles. As shown in Figure 4g–h, NW−$P_2W_{17}V$ exhibited a superior cycling stability with an initial transmittance variation of approximately 38.32%, wherein ~86% of the initial value was retained after 1000 cycles. This outstanding cycling stability should permit long-term application in real environments.

3.3. Energy-Storage Performance

The electrochemical performances of the thin films were then evaluated using CV and galvanostatic charge-discharge (GCD) tests. Figure 5a shows the CV curves of the NW−$P_2W_{17}V$ film measured at different scan rates, wherein it can be seen that upon increasing the scan rate from 50 to 150 mV s^{-1}, no obvious changes in shape were observed for the CV curves, although the peak potential moved slightly. The presence of characteristic symmetric reversible peaks for the NW−$P_2W_{17}V$ film also indicate its good capacitive behaviour upon ion insertion/extraction. Furthermore, the inset of Figure 5a shows a good linear relationship between the current density and the scan rate, indicating a fast electron transfer kinetic characteristic in these redox-active materials, which therefore represents a typical surface-controlled process. Figure 5b shows the CV curves of the NW−P$_2$W$_{17}$V and FTO−P$_2$W$_{17}$V films obtained using a three-electrode system at the same scan rate in a solution of HOAc-NaAc at pH 3.5. The composite film displayed three pairs of redox peaks, which can be attributed to the redox reaction between W^{VI} and W^V, indicating a typical faradic behaviour. The redox peaks of the NW−P$_2$W$_{17}$V film have higher peak current values than those of the FTO−P$_2$W$_{17}$V film, indicating the high conductivity and low internal resistance of the NW−P$_2$W$_{17}$V film. These increased peak current values can be attributed to the influence of faradaic reactions and to hydrogen ion (H$^+$) intercalation at the electrode/electrolyte interface.

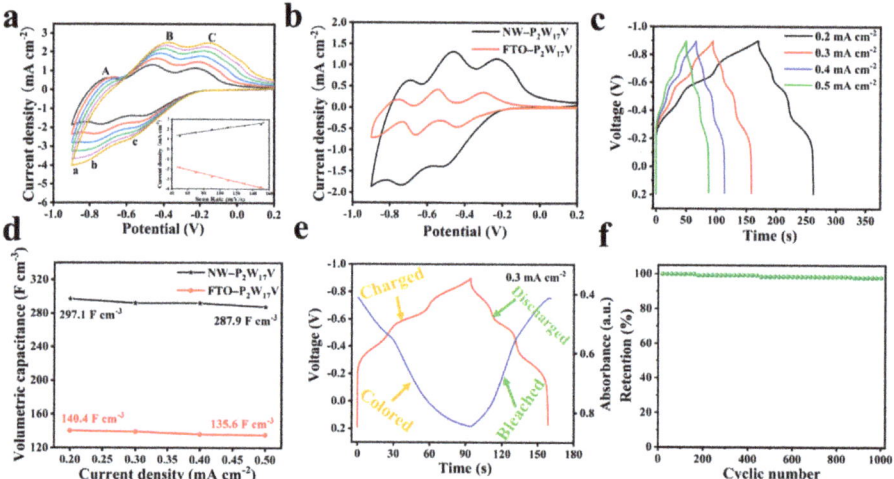

Figure 5. (**a**) CV for the NW−P$_2$W$_{17}$V film at different scan rates (from inner to outer): 50, 70, 90, 110, 130, and 150 mV s^{-1}. The inset shows plots of the anodic and the cathodic peak currents for C-c against scan rates; (**b**) CV for NW−P$_2$W$_{17}$V and FTO−P$_2$W$_{17}$V films at a scan rate of 50 mV/s; (**c**) Charge/discharge curves of NW−P$_2$W$_{17}$V film at various current densities; (**d**) Volumetric capacitance at various current densities of NW−P$_2$W$_{17}$V and FTO−P$_2$W$_{17}$V films; (**e**) In situ transmittance evolution at 580 nm with the charging and discharging process of the NW−P$_2$W$_{17}$V film; (**f**) Cycle performance of NW−P$_2$W$_{17}$V film measured under a current density of 0.2 mA cm^{-2}.

The diffusion coefficient of H$^+$ ions for insertion and extraction can be estimated based on the measured peak current, I$_p$ (A) [34,35]:

$$I_p = 2.69 \times 10^5 AC\sqrt{Dvn^3} \tag{3}$$

where I$_p$ is the peak current, A is the area of the film (cm^2), n is the number of electrons, D is the diffusion coefficient of the H$^+$ ions (cm^2 s^{-1}), C is the concentration of the H$^+$ ions in the electrolyte solution (mol cm^{-3}), and v is the scan rate (V s^{-1}). The diffusion rate of H$^+$ in NW−P$_2$W$_{17}$V was faster than that in FTO−P$_2$W$_{17}$V. This enhanced diffusion rate for NW−P$_2$W$_{17}$V therefore accounted for the superior electrical conductivity of this material.

Owing to their fast ion intercalation/deintercalation properties and excellent cycling stabilities, we envisaged that the composite films could have great potential for use in energy-storage applications. Thus, to further evaluate the capacitive behaviours of the composite films, a series of GCD measurements were carried out at different current densities. Figure 5c shows the potential responses of the NW−P$_2$W$_{17}$V film under different currents, in addition to the dependence of the volumetric capacitance of the composite film on the current density. The GCD curves collected under different current densities are displayed in Figure 5c, which shows that the shapes of the CD profiles were essentially retained for all the applied current ranges, demonstrating the superior charge/discharge reversibility of the sample [36]. Plateau regions are observed in the GCD curves, and the positions of the three plateaus are consistent with the CV curves, thereby indicating that the capacitance is mainly caused by the faradaic redox reaction, whereas the existence of plateaus in the curves illustrates a sound pseudocapacitive behaviour [37,38]. The calculated volumetric capacitance as a function of the current density is shown in Figure 5d. The volumetric capacitance gradually declines as the current density increases, mainly because the limited ion diffusion rate is inaccessible, and so adequate surface redox reactions of the active materials cannot be ensured at high current densities. Furthermore, the value obtained for the NW−$P_2W_{17}V$ film was higher than that of the FTO−$P_2W_{17}V$ film, which was ascribed to the interactions and synergistic effects between the $P_2W_{17}V$ and TiO_2 NW materials. Furthermore, the GCD curves at 0.3 mA cm^{-2} and the corresponding in situ transmittance at 580 nm were collected and plotted in Figure 5e. During the charging process, the NW−$P_2W_{17}V$ electrode gradually became coloured, and the decrease in transmittance was distinguishable. In contrast, the colour of the electrode was reversibly bleached during the discharge process.

The long-term cycling stability is another vital index for evaluating the properties of electrode materials [39,40]. As shown in Figure 5f, the NW−$P_2W_{17}V$ film revealed an excellent cyclic stability with its volumetric capacitance being almost fully maintained after 1000 cycles at 0.2 mA cm^{-2} in a voltage range of −0.5 to 0.2 V.

Subsequently, electrochemical impedance spectroscopy (EIS) was employed to investigate the inner resistances and capacitance properties of the thin films [30]. Figure 6a shows the Nyquist plots of the NW−$P_2W_{17}V$ and FTO−$P_2W_{17}V$ films with a frequency range of 0.01–100,000 Hz and a signal amplitude of ±5 mV. The electrode system can be described by a simple equivalent circuit (see the inset of Figure 6a), which was selected to fit the obtained impedance data for the NW−$P_2W_{17}V$ composite film. The high-frequency part of the semicircle in the EIS spectrum indicates the speed of the electron transfer process, and the diameter is closely related to the electron transfer resistance (Rct). The Rct of the FTO−$P_2W_{17}V$ film was significantly smaller than that of the NW−$P_2W_{17}V$ film, indicating the lower Rct and the higher electron transfer rate of NW−$P_2W_{17}V$ composite film. As outlined in Figure 6b, we constructed an EES device using $LiClO_4$/PC as the electrolyte, the NW−$P_2W_{17}V$ composite film as the negative electrode, and FTO as the positive electrode. Importantly, this EES device was capable of lighting a red LED (Figure 6c). After charging for 10 s, the device became dark blue in colour, and the system lit the red LED for a total of 20 s. These results indicate that the energy-storage states were directly reflected by the colour change. More specifically, as the charge stored inside the device increased, its colour deepened. Overall, these observations verify the potential practical application of our device in energy-storage smart windows and visual monitoring systems.

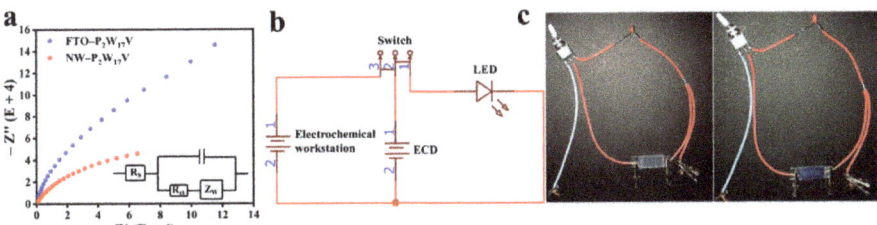

Figure 6. (a) The EIS figures of the films with different components FTO−$P_2W_{17}V$ and NW−$P_2W_{17}V$ film, the inset shows a simple equivalent circuit about the NW−$P_2W_{17}V$ electrode system; (b) Structural diagram of the solid-state EC device architecture used in this work; (c) Photo of a red LED lit up and out by a solid-state EC device.

4. Conclusions

In this work, a suitably designed nanocomposite film composed of vanadium-substituted Dawson-type POMs were fabricated on a TiO$_2$ nanowire array substrate. Compared with the dense packing structure, the core—shell nano structure exhibited enhanced EC and electrochemical properties with significant optical contrast (38.32% at 580 nm), short response time (1.65 and 1.64 s for colouring and bleaching, respectively), and satisfactory volumetric capacitance (297.1 F cm^{-3} at 0.2 mA cm^{-2}), which mainly originate from the unique three-dimensional structure of a nanocomposite with low tortuosity and a high specific surface area. TiO$_2$ NW not only provided a transparent substrate with greater adhesion, but it also shortened the electrons/ions diffusion pathway, resulting in uniform and fast reaction kinetic characteristics. A solid-state EES device was fabricated using the composite film as the cathode. In terms of its potential practical applications, the developed device was demonstrated to light up a red LED, and the energy-storage state of the device was easily monitored by observing its change in colour, so as to achieve the purpose of real-time monitoring, and avert the damage caused by overcharging and over-discharging to the supercapacitor. These results therefore confirm the promising features of POM-based EES devices and demonstrate their potential for use in a wide range of multifunctional supercapacitors, such as self-charging supercapacitors, smart energy storage windows, and electrochromic supercapacitors.

Supplementary Materials: The following supporting information can be downloaded at: https://www.mdpi.com/article/10.3390/molecules27134291/s1, Figure S1: The IR spectra of K$_7$[P$_2$W$_{17}$VO$_{62}$]·18H$_2$O; Figure S2: The UV-vis spectra of K$_7$[P$_2$W$_{17}$VO$_{62}$]·18H$_2$O; Figure S3: CV curve of K$_7$[P$_2$W$_{17}$VO$_{62}$]·18H$_2$O in HOAc-NaOAc solution (pH = 3.5); Figure S4: The SEM images of FTO−P$_2$W$_{17}$V (inset: the cross-sectional images of prepared films); Figure S5: The SEM images of TiO$_2$ NW (inset: the cross-sectional images of prepared films); Figure S6: 2D AFM images of (a) FTO−P$_2$W$_{17}$V and 3D AFM images of (b) FTO−P$_2$W$_{17}$V; Figure S7: High-resolution XPS spectra for C1s (a), V2p (b) and P2p (c) of FTO−P$_2$W$_{17}$ film; Figure S8: EDS elemental mapping patterns of P in the NW−P$_2$W$_{17}$V film.

Author Contributions: Conceptualization, X.Q. and Y.Y.; methodology, X.Q. and Y.Y.; validation, Y.F., D.C. and X.Q.; formal analysis, D.C. and L.Z.; investigation, L.Z., Z.L. and X.Y.; resources, X.Q. and Y.Y.; data curation, Y.F. and X.Q.; writing—original draft preparation, Y.F.; writing—review and editing, X.Q.; funding acquisition, X.Q. and X.Y. All authors have read and agreed to the published version of the manuscript.

Funding: This research was financially supported by the National Natural Science Foundation of China (22071080, 21902058) and the Natural Science Foundation of Jilin Province (YDZJ202101ZYTS175).

Institutional Review Board Statement: Not applicable.

Informed Consent Statement: Not applicable.

Data Availability Statement: Data are contained within the article and supplementary materials.

Conflicts of Interest: The authors declare no conflict of interest.

Sample Availability: Samples of the compounds are available from the authors.

References

1. Chang, P.; Mei, H.; Zhang, M.G.; Zhao, Y.; Wang, X.; Cheng, L.F.; Zhang, L.T. 3D printed electrochromic supercapacitors with ultrahigh mechanical strength and energy density. *Small* **2021**, *17*, 2102639. [CrossRef] [PubMed]
2. Ong, W.J.; Zheng, N.; Antonietti, M. Advanced nanomaterials for energy conversion and storage: Current status and future opportunities. *Nanoscale* **2021**, *13*, 9904–9907. [CrossRef]
3. Pomerantseva, E.; Bonaccorso, F.; Feng, X.L.; Cui, Y.; Gogotsi, Y. Energy storage: The future enabled by nanomaterials. *Science* **2019**, *366*, 8285. [CrossRef] [PubMed]
4. Xu, K.; Zhang, Q.Q.; Hao, Z.D.; Tang, Y.H.; Wang, H.; Liu, J.B.; Yan, H. Integrated electrochromic supercapacitors with visual energy levels boosted by coating onto carbon nanotube conductive networks. *Sol. Energy Mater. Sol. Cells* **2020**, *206*, 110330. [CrossRef]

5. Koo, B.R.; Jo, M.H.; Kim, K.H.; Ahn, H.J. Amorphous-quantized $WO_3 \cdot H_2O$ films as novel flexible electrode for advanced eletrchromic energy storage devices. *Chem. Eng. J.* **2021**, *424*, 130383. [CrossRef]
6. Ling, H.; Wu, J.C.; Su, F.Y.; Tian, Y.Q.; Liu, Y.J. Automatic light-adjusting electrochromic device powered by perovskite solar cell. *Nat. Commun.* **2021**, *12*, 1010. [CrossRef]
7. Koo, B.R.; Jo, M.H.; Kim, K.H.; Ahn, H.J. Multifunctional electrochromic energy storage devices by chemical cross-linking: Impact of a $WO_3 \cdot H_2O$ nanoparticle-embedded chitosan thin film on amorphous WO_3 films. *NPG Asia Mater.* **2020**, *12*, 10. [CrossRef]
8. Zhang, P.P.; Zhu, F.; Wang, F.X.; Wang, J.H.; Dong, R.H.; Zhuang, X.D.; Schmidt, O.G.; Feng, X.L. Stimulus-responsive microsupercapacitors with ultrahigh energy density and reversible electrochromic window. *Adv. Mater.* **2017**, *29*, 1604491. [CrossRef]
9. Xue, J.Y.; Li, W.J.; Song, Y.; Li, Y.; Zhao, J.P. Visualization electrochromic-supercapacitor device based on porous Co doped NiO films. *J. Alloys Compd.* **2021**, *857*, 158087. [CrossRef]
10. Rai, V.; Singh, R.S.; Blackwood, D.J.; Zhili, D. A review on recent advances in electrochromic devices: A material approach. *Adv. Eng. Mater.* **2020**, *22*, 2000082. [CrossRef]
11. Maity, S.; Vannathan, A.A.; Kumar, K.; Das, P.P.; Mal, S.S. Enhanced power density of graphene oxide phosphotetradecavanadate nanohybrid for supercapacitor electrode. *J. Mater. Eng. Perform.* **2021**, *30*, 1371–1377. [CrossRef]
12. Kim, J.; Rémond, M.; Kim, D.; Jang, H.; Kim, E. Electrochromic conjugated polymers for multifunctional smart windows with integrative functionalities. *Adv. Mater. Technol.* **2020**, *5*, 1900890. [CrossRef]
13. Wang, C.L.; Rong, S.; Zhao, Y.Q.; Wang, X.M.; Ma, H.Y. Three-dimensional supramolecular crystalline materials based on Kegin-based polyoxometalates and 1,2-Bis (4-pyridyl) ethylene for supercapacitor electrodes. *Transition Met. Chem.* **2021**, *46*, 335–343. [CrossRef]
14. Wang, S.M.; Wang, Y.H.; Wang, T.; Han, Z.B.; Cho, C.; Kim, E. Charge-Balancing Redox Mediators for High Color Contrast Electrochromism on Polyoxometalates. *Adv. Mater. Technol.* **2020**, *5*, 2000326. [CrossRef]
15. Fu, Z.J.; Qu, Z.Y.; Yu, T.; Bi, L.H. Study on electrochromic-fluorescence switching performance of film based on silicomolybdotungstate and silica nanoparticles doped with negative charged dye. *J. Electroanal. Chem.* **2019**, *855*, 113623. [CrossRef]
16. Li, Y.T.; Yan, L.T.; Zhang, J.; Xu, M.Y.; Zhu, Y.Y. Preparation of graphene/$W_{18}O_{49}$ nanorod composites and their application in electrochromic performance. *J. Mater. Sci. Mater. Electron.* **2019**, *30*, 20181–20188. [CrossRef]
17. Truong, Q.D.; Le, T.S.; Hoa, T.H. Ultrathin TiO_2 rutile nanowires enable reversible Mg-ion intercalation. *Mater. Lett.* **2019**, *254*, 357–360. [CrossRef]
18. Khanna, S.; Marathey, P.; Paneliya, S.; Chaudhari, R.; Vora, J. Fabrication of rutile—TiO_2 nanowire on shape memory alloy: A potential material for energy storage application. *Mater. Today Proc.* **2021**, *46*, 335–343. [CrossRef]
19. Ji, Q.; Hou, Y.F.; Wei, S.X.; Liu, Y.; Du, P.; Luo, L.H.; Li, W.P. Excellent energy storage performance in bilayer composites combining aligned TiO_2 nanoarray and random TiO_2 nanowires with Poly(vinylidene fluoride). *J. Phys. Chem. C* **2020**, *124*, 2864–2871. [CrossRef]
20. Lv, H.M.; Li, N.; Zhang, H.C.; Tian, Y.L.; Zhang, H.M.; Zhang, X.; Qu, H.Y.; Liu, C.; Jia, C.Y.; Zhao, J.P.; et al. Transferable TiO_2 nanotubes membranes formed via anodization and their application in transparent. *Sol. Energy Mater. Sol. Cells* **2016**, *150*, 57–64. [CrossRef]
21. Zhang, S.H.; Chen, S.; Yang, F.; Hu, F.; Yan, B.; Gu, Y.C.; Jiang, H.; Cao, Y.; Xiang, M. High-performance electrochromic device based on novel polyaniline nanofibers wrapped antimony-doped tin oxide/TiO_2 nanorods. *Org. Electron.* **2019**, *65*, 341–348. [CrossRef]
22. Qu, X.S.; Feng, H.; Liu, S.P.; Yang, Y.Y.; Ma, C. Enhanced electrochromic performance of nanocomposite film based on Preysslertype polyoxometalate and TiO_2 nanowires. *Inorg. Chem. Commun.* **2018**, *98*, 174–179. [CrossRef]
23. Qu, X.S.; Ma, C.; Fu, Y.; Liu, S.P.; Wang, J.; Yang, Y.Y. Construction of a vertically arrayed three-dimensional composite structure as a high coloration efficiency electrochromic film. *New J. Chem.* **2020**, *44*, 4177–4184. [CrossRef]
24. Qu, X.S.; Fu, Y.; Ma, C.; Yang, Y.Y.; Shi, D.; Chu, D.X.; Yu, X.Y. Bifunctional electrochromic-energy storage materials with ehanced performance obtained by hybridizing TiO_2 nanowires with POMs. *New J. Chem.* **2020**, *44*, 15475–15482. [CrossRef]
25. Harmalker, S.P.; Leparulo, M.A.; Pope, M.T. Mixed-valence chemistry of adjacent vanadium centers in heteropolytungstate anions synthesis and electronic structures of mono-, di-, and trisubstituted derivatives of α–$[P_2W_{18}O_{62}]^{6-}$. *J. Am. Chem. Soc.* **1983**, *105*, 4286–4292. [CrossRef]
26. Abbessi, M.; Contant, R.; Thouvenot, R.; Hervé, G. Dawson type heteropolyanions. 1. Multinuclear ($^{31}P, ^{51}V, ^{183}W$) NMR structural investigations of octadeca (molybdotungstovanado) diphosphates α-1,2,3-$[P_2MM'_2W_{15}O_{62}]^{n-}$ (M, M'=Mo, V, W): Syntheses of new related compounds. *Inorg. Chem.* **1991**, *30*, 1695–1702. [CrossRef]
27. Liu, S.P.; Qu, X.S. Construction of nanocomposite film of Dawson-type polyoxometalate and TiO_2 nanowires for electrochromic applications. *Appl. Surf. Sci.* **2017**, *412*, 189–195. [CrossRef]
28. Ahmed, I.; Wang, X.X.; Boualili, N.; Xu, H.L.; Farha, R.; Goldmann, M.; Ruhlmann, L. Photocatalytic synthesis of silver dendrites using electrostatic hybrid Films of porphyrin–polyoxometalate. *Appl. Catal. A Gen.* **2012**, *447*, 89–99. [CrossRef]
29. Walls, J.M.; Sagu, J.S.; Wijayantha, K.G. Upul. Microwave synthesised Pd–TiO_2 for photocatalytic ammonia production. *RSC Adv.* **2019**, *9*, 6387–6394. [CrossRef]
30. Zhang, D.; Ma, H.Y.; Chen, Y.Y.; Pang, H.J.; Yu, Y. Amperometric detection of nitrite based on Dawson-type vanodotungstophosphate and carbon nanotubes. *Anal. Chim. Acta* **2013**, *792*, 35–44. [CrossRef]

31. Sun, S.B.; Tang, C.J.; Jiang, Y.C.; Wang, D.S.; Chang, X.T.; Lei, Y.H.; Wang, N.N.; Zhu, Y.Q. Flexible and rechargeable electrochromic aluminium-ion battery based on tungsten oxide film electrode. *Sol. Energy Mater. Sol. Cells* **2020**, *207*, 110332. [CrossRef]
32. Cai, G.F.; Chen, J.W.; Xiong, J.Q.; Lee-Sie Eh, A.; Wang, J.X.; Higuchi, M.; Lee, P.S. Molecular level assembly for high-performance flexible electrochromic energy-storage devices. *ACS Energy Lett.* **2020**, *5*, 1159–1166. [CrossRef]
33. Sajitha, S.; Aparna, U.; Deb, B. Ultra-thin manganese dioxide-encrusted vanadium pentoxide nanowire mats for electrochromic energy storage applications. *Adv. Mater. Interfaces* **2019**, *6*, 2001928. [CrossRef]
34. Gu, X.; Liu, Y.B.; Wang, J.X.; Xiao, X.D.; Ca, X.S.; Sheng, G.Z. Synthesis of high-performance electrochromic thin films by a low-cost method. *Ceram. Int.* **2021**, *47*, 7837–7844.
35. Libansky, M.; Zima, J.; Barek, J.; Reznickova, A.; Svorcik, V.; Dejmkova, H. Basic electrochemical properties of sputtered gold film electrodes. *Electrochim. Acta* **2017**, *251*, 452–460. [CrossRef]
36. Shi, Y.D.; Sun, M.J.; Zhang, Y.; Cui, J.W.; Wu, Y.Q.; Ta, H.H.; Liu, J.Q.; Wu, Y.C. Structure modulated amorphous/crystalline WO_3 nanoporous arrays with superior electrochromic energy storage performance. *Sol. Energy Mater. Sol. Cells* **2020**, *212*, 110579. [CrossRef]
37. Wang, Y.Y.; Jia, X.T.; Zhu, M.H.; Liu, X.C.; Chao, D.M. Oligoaniline-functionalized polysiloxane/prussian blue composite towards bifunctional electrochromic supercapacitors. *New J. Chem.* **2020**, *44*, 8138–8147. [CrossRef]
38. Guo, Q.F.; Zhao, X.Q.; Li, Z.Y.; Wang, B.Y.; Wang, D.B.; Nie, G.M. High performance multicolor intelligent supercapacitor and its quantitative monitoring of energy storage level by electrochromic parameters. *ACS Appl. Energy Mater.* **2020**, *3*, 2727–2736. [CrossRef]
39. Zhou, S.Y.; Wang, S.; Zhou, S.J.; Xu, H.B.; Zhao, J.P.; Wang, J.; Li, Y. Electrochromic-supercapacitor based on MOF derived hierarchical-porous NiO film. *Nanoscale* **2020**, *12*, 8934–8941. [CrossRef]
40. Shi, Y.D.; Sun, M.J.; Chen, W.J.; Zhang, Y.; Shu, X.; Qin, Y.Q.; Zhang, X.R.; Shen, H.J.; Wu, Y.C. Rational construction of porous amorphous WO_3 nanostructures with high electrochromic energy storage performance: Effect of temperature. *J. Non-Cryst. Solids* **2020**, *549*, 120337. [CrossRef]

Article

Controllable Assembly of Vanadium-Containing Polyoxoniobate-Based Materials and Their Electrocatalytic Activity for Selective Benzyl Alcohol Oxidation

Xiaoxia Li, Ni Zhen, Chengpeng Liu, Di Zhang, Jing Dong, Yingnan Chi * and Changwen Hu *

Key Laboratory of Cluster Science of Ministry of Education, Beijing Key Laboratory of Photoelectronic/Electrophotonic Conversion Materials, School of Chemistry and Chemical Engineering, Beijing Institute of Technology, Beijing 102488, China; 3120191243@bit.edu.cn (X.L.); zhenni1220@163.com (N.Z.); laucouple@163.com (C.L.); dizhang_pub@163.com (D.Z.); 20200808@btbu.edu.cn (J.D.)
* Correspondence: chiyingnan7887@bit.edu.cn (Y.C.); cwhu@bit.edu.cn (C.H.)

Abstract: During the controllable synthesis of two vanadium-containing Keggin-type polyoxoniobates (PONbs), $[Ni(en)_2]_5[PNb_{12}O_{40}(VO)_5](OH)_5 \cdot 18H_2O$ (**1**) and $[Ni(en)_3]_5[PNb_{12}O_{40}(VO)_2] \cdot 17H_2O$ (**2**, en – ethylenediamine) are realized by changing the vanadium source and hydrothermal temperature. Compounds **1** and **2** have been thoroughly characterized by single-crystal X-ray diffraction analysis, FT-IR spectra, X-ray photoelectron spectrum (XPS), powder X-ray diffraction (PXRD), etc. Compound **1** contains a penta-capped Keggin-type polyoxoniobate $\{PNb_{12}O_{40}(VO)_5\}$, which is connected by adjacent $[Ni(en)_2]^{2+}$ units into a three-dimensional (3D) organic-inorganic framework, representing the first nickel complexes connected vanadoniobate-based 3D material. Compound **2** is a discrete di-capped Keggin-type polyoxoniobate $\{PNb_{12}O_{40}(VO)_2\}$ with $[Ni(en)_3]^{2+}$ units as counter cations. Compounds **1** and **2** have poor solubility in common solvents and can keep stable in the pH range of 4 to 14. Notably, both **1** and **2** as electrode materials are active for the selective oxidation of benzyl alcohol to benzaldehyde. Under ambient conditions without adding an alkaline additive, compound **1** as a noble metal free electrocatalyst can achieve 92% conversion of benzyl alcohol, giving a Faraday efficiency of 93%; comparatively, **2** converted 79% of the substrate with a Faraday efficiency of 84%. The control experiments indicate that both the alkaline polyoxoniobate cluster and the capped vanadium atoms play an important role during the electrocatalytic oxidation process.

Keywords: vanadium-containing polyoxoniobates; electrocatalysis; benzyl alcohol oxidation; nickel complexes

Citation: Li, X.; Zhen, N.; Liu, C.; Zhang, D.; Dong, J.; Chi, Y.; Hu, C. Controllable Assembly of Vanadium-Containing Polyoxoniobate-Based Materials and Their Electrocatalytic Activity for Selective Benzyl Alcohol Oxidation. *Molecules* **2022**, *27*, 2862. https://doi.org/10.3390/molecules27092862

Academic Editor: Xiaobing Cui

Received: 14 April 2022
Accepted: 29 April 2022
Published: 30 April 2022

Publisher's Note: MDPI stays neutral with regard to jurisdictional claims in published maps and institutional affiliations.

Copyright: © 2022 by the authors. Licensee MDPI, Basel, Switzerland. This article is an open access article distributed under the terms and conditions of the Creative Commons Attribution (CC BY) license (https://creativecommons.org/licenses/by/4.0/).

1. Introduction

Polyoxoniobates (PONbs), as a unique branch of polyoxometalates (POMs), have drawn widespread attention in the past few decades due to their diverse structures and multiple applications in catalysis, nuclear-waste treatment, and virology [1,2]. Nevertheless, compared with other POM members, such as polyoxotungstates, polyoxomolybdates, and polyoxovanadates, the development of PONbs is relatively slow due to the lack of soluble Nb precursors, and their low reactivity and narrow working pH range [3]. Recently, great progress has been made in isopolyoxoniobates and some large clusters, such as $\{Nb_{27}O_{76}\}$, $\{Nb_{32}O_{96}\}$, $\{Nb_{52}O_{150}\}$, $\{Nb_{81}O_{225}\}$, $\{Nb_{114}O_{316}\}$, and the highest nuclearity $\{Nb_{288}O_{768}(OH)_{48}(CO_3)_{12}\}$ have been reported [4–7]. In 2002, the first Keggin-type PONb $\{(Ti_2O_2)SiNb_{12}O_{40}\}$ was successfully synthesized by Nyman et al., marking the beginning of heteropolyoxoniobate chemistry [8]. After that, a series of heteropolyoxoniobates were reported and the Keggin-type $\{XNb_{12}O_{40}\}^{n-}$ (X = Si, Ge, P) are the most extensively studied [9,10].

In the periodic table, Nb and V are neighbors with similar ionic radius and electronegativity, and their hydrolysis and condensation can be performed under alkaline conditions. Inspired by these similarities, in 2011, we synthesized the first Keggin-type vanadium-containing PONb {VNb$_{12}$O$_{40}$(VO)$_2$} stabilized by Cu complex units [11]. Since then, a series of vanadium-containing Keggin-type PONbs and their derivatives have been reported, including {XNb$_{12}$O$_{40}$(VO)$_2$} (X = Si, Ge, P, V), {XNb$_{12}$O$_{40}$(VO)$_4$} (X = As, V), {PNb$_{10}$V$_2$O$_{40}$(VO)$_4$}, {XNb$_{12}$O$_{40}$(VO)$_6$} (X = P, V), {VNb$_{14}$O$_{42}$L$_2$} (L = CO$_3^{2-}$, NO$_3^-$), {XNb$_8$V$_4$O$_{40}$(VO)$_4$} (X = P, V, As), {AsNb$_9$V$_3$O$_{40}$(VO)$_4$}, {TeNb$_9$V$_2$O$_{37}$}, and {TeNb$_9$V$_3$O$_{39}$}, where the introduced V acts as central, capping, or/and substituted atoms [12–23]. Notably, most of the clusters were modified by metal-complex units. The use of the metal-complex unit not only contributes to the isolation of novel Keggin-type PONbs, but can also link the discrete PONb clusters into extended structures. In general, most transition metals tend to hydrolyze rapidly into precipitation under alkaline conditions, and thereby the coexistence of transition metal ions with basic PONbs is a challenge [3]. As the Cu ion can tolerate the alkaline synthesis conditions combining with its Jahn–Teller effect, Cu-complexes are the dominated metal organic units in the synthesis of PONb-based hybrids [24–28]. In contrast, Ni-complexes were seldom used and the extended structure based on V-containing PONb hybrids and a Ni-complex is rare.

Compared with other POM members, the catalytic properties of PONbs are not extensively explored. Due to their Brønsted basicity, PONbs have been used to promote the hydrolysis of chemical warfare agents [29,30]. The introduction of V endows basic PONb clusters with interesting redox properties. For example, a double-anion cluster {PNb$_{12}$O$_{40}$(VO)$_2$(V$_4$O$_{12}$)$_2$} was successfully prepared in our group, which can effectively promote the basic hydrolysis of the nerve agent simulant and the oxidative decontamination of the sulfur mustard simulant [31]. Then, we found that the organic-inorganic hybrids based on {PNb$_{12}$O$_{40}$(VO)$_2$} were active for the selective oxidation of benzyl-alkanes to ketone [14]. Our investigation indicates that the V atoms of {V$_5$Nb$_{23}$O$_{80}$} and {V$_6$Nb$_{23}$O$_{81}$} play a key role in the selective oxidation of the sulfur mustard simulant [32]. Owing to their fast and reversible electron transfer behavior, POMs are also a kind of promising electrocatalyst [33–35]. Recently, the covalent triazine framework immobilized {PMo$_{10}$V$_2$O$_{40}$} shows excellent activities in the electrocatalytic oxidation of benzyl alcohols and ethylbenzene [36,37]. However, the electrocatalytic activity of vanadium-containing PONbs is nearly unexplored.

Herein, we report the controllable synthesis and structural characterization of two vanadium-containing Keggin-type PONbs: [Ni(en)$_2$]$_5$[PNb$_{12}$O$_{40}$(VO)$_5$](OH)$_5$·18H$_2$O (**1**) and [Ni(en)$_3$]$_5$[PNb$_{12}$O$_{40}$(VO)$_2$]·17H$_2$O (**2**, en = ethylenediamine) (Scheme 1). Compound **1** contains a Keggin-type PONb capped by five vanadyl groups, which was further connected by [Ni(en)$_2$]$^{2+}$ units into a three-dimensional (3D) organic-inorganic framework. Compound **2** is a di-capped discrete vanadoniobate cluster with a [Ni(en)$_3$]$^{2+}$ unit as count cations. Interestingly, **1** and **2** as electrode materials can catalyze the selective oxidation of benzyl alcohol to benzaldehyde under alkaline additive free conditions and compound **1** is more active than **2**. The control experiments show that both the capped V atom and the basic PONb cluster contribute to the enhancement of electrocatalytic activity.

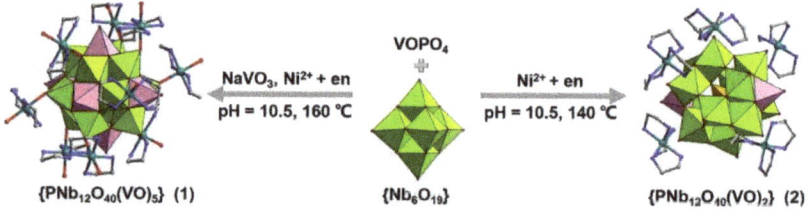

Scheme 1. Controllable synthesis of **1** and **2**.

2. Experimental Section

2.1. Materials and Methods

The hexaniobate precursor $K_7HNb_6O_{19} \cdot 13H_2O$, vanadyl phosphate $VOPO_4 \cdot 2H_2O$, $TMA_{10}H_5[PNb_{12}O_{40}] \cdot 30.5H_2O$, $K_6[V_{10}O_{28}] \cdot 10H_2O$, and $TMA_9[PNb_{12}V_2O_{42}] \cdot 19H_2O$ were prepared according to previous literature methods and identified by IR spectroscopy [10,38–41]. Other starting chemicals and solvents were purchased from commercial source and used without further purification. IR spectra in KBr pellets were collected in the range of 400–4000 cm^{-1} using a Bruker FT-IR spectrometer (Leipzig, Germany). Thermogravimetric analyses (TGA) of the compounds were performed using a LABSYS EVO device (Setaram Inc., Lyon, France) from room temperature to 800 °C under N_2 atmosphere. Elemental analyses (Nb, V, Ni, P) were measured on a Thermo ICP atomic emission spectrometer (Waltham, MA, USA); C, H, N were performed on an ElementarVario EL cube Elmer CHN elemental analyzer (Langenselbold, Germany). X-ray photoelectron spectrum (XPS) analysis were measured on a Thermo ESCALAB 250 spectrometer using Al Kα radiation as the X-ray source (1486.7 eV). Powder X-ray diffraction (PXRD) data were obtained on SHIMADZU XRD-6000 X-ray diffractometer (Kyoto, Japan) with Cu Kα radiation (λ = 1.54 Å; 2θ = 5–50°). Gas chromatograph analyses were detected on a Shimadzu GC-2014C instrument with an FID detector equipped with an HP 5 ms capillary column. The hydrogen was detected by a Techcomp GC-9700 gas chromatograph (Shanghai, China) with a 5 Å molecular sieve column (2 m × 2 mm) and a thermal conductivity detector (TCD). Temperature-programmed chemisorption of carbon dioxide (CO_2-TPD) was performed on a PCA-1200 temperature-programmed chemisorption instrument. The UV-vis spectra were measured on a UV-2600 (Builder, Beijing, China). Electrochemical surface area experiments were measured on a CHI660E electrochemical workstation (CH Instruments, Shanghai, China). Other electrochemical experiments were performed on an Ivium-OctoStat30 multi-channel electrochemical workstation (Eindhoven, The Netherlands).

2.2. Synthesis of $[Ni(en)_2]_5[PNb_{12}O_{40}(VO)_5](OH)_5 \cdot 18H_2O$ (1)

$Ni(CH_3COO)_2 \cdot 4H_2O$ (0.2240 g, 0.9 mmol) was dissolved in distilled water (0.83 mL), and to this en (0.13 mL) was added, obtaining a purple solution. Then the resulting solution was added dropwise to $K_7HNb_6O_{19} \cdot 13H_2O$ (0.1507 g, 0.11 mmol) aqueous solution (8 mL) containing $VOPO_4 \cdot 2H_2O$ (0.0534 g, 0.27 mmol) and $NaVO_3$ (0.0329 g, 0.27 mmol) under stirring. The pH value of the mixture was adjusted to 10.50 using 2 M NaOH. The mixture was transferred to a Teflon-lined autoclave (23 mL) and kept at 160 °C for 72 h, and then slowly cooled to room temperature after 24 h, and brown block crystals of **1** were isolated in about 17.5% yield (based on Nb). Anal. Calcd (%) for **1**: C, 7.02; H, 3.56; N, 8.18; Ni, 8.57; P, 0.90; V, 7.44; Nb, 32.56. Found: C, 7.57; H, 3.47; N, 7.83; Ni, 8.06; P, 0.48; V, 7.05; Nb, 32.34. FT-IR (cm^{-1}): 3451 (s), 2947 (w), 2883 (w), 1589 (s), 1460 (m), 1395 (w), 1330 (w), 1282 (s), 1233 (w), 1112 (w), 1032 (s), 942 (s), 870 (s), 805 (w), 700 (s), 627 (w), 498 (w), 473 (w).

2.3. Synthesis of $[Ni(en)_3]_5[PNb_{12}O_{40}(VO)_2] \cdot 17H_2O$ (2)

Compound **2** was synthesized by a similar procedure to that of **1**, but without adding $NaVO_3$. The pH value of the mixture was adjusted to 10.50 using 2 M NaOH and transferred to a Teflon-lined autoclave (23 mL), kept at 140 °C for 72 h, and then slowly cooled to room temperature. Brown block crystals of **2** were isolated in about 21.8% yield (based on Nb). Anal. Calcd (%) for **2**: C, 10.53; H, 4.54; N, 12.28; Ni, 8.58; P, 0.91; V, 2.98; Nb, 32.59. Found: C, 10.46; H, 4.42; N, 11.84; Ni, 8.31; P, 0.64; V, 2.77; Nb, 32.86. FT-IR (cm^{-1}): 3425 (s), 2945 (w), 2875 (w), 1593 (s), 1469 (m), 1363 (w), 1318 (w), 1275 (s), 1233 (w), 1106 (w), 1027 (s), 947 (s), 877 (s), 815 (w), 709 (s), 638 (w), 505 (w), 469 (w).

2.4. Preparation of Working Electrode

Grinded crystal samples of compound **1** or **2** (10 mg) and acetylene black (3 mg) were dispersed uniformly in isopropanol (0.5 mL) containing 5 wt% Nafion under ultrasonic conditions. A total of 50 μL of the suspension was drop-cast onto a piece of carbon cloth (1 cm^2) and then dried slowly at room temperature.

2.5. Cyclic Voltammetry Experiments

Cyclic voltammetry experiments were performed in acetonitrile (10 mL) containing supporting electrolyte LiClO$_4$ (1.0 mmol) and benzyl alcohol (0.5 mmol) under ambient conditions, and the scan rate was kept at 40 mV s^{-1}. The cyclic voltammetry tests of **1** and **2** were performed using a three-electrode setup: carbon cloth modified by **1** or **2** as the working electrode, platinum plate electrode as the counter electrode, and Ag/Ag$^+$ as the reference electrode.

2.6. Electrochemical Surface Area Experiments

Electrochemical surface areas of compounds **1** and **2** were estimated by the capacitance of the double layer C_{dl}, which were determined by cyclic voltammetry tests [42]. For the cyclic voltammetry tests of **1** and **2**, glassy carbon electrode (3 mm diameter) was dripped with 5 μL isopropanol suspension of **1** or **2**, and served as the working electrode. The potential window was 0.01–0.13 V vs. Ag/Ag$^+$, where no Faradaic processes occur. The scan rates were 10 mV s^{-1}, 20 mV s^{-1}, 30 mV s^{-1}, 40 mV s^{-1}, 50 mV s^{-1}, 60 mV s^{-1}, 70 mV s^{-1}, 80 mV s^{-1}, 90 mV s^{-1}, and 100 mV s^{-1}. The C_{dl} was calculated by plotting the relationship between Δj and scanning rate at 0.07 V ($\Delta j = j_a - j_c$, j_a, and j_c represent the current densities of the anode and cathode, respectively), and the slope of the image is twice that of C_{dl}.

2.7. Controlled Potential Electrolysis Experiments

Bulk electrolysis experiments were performed in an undivided cell using a three-electrode setup with carbon cloth modified by catalysts as the working electrode, platinum plate electrode as the counter electrode, and Ag/Ag$^+$ electrode as the reference electrode. A mixture of acetonitrile (10 mL) containing supporting electrolyte LiClO$_4$ (1.0 mmol) and benzyl alcohol (0.5 mmol) was added to the undivided cell with applied potential of 1.6 V vs. Ag/Ag$^+$. After the electrolysis experiment, biphenyl was added to the reaction solution as internal standard, and then the product was quantitatively detected by GC. For the recycle test, carbon cloth modified by compound **1** was washed three times with acetonitrile and ethyl alcohol, and dried for the next cycle.

3. Results and Discussion
3.1. Synthesis and Structure

Compound **1** was prepared by the hydrothermal reaction of K$_7$HNb$_6$O$_{19}$·13H$_2$O, VOPO$_4$·2H$_2$O, NaVO$_3$, Ni(CH$_3$COO)$_2$·4H$_2$O, and en at 160 °C. The single-crystal structural analysis (Table S1) reveals that **1** crystallizes in the monoclinic C2/c space group. Compound **1** is a 3D organic-inorganic framework constructed from [PNb$_{12}$O$_{40}$(VO)$_5$]$^{5-}$ ({PNb$_{12}$V$_5$}) clusters and ten [Ni(en)$_2$]$^{2+}$ linkers. The C$_{2h}$ symmetric {PNb$_{12}$V$_5$} cluster contains a typical Keggin-type [PNb$_{12}$O$_{40}$]$^{15-}$ ({PNb$_{12}$}) capped by five {VO} units (Figure 1a). In the {PNb$_{12}$} cluster, the centered heteroatom P connects with four edge-sharing {Nb$_3$O$_{13}$} subunits by sharing μ$_4$-O atoms; the P–O$_c$ (O$_c$ = central oxygen) bond lengths are in the range of 1.539–1.550 Å and the O–P–O angles are in the range of 108.7–110.6°. Each Nb center is six-coordinated with octahedral geometry and the bond lengths are 1.730–1.768 Å for Nb–O$_t$ (O$_t$ = terminal oxygen), 1.912–2.035 Å for Nb–O$_b$ (O$_b$ = bridge oxygen), and 2.539–2.583 Å for Nb–O$_c$. There are six square windows on {PNb$_{12}$}, which are capped by four 100% occupied {VO} and two 50% occupied {VO} (front ellipses style) (Figure 1a). The penta-capped Keggin-type {PNb$_{12}$O$_{40}$(VO)$_5$} in compound **1** presents a new V-containing PONb skeleton. All of the V centers are coordinated with four μ$_2$-O atoms from {PNb$_{12}$} and one terminal oxygen atom. The bond lengths of V–O$_b$ are 1.929–2.340 Å, and those of V–O$_t$ are 1.605–1.748 Å. According to the bond-valence sum (BVS) calculation, the five-capped V atoms are all in +4 oxidation state (Table S6). The oxidation state of the V atoms was further confirmed by XPS measurement. In the XPS spectrum of **1** (Figure S5), the peaks at 523.2 eV and 516.0 eV are attributable to V^{4+} $2p_{1/2}$ and V^{4+} $2p_{3/2}$, respectively.

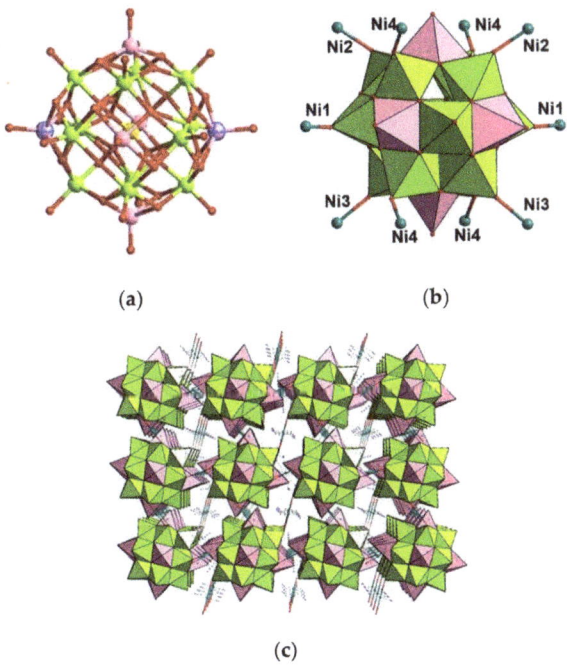

Figure 1. Crystal structure of **1**. (**a**) Ball-and-stick representation of {PNb$_{12}$O$_{40}$(VO)$_5$} in **1**. (**b**) Connections between {PNb$_{12}$O$_{40}$(VO)$_5$} and nickel ions. (**c**) The 3D framework of **1** viewed along the *b* direction. Color codes: Nb, green; V, pink; Ni, dark green; P, yellow; O, red; N, blue; C, gray.

There are four crystallographically independent Ni centers in **1** (Figure 1b); each one is coordinated with four N atoms from two en ligand and two terminal O atoms from two adjacent {PNb$_{12}$O$_{40}$(VO)$_5$}. The Ni–N distances are in the range of 2.084–2.118 Å and the Ni–O$_t$ distances are from 2.080 Å to 2.154 Å. Each {PNb$_{12}$O$_{40}$(VO)$_5$} cluster was connected by ten adjacent [Ni(en)$_2$]$^{2+}$ linkers to form a 3D framework (Figure 1c). To our knowledge, it represents the first extended V-containing PONb connected by Ni-complex units.

When a similar hydrothermal reaction was performed at 140 °C without adding NaVO$_3$, compound **2** was obtained. Compound **2** crystallizes in the orthorhombic *Pna2$_1$* space group. Compound **2** contains a C$_{2v}$ symmetric discrete [PNb$_{12}$O$_{40}$(VO)$_2$]$^{10-}$ ({PNb$_{12}$V$_2$}) cluster and five [Ni(en)$_3$]$^{2+}$ units as counter cations (Scheme 1). The polyanion {PNb$_{12}$V$_2$} features a {PNb$_{12}$} cluster capping two {VO} units (Figure 2). As with **1**, the centered heteroatom P is coordinated with four μ$_4$-O atoms, and all of the Nb centers are six-coordinated with octahedral geometry. The P–O$_c$ bond lengths are in the range of 1.549–1.555 Å, and the O–P–O angles are in the range of 108.4–111.3°. The bond lengths of Nb–O$_t$, Nb–O$_b$, and Nb–O$_c$ in {PNb$_{12}$V$_2$} are in the ranges 1.743–1.776 Å, 1.888–2.125 Å, and 2.498–2.604 Å, respectively. Two {VO} units are located on two opposite square windows on the surface of {PNb$_{12}$}, and each V center exhibits square pyramidal geometry. The bond lengths of V–O$_t$ and V–O$_b$ are in the ranges 1.620–1.623 Å and 1.853–1.979 Å, respectively. Five free [Ni(en)$_3$]$^{2+}$ are distributed around {PNb$_{12}$V$_2$} and the Ni–N bond lengths are in the range of 2.061–2.164 Å. The XPS spectrum (Figure S6) indicates that there are both V^{+4} and V^{+5} oxidation states in **2**. In the V 2p region of **2**, the peaks at 523.4 eV and 515.8 eV are assigned to +4 oxidation state, and the peaks at 524.8 eV and 517.3 eV are attributable to +5 oxidation state, which are consistent with the BVS values of V (3.98 for V1 and 4.64 for V2) (Table S7).

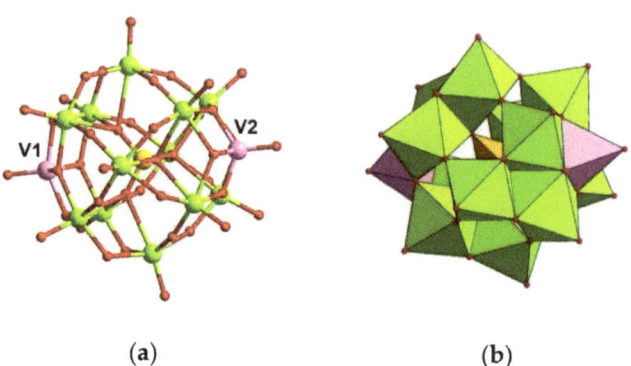

Figure 2. Crystal structure of polyanion in **2**. (**a**) Ball-and-stick representation of {PNb$_{12}$O$_{40}$(VO)$_2$} in **2**. (**b**) Polyhedral representation of {PNb$_{12}$O$_{40}$(VO)$_2$} in **2**. Color codes: Nb, green; V, pink; P, yellow; O, red.

We systematically explored the factors that influence the synthesis of **1** and **2**. It is found that the used en can effectively protect Ni^{2+} from hydrolysis under alkaline conditions. When 1,2-diaminopropane or 1,3-diaminopropane was used instead of en, the amount of precipitation was obtained. In addition, control experiments show that temperature and vanadium source play important roles in the synthesis of **1** and **2**. Following a procedure similar to that of **1**, compound **2** was obtained by removing NaVO$_3$ at 140 °C. In addition, compound **2** cannot be obtained by lowering the hydrothermal temperature or varying the ratio of VOPO$_4$ to NaVO$_3$ in the synthesis of **1**. Therefore, we speculate that the evaluated hydrothermal temperature and the use of NaVO$_3$ might increase the number of vanadium caps in the PONb cluster, and meanwhile the terminal O atoms of the Keggin-type {PNb$_{12}$} would be activated by introducing additional vanadyl caps. As a result, two-capped {PNb$_{12}$O$_{40}$(VO)$_2$} was isolated as a discrete cluster with the Ni-complex as counter cations, and five-capped {PNb$_{12}$O$_{40}$(VO)$_5$} gave rise to a 3D framework by using the Ni-complex as linker.

The IR spectra of **1** and **2** (Figure S2) were recorded from 4000 to 400 cm^{-1}. The terminal M = O$_t$ (M = Nb, V) vibrations are at 942 and 870 cm^{-1} for **1**, and at 947 and 877 cm^{-1} for **2**. The peaks at 805, 700, 627, 498, and 473 cm^{-1} of **1**, and at 815, 709, 638, 505, and 469 cm^{-1} of **2** are attributed to the bridging M–O–M vibrations. The peaks at 1032 cm^{-1} for **1**, and at 1027 cm^{-1} for **2**, are attributed to P–O$_c$ vibration. The peaks at 1642–1048 cm^{-1} for **1**, and at 1593–1035 cm^{-1} for **2** can be assigned to the en ligand. In addition, for the two compounds, the broad band above 3000 cm^{-1} is attributed to the O–H vibrations of water molecules and/or the hydroxyl groups.

The phase purity of **1** and **2** was confirmed by PXRD (Figure S3), where the collected diffraction peaks match well with the simulated ones. Compounds **1** and **2** are nearly insoluble in water and common organic solvents, such as CH$_2$Cl$_2$, THF, CH$_3$COCH$_3$, CH$_3$CN, and DMF (Figures S8 and S9). Therefore, we tested their pH stability in aqueous solution modified by PXRD (Figure 3) and IR spectra (Figure S7). As shown in Figure 3, **1** and **2** remained stable in the pH range of 4–14 after soaking for 24 h and began to decompose when the solution pH was 3. In addition, the crystals of **1** and **2** can keep their structure integrity after heating in organic solvent in the temperature range of 40–80 °C (Figures S10 and S11).

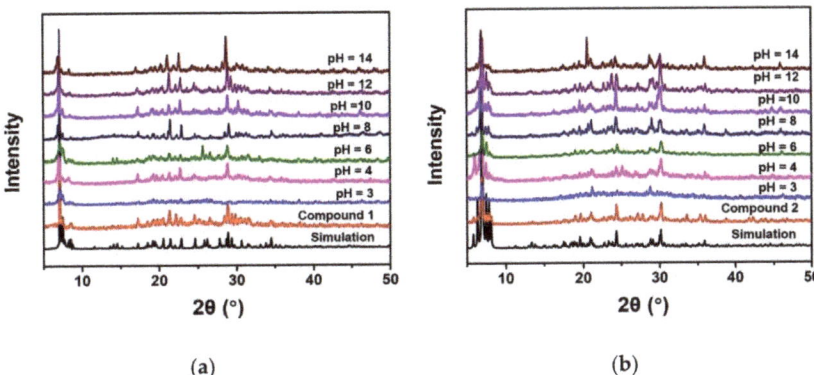

Figure 3. PXRD patterns of **1** (**a**) and **2** (**b**) after being soaked in aqueous solutions with different pH values for 24 h.

3.2. Electrocatalytic Selective Oxidation of Benzyl Alcohol

The selective oxidation of alcohols to aldehydes is one of the important organic transformations [43,44]. Compared with the traditional oxidation processes, the electrochemical oxidation provides an efficient and sustainable alternative [45,46]. Driven by electricity, alcohols can be oxidized on the anodic electrode under ambient conditions with hydrogen released from the cathodic electrode. Although some electrocatalysts have been developed in the anodic oxidation of alcohols, the selective oxidation of alcohols to the corresponding aldehydes remains a challenge, and in the reported system, the reaction activity significantly relies on the addition of alkaline additives [36,47,48]. Therefore, it is necessary to develop efficient and cost-effective electrode materials to realize the selective oxidation of alcohols under alkaline additive free conditions.

Considering that V-containing PONbs **1** and **2** both have both Brønsted basicity and redox activity, we investigate the electrocatalytic activities of **1** and **2** using the selective oxidation of benzyl alcohol (BA) to benzaldehyde as a model reaction. The electrocatalytic activity of **1** and **2** was first evaluated by the cyclic voltammetry (CV) method, which was performed in an acetonitrile solution containing LiClO$_4$ and BA with a carbon cloth modified by **1** or **2** as the working electrode. As shown in Figure 4, for **1** and **2**, the addition of BA leads to the significant increase in the anodic peak currents, indicating that the two compounds have a fast electrocatalytic response to the oxidation of BA. Notably, the anodic peak current of **1** is obviously higher than that of **2**, revealing that **1** has better electrocatalytic performance than **2**. The electrocatalytic activities of **1** and **2** were further verified by bulk electrolysis experiments performed in an undivided cell using **1** or **2** modified carbon cloth as the working electrode. As shown in Figure 5a and Table 1, both **1** and **2** are active for the selective oxidation of BA. Under ambient conditions, 92% of BA was converted by **1** in 6 h at the potential of 1.6 V vs. Ag/Ag$^+$ and the selectivity for benzaldehyde reached 95%, giving the Faradaic efficiency (FE) of 93% (Table 1, entry 2). In addition, a trace amount of N-benzylacetamide as the only by-product was detected (Figures S12 and S14). Under the otherwise identical conditions, the catalytic activity of **2** (conversion: 79%, selectivity: 90%, FE: 84%, Table 1, entry 3) is lower than that of **1**. During the electrolysis process, hydrogen was released on the counter electrode (Figure S13).

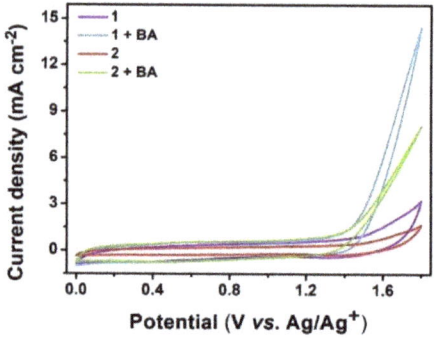

Figure 4. CV curves of **1** and **2** with BA (0.5 mmol) and without BA, obtained with a carbon cloth working electrode modified with **1** or **2** at a scan rate of 40 mV s^{-1}.

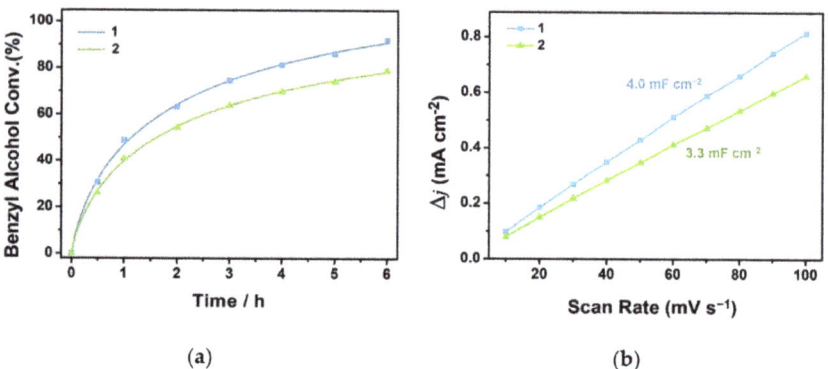

Figure 5. (a) Time profile for the electrocatalytic oxidation of BA by **1** and **2**; (b) The scan-rate dependence of current density at potential = 0.07 V vs. Ag/Ag$^+$ for **1** and **2**.

Table 1. Electrocatalytic selective oxidation of benzyl alcohol by different catalysts [a].

Entry	Cat.	Conv. (%) [b]	Sele. (%) [b]	FE (%)
1	-	42	73	66
2	Compound **1**	92	95	93
3	Compound **2**	79	90	84
4	Ni(en)$_3$Cl$_2$·H$_2$O	48	69	70
5	K$_7$H[Nb$_6$O$_{19}$]·13H$_2$O	72	79	76
6	[N(CH$_3$)$_4$]$_{10}$H$_5$[PNb$_{12}$O$_{40}$]·30.5H$_2$O	70	78	76
7	K$_6$[V$_{10}$O$_{28}$]·10H$_2$O	83	76	78
8	[N(CH$_3$)$_4$]$_9$[PNb$_{12}$O$_{40}$(VO)$_2$]·19H$_2$O	74	89	82

[a] Standard reaction conditions: CH$_3$CN (10 mL), LiClO$_4$ (1.0 mmol), BA (0.5 mmol), reaction time: 6 h, constant potential: 1.6 V vs. Ag/Ag$^+$. [b] The product conversion and selectivity were determined by GC analysis using biphenyl as internal standard.

To investigate the influence of the Ni-complex unit, PONb, and the capped V of **1** and **2** on the electrocatalytic selective oxidation of BA, the following control experiments were carried out (Table 1, entries 4–8). As shown in Table 1, entry 4, the electrocatalytic activity of the Ni-complex unit is negligible, because the catalytic performance of Ni(en)$_3$Cl$_2$ (conversion: 48%, selectivity: 69%, FE: 70%) is almost comparable to that of the blank test (conversion: 42%, selectivity: 73%, FE: 66%, Table 1, entry 1). When carbon cloth modified by K$_7$H[Nb$_6$O$_{19}$]·13H$_2$O was used as a working electrode, the conversion of BA (72%) was improved relative to the blank test, but the selectivity (79%) for benzaldehyde was still common (Table 1, entry 5). Moreover, a similar result was obtained by [N(CH$_3$)$_4$]$_{10}$H$_5$[PNb$_{12}$O$_{40}$] (conversion: 70%, selectivity: 78%, FE: 76%, Table 1, entry 6). The above results indicate that PONbs contribute to the conversion of BA because basic PONbs might facilitate the dehydrogenation oxidation of BA. Therefore, the temperature-programmed desorption of the carbon dioxide (CO$_2$-TPD) measurement for 1 and 2 was performed, where the desorption peaks at 152 °C for 1 and 148 °C for 2 corresponding to the weak base site were observed, respectively (Figure S15). In addition, polyoxovanadate, K$_6$[V$_{10}$O$_{28}$], can convert 83% of the substrate (Table 1, entry 7), but its selectivity (76%) is lower than that of the V-containing PONb 1 or 2. As shown in Table 1, entry 8, the catalytic activity of the bicapped Keggin-type [N(CH$_3$)$_4$]$_9$[PNb$_{12}$O$_{40}$(VO)$_2$] (conversion: 74%, selectivity: 89%) is similar to that of 2 (conversion: 79%, selectivity: 90%). The control experiments above show that both the PONb cluster and the V caps contribute to the enhancement of BA oxidation. Then, we speculate that the different catalytic activity of 1 and 2 is mainly caused by their different number of V caps. This is further confirmed by the electrochemical surface area (ECSA) measurement. The ECSA of 1 (4.0 mF·cm^{-2}) with five V caps is higher than that of 2 (3.3 mF·cm^{-2}) with two V caps (Figures 5b and S16).

To explore the optimal reaction conditions, we systematically investigated the influences of electrolyte, solvent, applied potential, and catalyst dosage on the electrocatalytic selective oxidation of BA by 1. As shown in Figure 6a, compared with other types of supporting electrolytes, 1 exhibits excellent catalytic performance by using LiClO$_4$. Meanwhile, it is found that acetonitrile with excellent conductivity exhibits a better performance than that of acetone, tetrahydrofuran, and N,N-dimethylformamide (Figure 6b). When the applied potential was increased from 1.4 to 1.6 V vs. Ag/Ag$^+$, the conversion of BA was increased from 16% to 92%, but when it reached 1.7 V vs. Ag/Ag$^+$, the selectivity decreased to 79%, although 98% of the BA was converted (Figure 6c). Therefore, 1.6 V vs. Ag/Ag$^+$ is the optimal potential. As shown in Figure 6d, the best catalytic performance was achieved by using 1.0 mg 1. After that, the catalytic activity was not further improved by increasing the catalyst dosage.

Moreover, the recyclability and stability of 1 were evaluated. As shown in Figure 7a, the catalytic activity of 1 is basically maintained after four cycles. There is no obvious change observed in the IR spectra of 1 before and after the reaction (Figure S17), revealing that compound 1 is basically stable after the recycle test. We compared the XPS spectra of 1 before and after the recycle. As shown in Figure 7b, in the V 2p region, the peaks at 523.2 eV and 516.0 eV are basically unchanged, indicating that the oxidation state of V remains +4 in 1. Meanwhile, to verify the heterogeneity of 1, the reaction solution was tested by ICP-OES (detection limit ca. 1 ppm) and no Nb, V, or Ni was detected, indicating that there is no catalyst leaching during the electrocatalytic process. In addition, no characteristic absorption of V-containing PONb is detected by the UV-vis spectrum (Figure S18).

In order to explore the possible mechanism of the electrocatalytic selective oxidation of BA by 1, we performed free radical trapping experiments. Oxygen radical scavenger, diphenylamine, and hydroxyl radical scavenger, *tert*-butanol, were added to the reaction system, respectively. As shown in Table S8, the oxidation of BA was significantly inhibited after the addition of diphenylamine. Therefore, we speculate that a free radical process was involved in the electrocatalytic oxidation of BA. Based on the experimental results, a plausible reaction mechanism was proposed (Figure S19). First, the electrocatalyst 1 in the reduced state (1-Red) is oxidized to its oxidized state (1-Ox) at the anode under constant potential. Then, the hydroxyl group of BA might be activated by the surface bridging O of the Keggin-type PONb cluster due to its Brønsted basicity [49,50]. After that, the BA is oxidized by 1-Ox through a $-1e^-/-1H^+$ process, generating the oxygen radical species (PhCH$_2$O$^\bullet$). PhCH$_2$O$^\bullet$ is further oxidized to benzaldehyde through another $-1e^-/-1H^+$

process, and meanwhile, 1-Ox is reduced to 1-Red, releasing protons to complete a catalytic cycle. The released protons are reduced at the cathode to produce H_2.

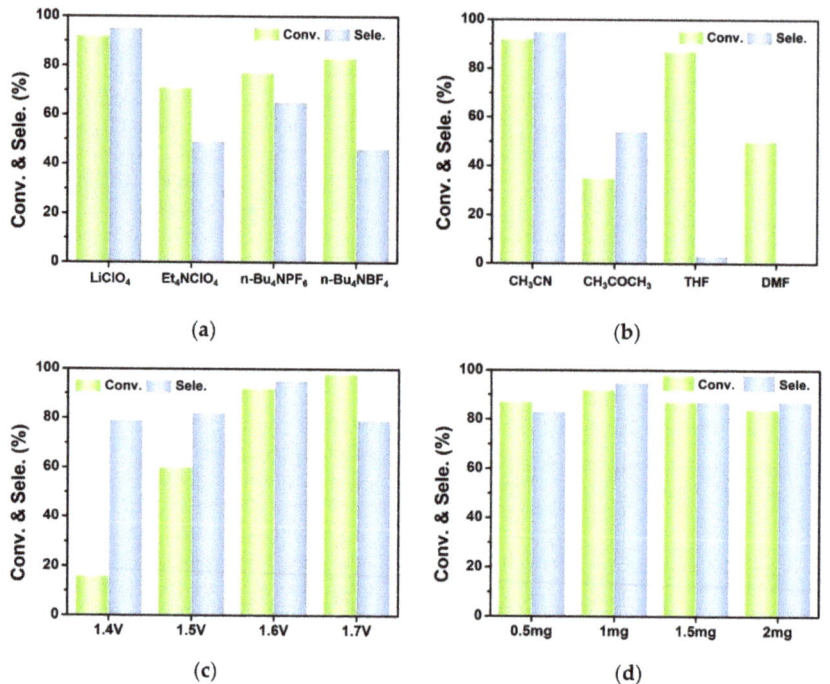

Figure 6. The influence of supporting electrolytes (**a**), solvent (**b**), applied potential (**c**), and catalyst dosage (**d**) on the electrocatalytic oxidation of BA by **1**.

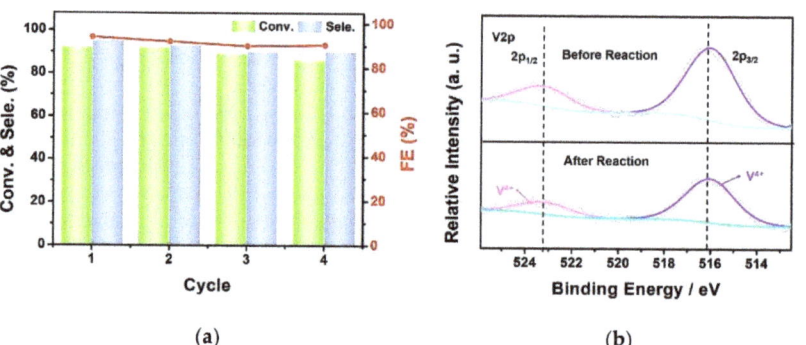

Figure 7. (**a**) Recycle test for the electrocatalytic selective oxidation of BA by **1**; (**b**) V 2p XPS spectra of **1** before and after the oxidation.

4. Conclusions

In summary, two novel vanadium-containing Keggin-type PONbs modified by a Ni-complex have been successfully synthesized by controlling temperature and the vanadium source. In **1**, the five-capped {$PNb_{12}O_{40}(VO)_5$} were connected by [$Ni(en)_2$]$^{2+}$ into a 3D framework; in **2**, the discrete bicapped {$PNb_{12}O_{40}(VO)_2$} was isolated with [$Ni(en)_3$]$^{2+}$ as counter cations. Compounds **1** and **2** as organic-inorganic hybrid materials exhibit good pH stability in aqueous solution and thermal stability in organic solvent. Importantly, under alkaline additive free conditions, compounds **1** and **2** are highly active for the selective oxidation of BA. The control experiments show that both the Brønsted basicity of PONb and the redox activity of the V caps play an important role in the electrocatalytic process. This study not only enriches the structure data base of vanadium-containing PONbs but also extends their catalytic application.

Supplementary Materials: CCDC 2164660 (**1**), 2164659 (**2**) contains the supplementary crystallographic data for this paper. These data can be obtained free of charge via http://www.ccdc.cam.ac.uk/conts/retrieving.html, accessed on 13 April 2022. The following supporting information can be downloaded at: https://www.mdpi.com/article/10.3390/molecules27092862/s1, Table S1: Crystal data and structure refinement for **1** and **2**; Tables S2–S5: Selected bond lengths and bond angles for **1** and **2**; Tables S6 and S7: BVS results of Ni and V atoms for **1** and **2**; Table S8: Electrocatalytic oxidation of benzyl alcohol (BA) catalyzed by **1** in the presence of radical scavengers; Figure S1: Digital photographs of **1** and **2**; Figure S2: the IR spectra of **1** and **2**; Figure S3: the PXRD patterns of **1** and **2**; Figure S4: The TG curves of **1** and **2**; Figure S5: The XPS spectra for Ni(2p) and V(2p) in **1**; Figure S6: The XPS spectra for Ni(2p) and V(2p) in **2**; Figure S7: The IR spectra of **1** and **2** after being soaked in aqueous solutions with different pH values for 24 h; Figures S8 and S9: The PXRD patterns and IR spectra of **1** and **2** after being soaked in different solvents; Figures S10 and S11: The PXRD patterns and IR spectra of **1** and **2** after heating in acetonitrile for 2 h; Figure S12: Gas chromatogram of the benzyl alcohol oxidation by **1**; Figure S13: The gas chromatograph of benzyl alcohol oxidation before and after reaction; Figure S14: GC-MS spectrum of the by-product; Figure S15: CO_2-TPD for **1** and **2**; Figure S16: CV curves of **1** and **2** at different scan rates; Figure S17: The IR spectra of **1** after the recycle test; Figure S18: UV-vis spectra of postreaction solution, BA, benzyl aldehyde, and $PNb_{12}V_2$; Figure S19: Proposed mechanism for the electrocatalytic selective oxidation of BA to benzaldehyde by **1**.

Author Contributions: Conceptualization, Y.C. and C.H.; methodology, Y.C. and C.H.; software, N.Z., C.L. and D.Z.; validation, X.L.; formal analysis, X.L., N.Z. and C.L.; investigation, X.L. and N.Z.; resources, Y.C., C.H. and J.D.; data curation, X.L., N.Z. and J.D.; writing—original draft preparation, X.L.; writing—review and editing, N.Z., C.L. and Y.C.; funding acquisition, Y.C. and C.H. All authors have read and agreed to the published version of the manuscript.

Funding: This research was financially supported by the National Natural Science Foundation of China (Nos. 21971010, 21871026, and 22101012).

Institutional Review Board Statement: Not applicable.

Informed Consent Statement: Not applicable.

Data Availability Statement: Data are contained within the article and supplementary materials.

Acknowledgments: The authors thank the Analysis and Testing Center, Beijing Institute of Technology, for device preparation and characterization.

Conflicts of Interest: The authors declare no conflict of interest.

Sample Availability: Samples of compounds **1** and **2** are available from the authors.

References

1. Nyman, M.; Powers, C.R.; Bonhomme, F.; Alam, T.M.; Maginn, E.J.; Hobbs, D.T. Ion-Exchange Behavior of One-Dimensional Linked Dodecaniobate Keggin Ion Materials. *Chem. Mater.* **2008**, *20*, 2513–2521. [CrossRef]
2. Zhao, H.; Li, Y.; Zhao, J.; Wang, L.; Yang, G. State-of-the-art Advances in the Structural Diversities and Catalytic Applications of Polyoxoniobate-based Materials. *Coord. Chem. Rev.* **2021**, *443*, 213966–213988. [CrossRef]
3. Nyman, M. Polyoxoniobate Chemistry in the 21st Century. *Dalton Trans.* **2011**, *40*, 8049–8058. [CrossRef]
4. Tsunashima, R.; Long, D.; Miras, H.N.; Gabb, D.; Pradeep, C.P.; Cronin, L. The Construction of High-nuclearity Isopolyoxoniobates with Pentagonal Building Blocks: $[HNb_{27}O_{76}]16-$ and $[H_{10}Nb_{31}O_{93}(CO_3)]^{23-}$. *Angew. Chem. Int. Ed.* **2010**, *49*, 113–116. [CrossRef] [PubMed]
5. Huang, P.; Qin, C.; Su, Z.; Xing, Y.; Wang, X.; Shao, K.; Lan, Y.; Wang, E. Self-assembly and Photocatalytic Properties of Polyoxoniobates: $\{Nb_{24}O_{72}\}$, $\{Nb_{32}O_{96}\}$, and $\{K_{12}Nb_{96}O_{288}\}$ Clusters. *J. Am. Chem. Soc.* **2012**, *134*, 14004–14010. [CrossRef] [PubMed]

6. Jin, L.; Zhu, Z.; Wu, Y.; Qi, Y.; Li, X.; Zheng, S. Record High-Nuclearity Polyoxoniobates: Discrete Nanoclusters {Nb_{114}}, {Nb_{81}}, and {Nb_{52}}, and Extended Frameworks Based on {Cu_3Nb_{78}} and {Cu_4Nb_{78}}. *Angew. Chem. Int. Ed.* **2017**, *56*, 16288–16292. [CrossRef] [PubMed]
7. Wu, Y.; Li, X.; Qi, Y.; Yu, H.; Jin, L.; Zheng, S. {$Nb_{288}O_{768}(OH)_{48}(CO_3)_{12}$}: A Macromolecular Polyoxometalate with Close to 300 Niobium Atoms. *Angew. Chem. Int. Ed.* **2018**, *57*, 8572–8576. [CrossRef]
8. Nyman, M.; Bonhomme, F.; Alam, T.M.; Rodriguez, M.A.; Cherry, B.R.; Krumhansl, J.K.; Nenoff, T.M.; Sattler, A.M. A General Synthetic Procedure for Heteropolyniobates. *Science* **2002**, *297*, 996–998. [CrossRef]
9. Zhang, Z.; Peng, J.; Shi, Z.; Zhou, W.; Khan, S.U.; Liu, H. Antimony-dependent Expansion for the Keggin Heteropolyniobate Family. *Chem. Commun.* **2015**, *51*, 3091–3093. [CrossRef]
10. Son, J.; Casey, W.H. Reversible Capping/Uncapping of Phosphorous-centered Keggin-type Polyoxoniobate Clusters. *Chem. Commun.* **2015**, *51*, 1436–1438. [CrossRef]
11. Guo, G.; Xu, Y.; Cao, J.; Hu, C. An Unprecedented Vanadoniobate Cluster with 'Trans-vanadium' Bicapped Keggin-type {$VNb_{12}O_{40}(VO)_2$}. *Chem. Commun.* **2011**, *47*, 9411–9413. [CrossRef]
12. Zhang, Y.; Shen, J.; Zheng, L.; Zhang, Z.; Li, Y.; Wang, E. Four Polyoxonibate-Based Inorganic–Organic Hybrids Assembly from Bicapped Heteropolyoxonibate with Effective Antitumor Activity. *Cryst. Growth Des.* **2013**, *14*, 110–116. [CrossRef]
13. Zhang, T.; Zhang, X.; Lü, Y.; Li, G.; Xiao, L.; Cui, X.; Xu, J. New Organic–inorganic Hybrid Compounds Based on [$SiNb_{12}V_2O_{42}$]$^{12-}$ with High Catalytic Activity for Styrene Epoxidation. *Inorg. Chem. Front.* **2017**, *4*, 1397–1404. [CrossRef]
14. Hu, J.; Dong, J.; Huang, X.; Chi, Y.; Lin, Z.; Li, J.; Yang, S.; Ma, H.; Hu, C. Immobilization of Keggin Polyoxovanadoniobate in Crystalline Solids to Produce Effective Heterogeneous Catalysts towards Selective Oxidation of Benzyl-alkanes. *Dalton Trans.* **2017**, *46*, 8245–8251. [CrossRef]
15. Li, S.; Ji, P.; Han, S.; Hao, Z.; Chen, X. Two Polyoxoniobates-based Ionic Crystals as Lewis Base Catalysts for Cyanosilylation. *Inorg. Chem. Commun.* **2020**, *111*, 107666. [CrossRef]
16. Li, N.; Liu, Y.; Lu, Y.; He, D.; Liu, S.; Wang, X.; Li, Y.; Liu, S. An Arsenicniobate-based 3D Framework with Selective Adsorption and Anion-exchange Properties. *New J. Chem.* **2016**, *40*, 2220–2224. [CrossRef]
17. Zhang, X.; Liu, S.; Li, S.; Wang, X.; Jin, H.; Cui, J.; Liu, Y. Hydrothermal Syntheses and Characterization of Two Multi-vanadium Capped Keggin Polyoxoniobates Derivatives. *Chem. J. Chin. Univ.* **2013**, *34*, 2046–2050. [CrossRef]
18. Hu, J.; Wang, Y.; Zhang, X.; Chi, Y.; Yang, S.; Li, J.; Hu, C. Controllable Assembly of Vanadium-Containing Polyoxoniobate-Based Three-Dimensional Organic-Inorganic Hybrid Compounds and Their Photocatalytic Properties. *Inorg. Chem.* **2016**, *55*, 7501–7507. [CrossRef]
19. Abramov, P.A.; Davletgildeeva, A.T.; Moroz, N.K.; Kompankov, N.B.; Santiago-Schubel, B.; Sokolov, M.N. Cation-Dependent Self-assembly of Vanadium Polyoxoniobates. *Inorg. Chem.* **2016**, *55*, 12807–12814. [CrossRef]
20. Huang, P.; Zhou, E.; Wang, X.; Sun, C.Y.; Wang, H.; Xing, Y.; Shao, K.; Su, Z. New Heteropolyniobates Based on a Bicapped Keggin-type {VNb_{14}} Cluster with Selective Adsorption and Photocatalytic Properties. *CrystEngComm* **2014**, *16*, 9582–9585. [CrossRef]
21. Shen, J.; Wu, Q.; Zhang, Y.; Zhang, Z.; Li, Y.; Lu, Y.; Wang, E. Unprecedented High-nuclear Transition-metal-cluster-substituted Heteropolyoxoniobates: Synthesis by {V_8} Ring Insertion into the POM Matrix and Antitumor Activities. *Chem. Eur. J.* **2014**, *20*, 2840–2848. [CrossRef] [PubMed]
22. Lan, Q.; Zhang, Z.; Li, Y.; Wang, E. Extended Structural Materials Composed of Transition-metal-substituted Arsenicniobates and Their Photocatalytic Activity. *RSC Adv.* **2015**, *5*, 44198–44203. [CrossRef]
23. Yang, Z.; Shang, J.; Yang, Y.; Ma, P.; Niu, J.; Wang, J. Synthesis, Structures and Stability of Three V-substituted Polyoxonibate Clusters Based on [$TeNb_9O_{33}$]$^{17-}$ Units. *Dalton Trans.* **2021**, *50*, 7610–7620. [CrossRef] [PubMed]
24. Wang, J.; Niu, H.; Niu, J. A Novel Lindqvist Type Polyoxoniobate Coordinated to Four Copper Complex Moieties: {$Nb_6O_{19}[Cu(2,2'-bipy)]_2[Cu(2,2'-bipy)_2]_2$}·19$H_2O$. *Inorg. Chem. Commun.* **2008**, *11*, 63–65. [CrossRef]
25. Wang, J.; Niu, H.; Niu, J. Preparation, Crystal structure, and Characterization of an Inorganic–organic Hybrid Polyoxoniobate [$Cu(en)_2$]$_3$[$Cu(en)_2(H_2O)$]$_{1.5}$[$K_{0.5}Nb_{24}O_{72}H_{14.5}$]$_2$·25$H_2O$. *J. Chem. Sci.* **2008**, *120*, 309–313. [CrossRef]
26. Guo, G.; Xu, Y.; Hu, C. A Polyoxoniobate Based on Dimeric Hexanionbate: [$Cu(en)_2$]$_4${[$Nb_6O_{19}H_2$]$K(H_2O)_5$}$_2$·(H_2en)·17H_2O. *J. Coord. Chem.* **2010**, *63*, 3137–3145. [CrossRef]
27. Liang, Z.; Qiao, Y.; Li, M.; Ma, P.; Niu, J.; Wang, J. Two Synthetic Routes Generate Two Isopolyoxoniobates Based on {Nb_{16}} and {Nb_{20}}. *Dalton Trans.* **2019**, *48*, 17709–17712. [CrossRef]
28. Li, C.; Zhao, D.; Li, N.; Ma, Y.; Wei, G.; Wang, G.; Zhang, D. Dimeric Dumbbell Architecture Based on PNb_{14} unit. *Inorg. Chem. Commun.* **2019**, *102*, 210–214. [CrossRef]
29. Kinnan, M.K.; Creasy, W.R.; Fullmer, L.B.; Schreuder-Gibson, H.L.; Nyman, M. Nerve Agent Degradation with Polyoxoniobates. *Eur. J. Inorg. Chem.* **2014**, *2014*, 2361–2367. [CrossRef]
30. Guo, W.; Lv, H.; Sullivan, K.P.; Gordon, W.O.; Balboa, A.; Wagner, G.W.; Musaev, D.G.; Bacsa, J.; Hill, C.L. Broad-Spectrum Liquid- and Gas-Phase Decontamination of Chemical Warfare Agents by One-Dimensional Heteropolyniobates. *Angew. Chem. Int. Ed.* **2016**, *55*, 7403–7407. [CrossRef]
31. Dong, J.; Hu, J.; Chi, Y.; Lin, Z.; Zou, B.; Yang, S.; Hill, C.L.; Hu, C. A Polyoxoniobate-Polyoxovanadate Double-Anion Catalyst for Simultaneous Oxidative and Hydrolytic Decontamination of Chemical Warfare Agent Simulants. *Angew. Chem. Int. Ed.* **2017**, *56*, 4473–4477. [CrossRef]

32. Zhen, N.; Dong, J.; Lin, Z.; Li, X.; Chi, Y.; Hu, C. Self-assembly of Polyoxovanadate-capped Polyoxoniobates and Their Catalytic Decontamination of Sulfur Mustard Simulants. *Chem. Commun.* **2020**, *56*, 13967–13970. [CrossRef]
33. Macht, J.; Janik, M.J.; Neurock, M.; Iglesia, E. Catalytic Consequences of Composition in Polyoxometalate Clusters with Keggin Structure. *Angew. Chem. Int. Ed.* **2007**, *46*, 7864–7868. [CrossRef]
34. Toma, F.M.; Sartorel, A.; Iurlo, M.; Carraro, M.; Parisse, P.; Maccato, C.; Rapino, S.; Gonzalez, B.R.; Amenitsch, H.; Da Ros, T.; et al. Efficient Water Oxidation at Carbon Nanotube-polyoxometalate Electrocatalytic Interfaces. *Nat. Chem.* **2010**, *2*, 826–831. [CrossRef]
35. Evtushok, V.Y.; Suboch, A.N.; Podyacheva, O.Y.; Stonkus, O.A.; Zaikovskii, V.I.; Chesalov, Y.A.; Kibis, L.S.; Kholdeeva, O.A. Highly Efficient Catalysts Based on Divanadium-Substituted Polyoxometalate and N-Doped Carbon Nanotubes for Selective Oxidation of Alkylphenols. *ACS Catal.* **2018**, *8*, 1297–1307. [CrossRef]
36. Li, Z.; Zhang, J.; Jing, X.; Dong, J.; Liu, H.; Lv, H.; Chi, Y.; Hu, C. A Polyoxometalate@covalent Triazine Framework as a Robust Electrocatalyst for Selective Benzyl Alcohol Oxidation Coupled with Hydrogen Production. *J. Mater. Chem. A* **2021**, *9*, 6152–6159. [CrossRef]
37. Li, Z.; Liu, C.; Geng, W.; Dong, J.; Chi, Y.; Hu, C. Electrocatalytic Ethylbenzene Valorization Using a Polyoxometalate@covalent Triazine Framework with Water as the Oxygen Source. *Chem. Commun.* **2021**, *57*, 7430–7433. [CrossRef]
38. Filowitz, M.; Ho, R.K.C.; Klemperer, W.G.; Shum, W. ^{17}O Nuclear Magnetic Resonance Spectroscopy of Polyoxometalates. 1. Sensitivity and Resolution. *Inorg. Chem.* **1979**, *18*, 93–103. [CrossRef]
39. Dietl, N.; Hockendorf, R.F.; Schlangen, M.; Lerch, M.; Beyer, M.K.; Schwarz, H. Generation, Reactivity towards Hydrocarbons, and Electronic Structure of Heteronuclear Vanadium Phosphorous Oxygen Cluster Ions. *Angew. Chem. Int. Ed.* **2011**, *50*, 1430–1434. [CrossRef]
40. Guilherme, L.R.; Massabni, A.C.; Dametto, A.C.; de Souza Correa, R.; De Araujo, A.S. Synthesis, Infrared Spectroscopy and Crystal Structure Determination of a New Decavanadate. *J. Chem. Crystallogr.* **2010**, *40*, 897–901. [CrossRef]
41. Son, J.H.; Ohlin, C.A.; Johnson, R.L.; Yu, P.; Casey, W.H. A Soluble Phosphorus-centered Keggin Polyoxoniobate with Bicapping Vanadyl Groups. *Chem. Eur. J.* **2013**, *19*, 5191–5197. [CrossRef]
42. Gao, S.; Lin, Y.; Jiao, X.; Sun, Y.; Luo, Q.; Zhang, W.; Li, D.; Yang, J.; Xie, Y. Partially Oxidized Atomic Cobalt Layers for Carbon Dioxide Electroreduction to Liquid Fuel. *Nature* **2016**, *529*, 68–71. [CrossRef]
43. Zhong, W.; Liu, H.; Bai, C.; Liao, S.; Li, Y. Base-Free Oxidation of Alcohols to Esters at Room Temperature and Atmospheric Conditions using Nanoscale Co-Based Catalysts. *ACS Catal.* **2015**, *5*, 1850–1856. [CrossRef]
44. Guo, Z.; Liu, B.; Zhang, Q.; Deng, W.; Wang, Y.; Yang, Y. Recent Advances in Heterogeneous Selective Oxidation Catalysis for Sustainable Chemistry. *Chem. Soc. Rev.* **2014**, *43*, 3480–3524. [CrossRef] [PubMed]
45. Horn, E.J.; Rosen, B.R.; Chen, Y.; Tang, J.; Chen, K.; Eastgate, M.D.; Baran, P.S. Scalable and Sustainable Electrochemical Allylic C-H Oxidation. *Nature* **2016**, *533*, 77–81. [CrossRef] [PubMed]
46. Yin, Z.; Zheng, Y.; Wang, H.; Li, J.; Zhu, Q.; Wang, Y.; Ma, N.; Hu, G.; He, B.; Knop-Gericke, A.; et al. Engineering Interface with One-Dimensional Co_3O_4 Nanostructure in Catalytic Membrane Electrode: Toward an Advanced Electrocatalyst for Alcohol Oxidation. *ACS Nano* **2017**, *11*, 12365–12377. [CrossRef]
47. Zheng, J.; Chen, X.; Zhong, X.; Li, S.; Liu, T.; Zhuang, G.; Li, X.; Deng, S.; Mei, D.; Wang, J. Hierarchical Porous NC@CuCo Nitride Nanosheet Networks: Highly Efficient Bifunctional Electrocatalyst for Overall Water Splitting and Selective Electrooxidation of Benzyl Alcohol. *Adv. Funct. Mater.* **2017**, *27*, 1704169–1704179. [CrossRef]
48. Rafiee, M.; Konz, Z.M.; Graaf, M.D.; Koolman, H.F.; Stahl, S.S. Electrochemical Oxidation of Alcohols and Aldehydes to Carboxylic Acids Catalyzed by 4-Acetamido-TEMPO: An Alternative to "Anelli" and "Pinnick" Oxidations. *ACS Catal.* **2018**, *8*, 6738–6744. [CrossRef]
49. Hayashi, S.; Sasaki, N.; Yamazoe, S.; Tsukuda, T. Superior Base Catalysis of Group 5 Hexametalates $[M_6O_{19}]^{8-}$ (M = Ta, Nb) over Group 6 Hexametalates $[M_6O_{19}]^{2-}$ (M = Mo, W). *J. Phys. Chem. C* **2018**, *122*, 29398–29404. [CrossRef]
50. Hayashi, S.; Yamazoe, S.; Tsukuda, T. Base Catalytic Activity of $[Nb_{10}O_{28}]^{6-}$: Effect of Countercations. *J. Phys. Chem. C* **2020**, *124*, 10975–10980. [CrossRef]

Article

First Organic–Inorganic Hybrid Compounds Formed by Ge-V-O Clusters and Transition Metal Complexes of Aromatic Organic Ligands

Hai-Yang Guo [1,2], Hui Qi [3], Xiao Zhang [4] and Xiao-Bing Cui [1,*]

[1] State Key Laboratory of Inorganic Synthesis and Preparative Chemistry and College of Chemistry, Jilin University, Changchun 130021, China; guohy@zjxu.edu.cn
[2] College of Biological, Chemical Science and Engineering, Jiaxing University, Jiaxing 314001, China
[3] The Second Hospital of Jilin University, Changchun 130021, China; qihui1977@sohu.com
[4] MIIT Key Laboratory of Critical Materials Technology for New Energy Conversion and Storage, School of Chemistry and Chemical Engineering, Harbin Institute of Technology, Harbin 150001, China; zhangx@hit.edu.cn
* Correspondence: cuixb@mail.jlu.edu.cn

Abstract: Three compounds based on Ge-V-O clusters were hydrothermally synthesized and characterized by IR, UV-Vis, XRD, ESR, elemental analysis and X-ray crystal structural analysis. Both [Cd(phen)(en)]$_2$[Cd$_2$(phen)$_2$V$_{12}$O$_{40}$Ge$_8$(OH)$_8$(H$_2$O)]·12.5H$_2$O (**1**) and [Cd(DETA)]$_2$[Cd(DETA)$_2$]$_{0.5}$[Cd$_2$(phen)$_2$V$_{12}$O$_{41}$Ge$_8$(OH)$_7$(0.5H$_2$O)]·7.5H$_2$O (**2**) (1,10-phen = 1,10-phenanthroline, en = ethylenediamine, DETA = diethylenetriamine) are the first Ge-V-O cluster compounds containing aromatic organic ligands. Compound **1** is the first dimer of Ge-V-O clusters, which is linked by a double bridge of two [Cd(phen)(en)]$^{2+}$. Compound **2** exhibits an unprecedented 1-D chain structure formed by Ge-V-O clusters and [Cd$_2$(DETA)$_2$]$^{4+}$ transition metal complexes (TMCs). [Cd(en)$_3$]{[Cd(η$_2$-en)$_2$]$_3$[Cd(η$_2$-en)(η$_2$-μ$_2$-en)(η$_2$-en)Cd][Ge$_6$V$_{15}$O$_{48}$(H$_2$O)]}·5.5H$_2$O (**3**) is a novel 3-D structure which is constructed from [Ge$_6$V$_{15}$O$_{48}$(H$_2$O)]$^{12-}$ and four different types of TMCs. We also synthesized [Zn$_2$(enMe)$_3$][Zn(enMe)]$_2$[Zn(enMe)$_2$(H$_2$O)]$_2$[Ge$_6$V$_{15}$O$_{48}$(H$_2$O)]·3H$_2$O (**4**) and [Cd(en)$_2$]$_2${H$_8$[Cd(en)]$_2$Ge$_8$V$_{12}$O$_{48}$(H$_2$O)}·6H$_2$O (**5**) (enMe = 1,2-propanediamine), which have been reported previously. In addition, the catalytic properties of these five compounds for styrene epoxidation have been assessed.

Keywords: polyoxometalates; vanadogermanate; secondary transition metal substituted Ge-V-O clusters

Citation: Guo, H.-Y.; Qi, H.; Zhang, X.; Cui, X.-B. First Organic–Inorganic Hybrid Compounds Formed by Ge-V-O Clusters and Transition Metal Complexes of Aromatic Organic Ligands. *Molecules* **2022**, *27*, 4424. https://doi.org/10.3390/molecules27144424

Academic Editor: Santiago Reinoso

Received: 17 June 2022
Accepted: 6 July 2022
Published: 11 July 2022

Publisher's Note: MDPI stays neutral with regard to jurisdictional claims in published maps and institutional affiliations.

Copyright: © 2022 by the authors. Licensee MDPI, Basel, Switzerland. This article is an open access article distributed under the terms and conditions of the Creative Commons Attribution (CC BY) license (https://creativecommons.org/licenses/by/4.0/).

1. Introduction

Several metallic materials have extensive uses, such as sensors, catalysis, fluids, regulated drug delivery and pigments [1–5]. Integration of certain metals to form polyoxometalates (POM) is a feasible and promising strategy for making new heteropolyoxometalates, which are of great interest due to their abundant structures and conceivable applications in magnetism, catalysis, medicine and electrochemistry [6–15]. Many different elements have been reported as compositions of heteropolyanions [16–18], and POMs containing different elements inspired an enormous amount of new research due to their range of intriguing applications [19–24]. Vanadium is of particular interest since it shows flexible coordination geometries as well as a variety of chemical valence states. During the past years, significant progress has been made in the syntheses of polyoxovanadates by incorporating group 15 elements (As[III]/Sb[III], vanadoarsenates [25–38] and vanadoantimonates [39–52]) into the well-known {V$_{18}$O$_{42}$} shell. In 2015, Monakhov, Bensch and Kögerler published a milestone review on derivatives of polyoxovanadates [53], in which the syntheses and structures of vanadoarsenates, vanadoantimonates and vanadogermanates were systematically reviewed. In addition, we have focused on preparations of vanadoarsenates [25–30], vanadoantimonates [40,41] and secondary transition metal substituted As-V-O clusters [54] for years. Here, we further extended our interest in Ge-V-O [55–65] and secondary transition metal substituted Ge-V-O clusters [6,66] based on two considerations. Firstly, the {As$^{III}_2$O$_5$} of the As-V-O cluster is not favorable for forming extended structures because the arsenic center did not have terminal oxygens, which can further interact with other bridging metal centers, whereas the {Ge$^{IV}_2$O$_7$} of the Ge-V-O cluster has two additional terminal oxygens, which can provide opportunities for forming extended structures via metal-oxygen covalent and dative bonds. Secondly, like the As-V-O cluster, some vanadiums of the Ge-V-O cluster can also be substituted by secondary transition metals to yield new organic–inorganic hybrid clusters [54]. Vanadogermanates can significantly expand the area of polyoxovanadate chemistry due to the introduction of a different functionality compared to the As-containing congeners. In 2003, A. J. Jacobson [57], A. Clearfield [60] and Lin [56] respectively reported the preparations of a series of Ge-V-O compounds, and then W. Bensch reported several Ge-V-O compounds in 2006, 2010 and 2013 [59,61,64]. In 2010 and 2014, Yang reported the syntheses of several secondary transition metal-substituted Ge-V-O clusters [6,66]. However, compared with vanadoarsenates, the number of vanadogermanates is still far too small, and especially the secondary transition metal substituted Ge-V-O clusters. It is still a great challenge for chemists to synthesize new vanadogermanates.

We found that all the previously reported Ge-V-O compounds were totally based on aliphatic organic ligands [6,39–52,66], while no Ge-V-O compounds constructed out of aromatic organic ligands were reported. The reason only aliphatic-ligand involving Ge-V clusters were reported can be listed as below: (1) GeO$_2$ is inert in neutral and acidic aqueous solutions; (2) the aqueous solution of the aromatic nitrogen-containing organic ligands is neutral. It is not favorable for the aggregation of Ge-V clusters. Therefore, it is very difficult to prepare aromatic-ligand-containing Ge-V clusters. The first Ge-V clusters were reported in 2003 [60], and no aromatic-ligand-containing Ge-V clusters have been prepared. On the other hand, the introduction of aromatic organic ligands can not only can enrich the structures of this kind of compound but can also ameliorate their polar, electricity, acid and redox properties [67–70]. The introduction of aromatic organic ligands may thereby lead to compounds with more interesting structures, topologies and properties (It is well known that the robustness of almost all MOFs is derived from the aromatic organic ligands [71]). An example: recently, S. K. Das reported an aromatic-ligand-containing polyoxometalate that can be used as an efficient electrocatalyst for water oxidation [72], but the aliphatic analog did not exhibit such an excellent electrocatalytic property. Based on aforementioned points, we then chose phen as the aromatic organic ligand to prepare Ge-V-O compounds. Fortunately, we successfully synthesized [Cd(phen)(en)]$_2$[Cd$_2$(phen)$_2$V$_{12}$O$_{48}$Ge$_8$(OH)$_8$(H$_2$O)]·12.5H$_2$O (**1**), [Cd(DETA)]$_2$[Cd(DETA)$_2$]$_{0.5}$[Cd$_2$(phen)$_2$V$_{12}$O$_{41}$Ge$_8$(OH)$_7$(0.5H$_2$O)]·7.5H$_2$O (**2**) and [Cd(en)$_3$]{[Cd(η$_2$-en)$_2$]$_3$[Cd(η$_2$-en)(η$_2$-en)(η$_2$-μ$_2$-en)Cd][Ge$_6$V$_{15}$O$_{48}$(H$_2$O)]}·5.5H$_2$O (**3**), of which compounds **1** and **2** are the first Ge-V-O compounds based on aromatic organic ligands. Compound **1** is the first dimer of Ge-V-O compound, of which Ge-V clusters are linked by a double bridge of [Cd(phen)(en)]$^{2+}$. Compound **2** exhibits a novel 1-D chain structure of which Ge-V-O clusters are fused by [Cd$_2$(DETA)$_2$]$^{4+}$ TMCs. Compound **3** is a novel 3-D structure which is constructed out of [Ge$_6$V$_{15}$O$_{48}$(H$_2$O)]$^{12-}$ clusters and five different types of TMCs. We also synthesized [Zn$_2$(enMe)$_3$][Zn(enMe)]$_2$[Zn(enMe)$_2$(H$_2$O)]$_2$[Ge$_6$V$_{15}$O$_{48}$(H$_2$O)]·3H$_2$O (**4**) [6] and [Cd(en)$_2$]$_2${H$_8$[Cd(en)]$_2$Ge$_8$V$_{12}$O$_{48}$(H$_2$O)}·6H$_2$O (**5**), which have been reported previously [54]. In addition, the catalytic properties of these five compounds have been investigated.

2. Experimental Section

2.1. Chemicals and Data Analysis

All the chemicals used were of reagent grade without further purification. C, H, N elemental analyses were carried out on a Perkin-Elmer 2400 CHN elemental analyser (Shanghai, China). Infrared spectra were recorded as KBr pellets on a Perkin-Elmer SPECTRUM ONE FTIR spectrophotometer. UV-vis spectra were recorded on a Shimadzu UV-

3100 spectrophotometer. Powder XRD patterns were obtained with a Scintag X1 powder diffractometer system using Cu Kα radiation with a variable divergent slit and a solid-state detector. Electron spin resonance (ESR) spectra were performed on a JEOL JES-FA200 spectrometer(Guangzhou, China) operating in the X-band mode. The g value was calculated by comparison with the spectrum of 1,1-diphenyl-2-picrylhydrazyl (DPPH), whereas the spin concentrations were determined by comparing the recorded spectra with that of an Mn marker and DPPH, using the built-in software of the spectrometer.

2.2. Syntheses of Compounds Based on Ge-V-O Clusters

2.2.1. $[Cd(phen)(en)]_2[Cd_2(phen)_2V_{12}O_{40}Ge_8(OH)_8(H_2O)]\cdot 12.5H_2O$ (1)

V_2O_5 (0.061 g, 0.33 mmol), GeO_2 (0.069 g, 0.67 mmol) and TMAH (TMAH = tetramethylammonium hydroxide) (0.10 mL) were added to H_2O (3.00 mL) solution with stirring for a half-hour. Then, $CdCl_2$ (0.061 g, 0.33 mmol), phen (0.066 g, 0.33 mmol) and 2,2'-bpy (2,2'-bpy = 2,2'-bipyridine, 0.052 g, 0.33 mmol) were added, the resulting suspension was further stirred for 4 h, the pH of the mixture was 5.0. Finally, the pH of the mixture was adjusted to 9.5 with en, which was stirred for another 0.5 h and then was sealed in a Teflon-lined stainless bomb and heated at 170 °C for 5 days. Brown rectangle crystals were collected by filtration and washed with water (Yield: 0.149 g, 51.60% based on GeO_2). Compound **1** can also be prepared by adjusting the pH to 10.0. When the pH of the mixture was adjusted to 9.5, more crystals were obtained, and the crystal quality was better. Anal. Calcd for $C_{52}H_{83}Cd_4Ge_8N_{12}O_{61.5}V_{12}$: C, 17.83; H, 2.39; N, 4.80%. Found: C, 17.71; H, 2.28; N, 4.83%.

2.2.2. $[Cd(DETA)]_2[Cd(DETA)_2]_{0.5}[Cd_2(phen)_2V_{12}O_{41}Ge_8(OH)_7(0.5H_2O)]\cdot 7.5H_2O$ (2)

V_2O_5 (0.067 g, 0.36 mmol), GeO_2 (0.069 g, 0.67 mmol) and TMAH (0.10 mL) were added to the H_2O (3.00 mL) solution with stirring for a half-hour. Then, $CdCl_2$ (0.183 g, 1 mmol) and phen (0.066 g, 0.33 mmol) were added, and the resulting suspension was further stirred for 4 h; the pH of the mixture was 5.0. Finally, the pH of the mixture was adjusted to 9.5 with DETA solution, which was stirred for 0.5 h, and then was sealed in a Teflon-lined stainless bomb and heated at 170 °C for 5 days. Black needle crystals were collected by filtration and washed with water (Yield: 0.086 g, 31.30% based on GeO_2). Anal. Calcd for $C_{36}H_{78}Cd_{4.5}Ge_8N_{13}O_{56}V_{12}$: C, 13.15; H, 2.39; N, 5.54%. Found: C, 12.91; H, 2.40; N, 5.33%.

2.2.3. $[Cd(en)_3]\{[Cd(\eta_2\text{-}en)_2]_3[Cd(\eta_2\text{-}en)(\eta_1\text{-}en)(\eta_2\text{-}en)Cd][Ge_6V_{15}O_{48}(H_2O)]\}\cdot 5.5H_2O$ (3)

GeO_2 (0.104 g, 1.00 mmol), NH_4VO_3 (0.2323 g, 2.00 mmol) and $CdCl_2$ (0.1831 g, 1.00 mmol) were added to a 25% aqueous solution of en (6.00 mL). The resulting suspension was further stirred for 12 h, then 2,2'-bpy (0.156 g, 1.0 mmol) were added, the final mixture (pH 9.7–10) was moved to a 35 mL Teflon-lined autoclave, sealed and kept at 170 °C for 5 days and then it was cooled to ambient temperature. Black square crystals were collected by filtration and washed with water (Yield: 0.413 g, 71.20% based on GeO_2). Anal. Calcd for $C_{24}H_{109}Cd_6Ge_6N_{24}O_{54.5}V_{15}$: C, 8.28; H, 3.16; N, 9.66%. Found: C, 8.19; H, 3.00; N, 9.63%.

2.3. X-ray Crystallography

The crystal data for compound **1** were measured on a Bruker Apex II diffractometer with graphite monochromated Mo Kα (λ = 0.71073 Å) radiation. The data for compounds **2** were measured on a Rigaku R-AXIS RAPID diffractometer with graphite monochromated Mo Kα (λ = 0.71073 Å) radiation, while the data for compound **3** were measured on an Agilent Technology SuperNova Eos Dual system with a Mo Kα (λ = 0.71073 Å) microfocus source and focusing multilayer mirror optics. None of the crystals showed evidence of crystal decay during the data collections. Refinements were carried out with SHELXS-2014/7 [73] and SHELXL-2014/7 [73] using Olex 2.0 interface via the full matrix least-squares on F2 method. In the final refinements, all atoms were refined anisotropically in compounds **1**–**3**. The hydrogen atoms of en, phen, DETA and enMe in the three compounds were placed in calculated positions and included in the structure factor calculations but not refined. In these heavy-atom structures with reflection data from poor-quality crystals it was not possible to see clear electron-density peaks in difference maps which would correspond with acceptable locations for the various H atoms bonded to water oxygen atoms. The refinements were then completed with no allowance for these water H atoms in the models; the CCDC number: 1,525,920 for 1, 2,024,572 for 2 and 1,525,922 for 3. The reflection intensity data for compounds **4** and **5** were also measured on a Rigaku R-AXIS RAPID diffractometer with graphite monochromated Mo Kα (λ = 0.71073 Å) radiation, and the results show that the two compounds have already been reported previously [6]. A summary of the crystallographic data and structure refinements for compounds **1–3** is given in Table 1.

Table 1. Crystal data and structure refinements for compounds **1–3**.

Head 1	Compound 1	Compound 2	Compound 3
Empirical formula	$C_{52}H_{83}Cd_4Ge_8N_{12}O_{61.5}V_{12}$	$C_{36}H_{78}Cd_{4.5}Ge_8N_{13}O_{56}V_{12}$	$C_{24}H_{109}Cd_6Ge_6N_{24}O_{54.5}V_{15}$
Formula weight	3501.90	3286.91	3480.39
Crystal system	Triclinic	Monoclinic	Monoclinic
space group	P-1	C 2/c	$P2_1/n$
a (Å)	14.5034(8)	17.193(3)	17.9913(3)
b (Å)	16.5920(9)	23.511(5)	23.6117(4)
c (Å)	23.0440(13)	26.373(5)	23.9327(4)
α (°)	71.648(4)	90	90
β (°)	84.130(4)	100.15(3)	91.7290(13)
γ (°)	75.454(4)	90	90
Volume (Å^3)	5093.0(5)	10,494(4)	10,162.1(3)
Z	2	4	4
D_C (Mg·m^{-3})	2.284	2.080	2.275
μ (mm^{-1})	4.282	4.242	4.367
F(000)	3390	6324	6728
θ for data collection	1.375–25.032	3.025–27.466	3.083–29.145
Reflections collected	28,997	45,848	54,419
Reflections unique	17,941	11,814	23,431
R(int)	0.1263	0.1080	0.0437
Completeness to θ	99.6	99.1	99.6
parameters	1360	662	1207
GOF on F^2	1.030	1.042	1.027
R [a] [$I > 2\sigma(I)$]	R1 = 0.0621	R1 = 0.0822	R1 = 0.0780
R [b] (all data)	ωR2 = 0.1660	ωR2 = 0.2629	ωR2 = 0.2417

[a] $R_1 = \sum ||F_o| - |F_c|| / \sum |F_o|$. [b] $\omega R_2 = \{\sum [w(F_o^2 - F_c^2)^2] / \sum [w(F_o^2)^2]\}^{1/2}$.

3. Results and Discussion

3.1. Synthesis Description

Compounds **1** and **2** are all based on $Cd_2Ge_8V_{12}$ and compounds **3** and **4** are based on Ge_6V_{15}. The alkalinity (pH > 9) and the stirring time of the reaction mixture are important for the formation of Ge_6V_{15} in compounds **3** and **4**. We have a relatively clear grasp of the synthetic conditions of the two different clusters. The molar ratio of V_2O_5 to GeO_2 for compounds **1** and **2** is about 1:2, and the molar ratios of NH_4VO_3 to GeO_2 for compounds **3** and **4** are 2:1. The addition of 2, 2′-bpy is important for the preparations of compounds **2** and **3**. Though it is absent in the products, 2, 2′-bpy is required for the syntheses of compounds **2** and **3**. It should be noted that such a phenomenon is not unusual in hydrothermal preparations [74].

3.2. Description of Crystal Structures

3.2.1. [Cd(phen)(en)]$_2$[Cd$_2$(phen)$_2$V$_{12}$O$_{40}$Ge$_8$(OH)$_8$(H$_2$O)]·12.5H$_2$O (1)

The asymmetric unit of **1** consists of a di-Cd-substituted Ge-V-O cluster [H$_8$Cd$_2$(phen)$_2$Ge$_8$V$_{12}$O$_{48}$(H$_2$O)]$^{4-}$ (Cd$_2$Ge$_8$V$_{12}$), two [Cd(phen)(en)]$^{2+}$ and 12.5 water molecules. As shown in Figure 1, an unusual feature of **1** is that two [Cd(phen)]$^{2+}$ take the place of the two VO^{2+} fragments located at the two opposite positions of {Ge$_8$V$_{14}$O$_{50}$} [59], forming Cd$_2$Ge$_8$V$_{12}$. The two substituted cadmiums each is coordinated by four oxygens from two {Ge$_2$O$_7$} units with Cd-O distances of 2.290(5)–2.366(5) Å, and two nitrogens from a phen ligand with Cd-N distances of 2.353(6)–2.427(7) Å. That is to say, two phen ligands were decorated onto the surface of Cd$_2$Ge$_8$V$_{12}$. The two phen located at the two sides of Cd$_2$Ge$_8$V$_{12}$ are not parallel to each other. There is a dihedral angle of 36.116° between the two phenanthroline-planes. All the bond distances in **1** are comparable to those of previously reported compounds [6,55–66]. Bond valence sum (BVS) calculations for Ge and V indicate that both Ge and V exist in the +4 oxidation-state (Table S1). BVS calculations were also conducted for the cadmium and oxygen atoms in compound **1** to determine the locations of the hydrogen atoms in compound **1** (see supporting information and discussions in "BVS calculations to determine the locations of hydrogen atoms for compounds **1–3**") [75].

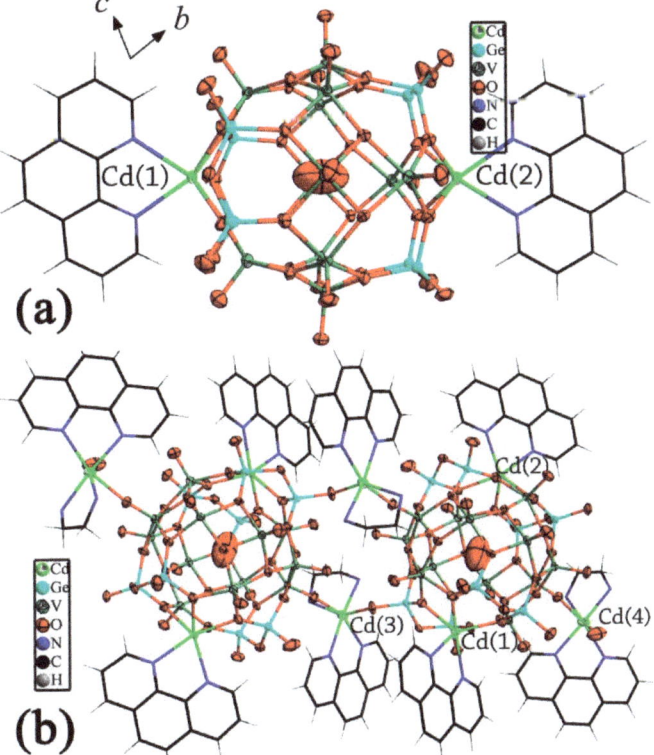

Figure 1. Ball-and-stick and wire representation of the di-Cd-substituted Ge-V-O cluster (**a**) and the dimer in compound **1** (**b**).

Except for [Cd(phen)]$^{2+}$, there are two [Cd(phen)(en)]$^{2+}$. It should be noted that the two [Cd(phen)(en)]$^{2+}$ are different from each other. Cadmium of [Cd(4)(phen)(en)]$^{2+}$ of the two is bonded to four nitrogens from a phen and an en with Cd-N distances of 2.250(8)–2.320(7) Å, a terminal oxygen from Cd$_2$Ge$_8$V$_{12}$ with the Cd-O distance of 2.385(5) Å and a water molecule with the Cd-O distance of 2.583(9) Å, exhibiting a cis-octahedral geometry. Therefore, the cluster acts as a monodentate inorganic ligand coordinating with Cd(4), forming a cluster supported transition metal complex (TMC). Cadmium of [Cd(3)(phen)(en)]$^{2+}$ of the two receives contributions from four nitrogens from a phen and an en with Cd-N distances of 2.290(7)–2.367(7) Å and two terminal oxygens from two

$Cd_2Ge_8V_{12}$ with Cd-O distances of 2.284(5)–2.443(5) Å. It should be noted that the two terminal oxygens involving Cd-O bonds are distinct: one is from a $\{Ge_2O_7\}$, but the other comes from a $\{VO_4\}$. Thus, Cd(3) TMC acts as a bridge linking two $Cd_2Ge_8V_{12}$ to construct a novel cluster dimer. It should be noted that there are two Cd(3) TMCs acting as a double bridge linking two $Cd_2Ge_8V_{12}$. The dimer further supports two Cd(4) TMCs at the two sides of the dimer. That is to say, Cd(3) TMCs act as bridges joining $Cd_2Ge_8V_{12}$, but Cd(4) TMCs terminate the connection of the clusters by the terminating water molecule.

Distances between the central water molecule of the $Cd_2Ge_8V_{12}$ and Cd(3) and Cd(4) are 7.886–7.888 Å, and the angle of Cd(3)-O1w-Cd(4) is 109.754(1)°.

The dimer of clusters was reported by our group in 2002 [76] and very recently [77]; the first one was based on the Mo_8V_6 cluster, and the second one was based on the $V_{15}O_{36}$ cluster. However, compound **1** here is the most complex one of the three, which is the first example of dimer of substituted clusters. The other two reported compounds are both based on traditional clusters but not the substituted one.

3.2.2. Cd(DETA)]$_2$[Cd(DETA)$_2$]$_{0.5}$[Cd$_2$(phen)$_2$V$_{12}$O$_{41}$Ge$_8$(OH)$_7$(0.5H$_2$O)]·7.5H$_2$O (**2**)

The building block $[H_7Cd_2(phen)_2Ge_8V_{12}O_{48}(0.5H_2O)]^{5-}$ ($Cd_2Ge_8V_{12}$) of **2** is almost identical to that of **1**, which is also a cadmium di-substituted Ge-V-O cluster; each substituted cadmium is also coordinated by a phen ligand. The main difference between the building blocks of compounds **2** and **1** is the number of the attached hydrogen atoms. There are only slight differences between the bond lengths and angles in compounds **2** and **1**. Bond valence sum calculations for Ge and V also indicate that Ge and V are in the +4 oxidation-state (Table S1).

Except for $[Cd(phen)]^{2+}$ TMCs, there are two different TMCs which are $[Cd(DETA)_2]^{2+}$ and $[Cd(DETA)]^{2+}$ (Figure 2). The two TMCs are thoroughly different from those in **1**. Cadmium of $[Cd(DETA)_2]^{2+}$ is bound to six nitrogens from two DETA ligands and a terminal oxygen from $Cd_2Ge_8V_{12}$ with Cd-O and Cd-N distances of 2.46(1) and 2.38(3)–2.52(3) Å. $[Cd(DETA)_2]^{2+}$, performing a similar role as Cd(4) TMC in compound **1**, serves as a TMC supported by $Cd_2Ge_8V_{12}$. Cd of $[Cd(DETA)]^{2+}$ is bonded to three nitrogens from a DETA with Cd-N distances of 2.26(1)–2.40(1) Å and two terminal oxygens from two $\{Ge_2O_7\}$ from two adjoining $Cd_2Ge_8V_{12}$ with Cd-O distances of 2.234(8)–2.240(8) Å, exhibiting a five-coordinated trigonal bipyramidal geometry. Cadmium of $[Cd(DETA)]^{2+}$ serves as a bridge connecting the two $Cd_2Ge_8V_{12}$. It should be noted that the two terminal oxygens was shared by the two $[Cd(DETA)]^{2+}$, meaning that two terminal oxygens simultaneously connect two $[Cd(DETA)]^{2+}$ to form a novel dimer $[Cd_2(DETA)_2O_2]$. The role of $[Cd_2(DETA)_2O_2]$ in compound **2** is only partly similar to that of Cd(3) TMC in compound **1**. Two Cd(3) TMCs serving as a double bridge links two $Cd_2Ge_8V_{12}$ to form a dimer in compound **1**, but $[Cd_2(DETA)_2O_2]$ in compound **2** acting as a single bridge connects two $Cd_2Ge_8V_{12}$, and for its two components $[Cd(DETA)]^{2+}$, is also joined by the two terminal oxygens to form a single building unit. Most importantly, $[Cd_2(DETA)_2O_2]$ in compound **2** connects $Cd_2Ge_8V_{12}$ to form a novel 1-D extended chain structure. It should be noted that the neighboring $Cd_2Ge_8V_{12}$ in the extended chain are oriented up and down, as shown in Figure 2. To our knowledge, compound **2** is the first extended structure based on a metal-substituted Ge-V-O cluster of aromatic organic ligands. Yang et. al. also reported a 1-D chain structure formed by similar substituted Ge-V-O clusters and coordination fragments [54]. However, Yang's cluster is based on aliphatic organic ligands but not aromatic organic ones. Secondly, Yang's coordination fragment is formed by en ligands rather than DETA ligands. Finally, the 1-D chain of Yang's compound is sinusoidal, but the one here is linear.

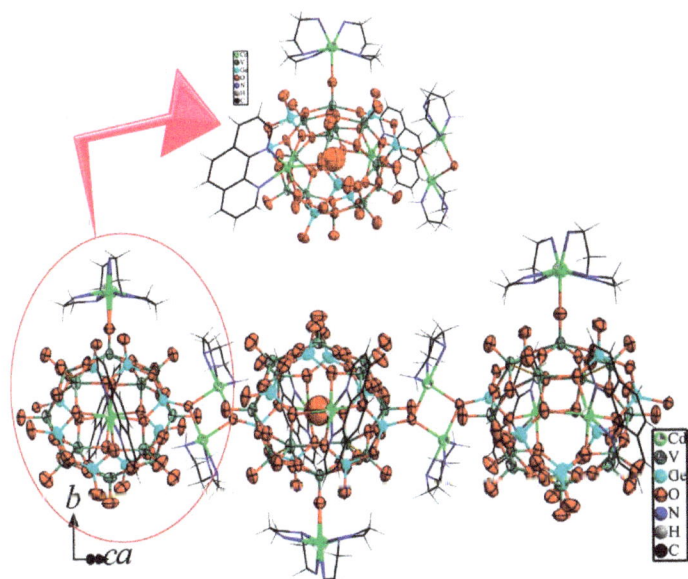

Figure 2. Ball-and-stick and wire representation of the building unit in the 1-D chain structure (**upper**) and the 1-D chain structure formed by Ge-V-O clusters and [Cd$_2$(DETA)$_2$O$_2$] (**lower**).

3.2.3. [Cd(en)$_3$]{[Cd(η_2-en)$_2$]$_3$[Cd(η_2-en)(η_2-μ_2-en)(η_2-en)Cd][Ge$_6$V$_{15}$O$_{48}$(H$_2$O)]}·5.5H$_2$O (**3**)

The asymmetric unit of compound **3** is composed of [Ge$_6$V$_{15}$O$_{48}$(H$_2$O)]$^{12-}$ (Ge$_6$V$_{15}$), [Cd(η_2-en)$_2$]$^{2+}$, [Cd(η_2-en)(η_1-en)]$^{2+}$, [Cd(η_2-en)$_3$]$^{2+}$ and 5.5 water molecules. The framework of the cluster in compound **3** is similar to those of {As$_6$V$_{15}$O$_{42}$} [16–18] and {Sb$_6$V$_{15}$O$_{42}$} [19–24], with {Ge$_2$O$_7$} displacing {As$_2$O$_5$} and {Sb$_2$O$_5$} in {As$_6$V$_{15}$O$_{42}$} and {Sb$_6$V$_{15}$O$_{42}$}. Although the oxo-cluster in compound **3** is thoroughly different from those in compounds **1** and **2**, the bond lengths and angles in compound **3** are comparable to those in compounds **1** and **2**. Bond valence sum calculations for Ge and V reveal that oxidation states of both Ge and V are +4 (Table S1).

It should be noted that [Cd(η_2-en)$_2$]$^{2+}$ of the five has two different configurations (Figure 3a). Cd(3) of [Cd(η_2-en)$_2$]$^{2+}$, which exhibits a trans-octahedral geometry, is bonded to four nitrogens from two en and two oxygens from two Ge$_6$V$_{15}$ with Cd-N and Cd-O distances in the range of 2.26(1)–2.31(2) Å and 2.229(9)–2.242(9) Å. Therefore, the trans-octahedral Cd(3) TMC joins two Ge$_6$V$_{15}$. Cd(5) of [Cd(η_2-en)$_2$]$^{2+}$ has a cis-octahedral geometry, which is coordinated by four nitrogens from two en with Cd-N distances of 2.33(2)–2.44(2) Å and two terminal oxygens in two cis-positions from two Ge$_6$V$_{15}$ with Cd-O distances of 2.228(7)–2.337(7) Å. Thus, the cis-octahedral Cd(5) TMC also connects two Ge$_6$V$_{15}$. Although Cd(3) and Cd(5) TMCs show different configurations, both their terminal oxygen atoms come from {Ge$_2$O$_7$} units of Ge$_6$V$_{15}$.

There are also two different [Cd(η_2-en)(η_2-μ_2-en)]$^{2+}$ TMCs in compound **3**. [Cd(6)(η_2-en)(η_2-μ_2-en)]$^{2+}$ presents a six-coordinated octahedral geometry with two nitrogens from a η_2-en, one nitrogen from a η_2-μ_2-en and three oxygens from two Ge$_6$V$_{15}$ with Cd-N and Cd-O distances of 2.30(1)–2.37(1) Å and 2.235(8)–2.610(8) Å (the first one oxygen is from one Ge$_6$V$_{15}$ and the remaining two oxygens are from the other Ge$_6$V$_{15}$). Cd(6) also serves as a bridge linking two Ge$_6$V$_{15}$. It should be noted that two oxygens of Cd(6) octahedron from two Ge$_6$V$_{15}$ are shared by Cd(5) octahedron. [Cd(4)(η_2-en)(η_2-μ_2-en)]$^{2+}$ is only five-coordinated by two nitrogens from a η_2-en, one nitrogen from a η_2-μ_2-en, and two oxygen atoms from two Ge$_6$V$_{15}$ with Cd-N and Cd-O distances of 2.29(1)–2.38(1) Å and 2.228(8)–2.238(8) Å, exhibiting a square pyramidal geometry. Cd(4) and Cd(6) are linked by η_2-μ_2-en to form a dumbbell-like dimer [Cd(η_2-en)(η_2-μ_2-en)(η_2-en)Cd]$^{4+}$. All five TMCs serve as bridges linking their neighboring clusters to form a novel 3-D framework structure. It should be noted that two terminal oxygens of Cd(1) octahedron are also shared by Cd(4) pyramid.

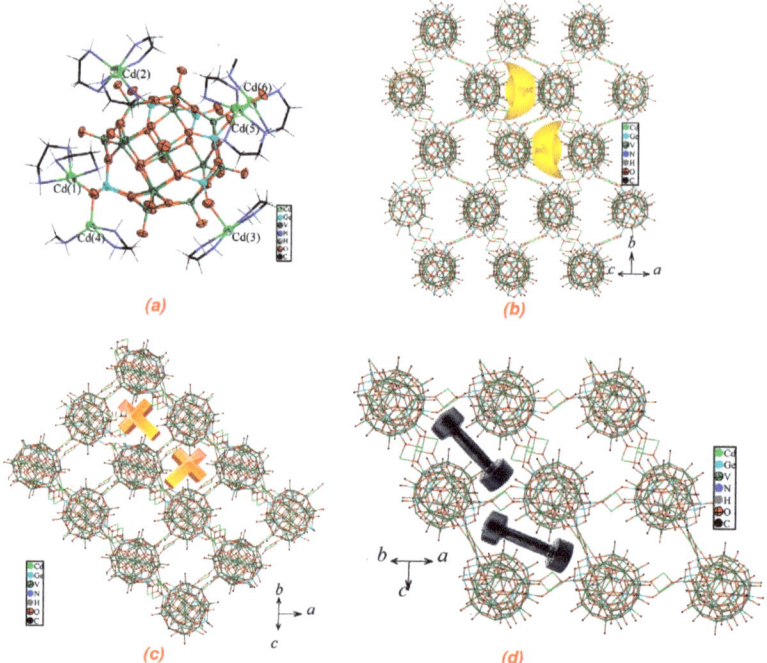

Figure 3. (**a**) Ball-and-stick and wire representation of the $[Ge_6V_{15}O_{48}]^{12-}$ cluster and five different types of TMCs in compound **3**; (**b**) the framework structure viewed along [101]; (**c**) the framework structure viewed along [011]; (**d**) the framework structure viewed along [110].

With the exception of the four different TMCs, there is a dissociated TMC $[Cd(\eta_2\text{-}en)_3]^{2+}$. Cd(2) of $[Cd(\eta_2\text{-}en)_3]^{2+}$ is chelated by three en with Cd-N distances in the range of 2.36(1)–2.40(1)Å. $[Cd(\mu_2\text{-}en)_3]^{2+}$ did not interact with any Ge_6V_{15}, which only serves as the space-filling agent and counterion.

In conclusion, there are five types of TMCs in compound **3**. To the best of our knowledge, compound **3** contains the largest number of TMC types.

The TMCs and the Ge_6V_{15} clusters are fused to form a novel 3-D framework structure via Cd-O covalent interactions, and the framework exhibits channels running along the [101], [110] and [011] directions. As shown in Figure 3, the framework exhibits gold ingot-shaped pores along the [101] direction. It should be noted that there are two kinds of such pores with different orientations. The framework exhibits dumbbell-shaped pores along the [110] direction; there are also two kinds of such pores with different orientations. The framework exhibits cross-shaped pores along the [011] direction; the pores here exhibit two orientations as well. The three kinds of channels intersect one another. Yang et al. also reported a 3-D structure formed by similar Ge-V-O clusters and coordination fragments [6]. However, there are several significant differences between our compound and Yang's compound. Firstly, and most importantly, the Ge-V-O cluster of Yang's compound is Ge_4V_{16}, but the corresponding cluster of our compound is Ge_6V_{15}. Secondly, Yang's compound is based on diethylenetriamine ligands but not en in our compound. Finally, Yang's compound did not exhibit various channels that were found in our compound.

3.3. BVS Calculations to Determine the Locations of Hydrogen Atoms of Compounds 1–3

Single crystal X-ray diffraction cannot exactly determine the positions of the hydrogen atoms from the Fourier maps. For further verifying the correctness of the formula of the three compounds, BVS calculations [75] were carried out to determine the positions of the hydrogen atoms for all the three compounds. As for compound **1**, the oxygens can be classified into eight groups: (1) seven Ge-O$_t$ terminal oxygens; (2) one Ge-O$_t$-Cd μ$_2$-oxygen; (3) eleven V-O$_t$ terminal oxygens; (4) one V-O$_t$-Cd μ$_2$-oxygen; (5) eight μ$_3$-oxygens located between two vanadiums and one germanium; (6) eight μ$_3$-oxygens located between three vanadiums; (7) eight μ$_3$-oxygens between a vanadium, cadmium and germanium; and (8) four μ$_2$-oxygens between two germaniums. All the atoms of the eight groups except groups (1) and (2) can be assigned to the −2 valence state, with BVS calculation results in the range of 1.56–2.16. With respect to the group (1) oxygens, all seven oxygens exist in the -1 valence state, with BVS results ranging from 1.01–1.04, indicating that all seven terminal Ge-O$_t$ oxygens are mono-protonated. The BVS value of the group (2) oxygen is 1.38, meaning that although this oxygen is coordinated by both one cadmium and one germanium, it exists in the −1 valence state. Therefore, the cluster in compound **1** is attached by eight hydrogens, and all eight hydrogens are attached on the eight Ge-O terminal oxygens.

As for compound **2**, the oxygens can also be divided into eight groups. Seven of the eight groups are similar to the corresponding groups in compound 1. Only the eighth one is not found in compound **1**: it is a μ$_3$-oxygen between two cadmiums and a germanium. This μ$_3$-oxygen is a terminal oxygen from a {Ge$_2$O$_7$} simultaneously interacting with two cadmiums and one germanium. Therefore, its valence state is not −1 but −2, with the BVS result of 1.85. In conclusion, only six of the eight terminal Ge-O$_t$ oxygens are mono-protonated. Thus, there is still one hydrogen atom whose position cannot be determined. We think this hydrogen should be disorderedly distributed on the surface of the cluster.

There are also seven groups of oxygens in compound **3**. However, only five of the seven have corresponding groups in compound 1. The five groups are: (1) V-O$_t$ terminal oxygens; (2) μ$_3$-oxygens between two vanadiums and one germanium; (3) μ$_3$-oxygens between three vanadiums; (4) μ$_2$-oxygens between two germaniums; and (5) μ$_2$-oxygen between one terminal vanadium and one cadmium. The remaining two groups are: (6) μ$_3$-oxygen between two cadmiums and one germanium, which has the corresponding group in compound 2; and (7) μ$_2$-oxygen between one cadmium and one germanium, which is only observed in compound 3. Compound **3** did not contain Ge-O$_t$ terminal oxygens, and all the Ge-O$_t$ terminal oxygens simultaneously interact with one or two cadmiums and finally form the group (6) and (7) oxygens. For the contributions of the cadmiums of group (6) and (7) oxygens, the BVS values of these oxygens are in the range of 1.56–2.01, indicating that there are no hydrogens attached on the cluster in compound **3**.

3.4. IR Spectrophotometry

The IR spectra of compounds **1–4** were recorded in the regions between 4000 and 200 cm^{-1} (Figure S1, Supporting Information). The strong peak at 984 cm^{-1} of compound **1** can be attributed to the stretching vibration of V=O. The patterns of the bands in the region characteristic of ν(V=O$_t$) indicate the presence of V^{IV} sites: clusters which contain exclusively V^{IV} generally possess ν(V=O$_t$) bands in the range of 970–1000 cm^{-1}, while bands in the region 940–960 cm^{-1} are characteristic of V^V. The observation of a strong absorbance in the 970–1000 cm^{-1} region provides a useful diagnostic for the presence of V^{4+} centers [78]. The strong peaks at 793 and 821 cm^{-1} of compound **1** may be due to asymmetric Ge-O stretching vibrations of {GeO$_4$}. The infrared spectrum of compound **2** is very similar to that of compound **1**. It also shows characteristic peaks at 983 cm^{-1} and 788 cm^{-1}, which should be ascribed to V=O$_t$ and Ge-O vibrations in compound **2**.

Compounds **3** and **4** are based on Ge$_6$V$_{15}$, which is different from that of compounds **1** and **2**. However, it should be noted that Ge$_6$V$_{15}$ is also formed by {GeO$_4$} and {V^{IV}O$_5$}; thus, the IR spectra of compounds **3** and **4** are very similar to those of compounds **1** and **2**.

The IR spectra of compounds **3** and **4** present characteristic peaks at 979, 801 cm^{-1} and 982, 800 cm^{-1}, respectively, which correspond to V=O$_t$ and Ge-O vibrations in compounds **3** and **4**. The main difference between the IR spectra of compounds **1** and **2** and **3** and **4** is that the bands at 667 and 660 cm^{-1} of compounds **1** and **2** are weak, but the corresponding bands at 691 and 692 cm^{-1} for compounds **3** and **4** are much stronger. Bands of 667–692 cm^{-1} can be ascribed to V-O-V vibrations.

3.5. XRD Powder Diffractometer

The powder X-ray diffraction patterns for compounds **1–4** are all in good agreement with the ones simulated based on the data of the single-crystal structures, indicating the purity of the as-synthesized products (Figure S2). The differences in the reflection intensity are probably due to preferred orientations in the powder samples of compounds **1–4**.

3.6. UV-Vis Spectrophotometry

The UV-vis spectra of compounds **1–4**, in the range of 250–600 nm, are presented in Figure S3. The UV-Vis spectrum of compound **1** displays an intense absorption sharp peak centered at about 266 nm, a shoulder peak at 294 nm and a peak tailing to the longer wavelength side (to about 450 nm), which can be assigned to O→V charge transfer, n→π* transitions of phen ligands and d→d transitions of complexes in compound **1**. The UV-Vis spectrum of compounds **2** displays an intense absorption peak at about 265 nm assigned to the O→V charge transfer in the polyoxoanion structure of compound **2**. The peak corresponding to the n→π* transitions of phen ligands was overlapped by the O→V charge transfer and cannot be separated.

The UV spectra of compounds **3** and **4** are similar to each other, but are different from those of compounds **1** and **2**, which exhibit absorption peaks at about 254 and 255 nm due to the O→V charge transfer in compounds **3** and **4**. The difference in the UV-Vis spectra between compounds **3–4** and compounds **1–2** may be due to the difference in their clusters.

3.7. ESR Spectrophotometry

The ESR spectra of compounds **1–4** were studied at room temperature (Figure S4). The ESR spectra of compounds **1–4** are very similar to one another, which show Lorentzian shapes accompanied by signals at g = 1.968, 1.968, 1.912 and 1.941, respectively, indicating that the vanadium atoms in compounds **1–4** are in a +4 oxidation-state. The ESR spectra further confirm the results of the bond valence sum calculations for compounds **1–4**.

3.8. Catalytic Activity

Epoxidation is an important industrial reaction, and epoxides are key intermediates in the manufacture of a wide variety of valuable products [79–81]. The epoxidation of styrene to styrene oxide with aqueous tertbutyl hydroperoxide (TBHP) using compound **1**, **2**, **3**, **4** or **5** as the catalyst was carried out in a batch reactor. In a typical run, the catalyst (compound **1** (2 mg, 0.57 μmol), compound **2** (2 mg, 0.60 μmol), compound **3** (2 mg, 0.58 μmol), compound **4** (2 mg, 0.62 μmol), compound **5** (2 mg, 0.70 μmol), 0.114 mL (1 mmol) of styrene and 2 mL of CH$_3$CN were added to a 10-mL two-neck flask equipped with a stirrer and a reflux condenser. The mixture was heated to 80 °C and then 2 mmol of TBHP was injected into the solution to start the reaction. The liquid organic products were quantified using a gas chromatograph (Shimadzu, GC-8A, Beijing, China) equipped with a flame detector and an HP-5 capillary column and identified by comparison with authentic samples and GC-MS coupling. In a blank experiment carried out in the absence of catalyst, no products were observed. Also, the styrene epoxidation reactions in the presence of GeO$_2$ (2 mg, 19.1 μmol) and V$_2$O$_5$ (2 mg, 11.0 μmol) were carried out respectively, and the activities are 24.8% and 71.2%, respectively, after 8 h.

Table 2 shows the catalytic reaction results of TBHP oxidation of styrene over various catalysts. As expected, all the catalysts are active for the oxidation of styrene. Compound **1** as a catalyst shows a performance with 50.1% conversion and 62.8% selectivity to styrene oxide after 8 h. Compound **2** shows the highest activity among the five with 96.3% conversion and 71.6% selectivity to styrene oxide. Compound **3** shows a catalytic performance with 81.4% conversion and 63.0% selectivity. The performance of compound **4** is similar to that of compound **3** with 84.1% conversion and 55.5% selectivity. The activity and selectivity of compound **5** are 41.7% and 67.1%, respectively. Compounds **3** and **4** are based on Ge_6V_{15}, group 12 metals (Cd and Zn) and similar organic ligands (en and enMe), and both exhibit extended framework structures (3-D and 2-D). Therefore, the catalytic activities of the two are similar. The structures of compounds **2** and **5** are more similar to each other. Compounds **2** and **5** are based on similar $Cd_2Ge_8V_{12}$ clusters and similar cadmium complexes, and both exhibit similar 1-D extended structures. The significant difference between compounds **2** and **5** is that compound **2** contains aromatic organic ligands but compound **5** does not; however, the catalytic activities of the two are thoroughly different from each other. To further understand the catalytic mechanism, we still need not only more Ge-V-O crystals but also more catalytic experimental results of the synthesized crystals. Although there have been no investigations on Ge-V-O metal-oxo-clusters as catalysts, there are some similar catalysis studies using catalysts formed by other POMs. The comparisons of the catalytic oxidation of styrene for compounds **1–5** and other reported POMs have been summarized in Table S2.

Table 2. Catalytic activity and product distribution.

Catalyst	Styrene Conversion [a] (%)	Product Selectivity [b] (mol%)		
		S	Bza	Others
GeO_2	24.8	58.6	39.8	1.7
V_2O_5	71.2	67.6	28.6	3.7
Compound 1	50.1	62.8	34.0	3.2
Compound 2	96.3	71.6	16.1	12.3
Compound 3	81.4	63.0	34.8	2.2
Compound 4	84.1	55.5	39.3	5.1
Compound 5	41.7	67.1	32.9	0.0

[a] Reaction conditions: catalyst 2 mg, styrene 0.114 mL (1 mmol), CH_3CN 2 mL, TBHP (2 mmol), temperature 80 °C and time 8 h. [b] So: Styrene oxide, Bza: benzaldehyde; Others: including benzoic acid and phenylacetaldehyde.

The recyclability and reusability of compound **3**, including the conversion and catalyst recovery in three cycles, were studied (Table 3). The same experimental conditions were used. Generally, when using soluble heteropolyacid (e.g., $H_3[PW_{12}O_{40}]$) as the catalyst, the used catalyst was recovered by precipitation and ion exchange [82]. In comparison, it was easy to separate (centrifugation) and recycle compound **3**. The process of recovery possibly resulted in the loss of approximately 40 wt.% after each cycle. The conversion dropped from 81.4% to 44.0% after three cycles.

Table 3. Recyclability and reusability of compound **3**.

Compound 3	Styrene Conversion [a] (%)	Product Selectivity [b] (mol%)		
		S	Bza	Others
1st run	81.4	63.0	34.8	2.2
2nd run	54.3	59.3	37.8	2.9
3rd run	44.0	43.9	53.0	3.1

[a] Reaction conditions: catalyst 2 mg, styrene 0.114 mL (1 mmol), CH_3CN 2 mL, TBHP (2 mmol), temperature 80 °C and time 8 h. [b] So: Styrene oxide, Bza: benzaldehyde; Others: including benzoic acid and phenylacetaldehyde.

Recovery experiments showed that compound **3** suffered significant activity losses after three cycles. However, the residual catalyst of compound **3** and the as-synthesized crystals used for X-ray analysis can still be considered homogeneous (Figure S5). The FT-IR spectra of compound **3** after the three cycles also remain identical to the one before the reaction (Figure S6).

4. Conclusions

The synthesis of Ge-V-O clusters, especially secondary metal substituted Ge-V-O clusters is still a great challenge for chemists. In this manuscript, we synthesized compounds **1** and **2**, which are the first examples formed by Ge-V-O clusters and transition metal complexes of aromatic organic ligands. Compounds **1** and **2** are also the first secondary metal substituted Ge-V-O clusters of aromatic organic ligands. Compound **3** is a novel 3-D framework with interesting channel structure. The catalytic properties of these compounds and two previously reported compounds have been investigated. We plan to apply these compounds in other oxidation catalytic reactions and hope to find applications of them in electrochemistry as well.

Supplementary Materials: The following supporting information can be downloaded at: https://www.mdpi.com/article/10.3390/molecules27144424/s1; Figure S1: IR spectra of compounds **1–4**; Figure S2: Simulated and experimental XRD patterns of compounds **1–4**; Figure S3: UV-Vis spectra of compounds **1–4**; Figure S4: EPR spectra of compounds **1–4**; Figure S5: Simulated, experimental XRD patterns and XRD patterns after three cycles of compounds **3**; Figure S6: FT-IR spectrum of compound **3** and FT-IR spectra of compound **3** after three cycles; Table S1: Bond valence sum calculations for Ge, V and O in compounds **1–3**; Table S2: Comparison of the catalytic performances of our compounds and other reported POMs [83,84]. References [85,86] are mentioned in Supplementary Materials.

Author Contributions: X.-B.C. and H.-Y.G. conceived the research and designed the experiments. H.-Y.G. performed the experiments. X.-B.C. and X.Z. analyzed and interpreted the data. X.-B.C. and H.Q. wrote and revised the article. All authors contributed to the final approval of the article. All authors have read and agreed to the published version of the manuscript.

Funding: This research was funded by the Jilin Provincial Department of Science and Technology (Grant 20190802027ZG) and A Project Supported by the Scientific Research Fund of Zhejiang Provincial Education Department, grant number 00321JYT01AL.

Institutional Review Board Statement: Not applicable.

Informed Consent Statement: Not applicable.

Data Availability Statement: Date is available form Corresponding author.

Conflicts of Interest: The authors declare no conflict of interest.

Sample Availability: Samples of the compounds are not available from the authors.

References

1. Kannan, K.; Radhika, D.; Sadasivuni, K.K.; Reddy, K.R.; Raghu, A.V. Nanostructured metal oxides and its hybrids for photocatalytic and biomedical applications. *Adv. Colloid. Interface. Sci.* **2020**, *281*, 102178. [CrossRef]
2. Vidyavathi, G.T.; Kumar, B.V.; Raghu, A.V.; Aravinda, T.; Hani, U.; Ananda-Murthy, H.C.; Shridhar, A.H. *Punica granatum* pericarp extract catalyzed green chemistry approach for synthesizing novel ligand and its metal(II) complexes: Molecular docking/DNA interactions. *J. Mol. Struct.* **2022**, *1249*, 131656. [CrossRef]
3. Basavarajappa, P.S.; Patil, S.B.; Ganganagappa, N.; Reddy, K.R.; Raghu, A.V.; Reddy, C.V. Recent progress in metal-doped TiO_2, non-metal doped/codoped TiO_2 and TiO_2 nanostructured hybrids for enhanced photocatalysis. *Int. J. Hydrogen Energy* **2020**, *45*, 7764–7778. [CrossRef]
4. Dakshayini, B.S.; Reddy, K.R.; Mishra, A.; Shetti, N.P.; Malode, S.J.; Basu, S.; Naveen, S.; Raghu, A.V. Role of conducting polymer and metal oxide-based hybrids for applications in ampereometric sensors and biosensors. *Microchem. J.* **2019**, *147*, 7–24. [CrossRef]
5. Kumar, S.; Reddy, K.R.; Reddy, C.V.; Shetti, N.P.; Sadhu, V.; Shankar, M.V.; Reddy, V.G.; Raghu, A.V.; Aminabhavi, T.M. *Nanostructured Materials for Environmental Applications*; Balakumar, S., Keller, V., Shankar, M.V., Eds.; Springer: Berlin/Heidelberg, Germany, 2021; p. 485.
6. Zhou, J.; Zhang, J.; Fang, W.H.; Yang, G.Y. A Series of Vanadogermanates from 1D Chain to 3D Framework Built by Ge–V–O Clusters and Transition-Metal-Complex Bridges. *Chem. Eur. J.* **2010**, *16*, 13253. [CrossRef]
7. Kikukawa, Y.; Yamaguchi, S.; Nakagawa, Y.; Uehara, K.; Uchida, S.; Yamaguchi, K.; Mizuno, N. Synthesis of a Dialuminum-Substituted Silicotungstate and the Diastereoselective Cyclization of Citronellal Derivatives. *J. Am. Chem. Soc.* **2008**, *130*, 15872–15878. [CrossRef]
8. Nyman, M.; Bonhomme, F.; Alam, T.M.; Bodriguez, M.A.; Cherry, B.R.; Krumhansl, J.L.; Nenoff, T.M.; Sattler, A.M. A General Synthetic Procedure for Heteropolyniobates. *Science* **2002**, *297*, 996–998. [CrossRef] [PubMed]

9. Kögerler, P.; Cronin, L. Polyoxometalate Nanostructures, Superclusters, and Colloids: From Functional Clusters to Chemical Aesthetics. *Angew. Chem. Int. Ed.* **2005**, *44*, 844–846. [CrossRef]
10. Rüther, T.; Hultgren, V.M.; Timko, B.P.; Bond, A.M.; Jackson, W.R.; Wedd, A.G. Electrochemical Investigation of Photooxidation Processes Promoted by Sulfo-polyoxometalates: Coupling of Photochemical and Electrochemical Processes into an Effective Catalytic Cycle. *J. Am. Chem. Soc.* **2003**, *125*, 10133–10143. [CrossRef]
11. Zheng, S.T.; Zhang, J.; Yang, G.Y. Designed Synthesis of POM–Organic Frameworks from {Ni_6PW_9} Building Blocks under Hydrothermal Conditions. *Angew. Chem. Int. Ed. Engl.* **2008**, *47*, 3909–3913. [CrossRef]
12. Zheng, S.T.; Yuan, D.Q.; Jia, H.P.; Zhang, J.; Yang, G.Y. Combination between lacunary polyoxometalates and high-nuclear transition metal clusters under hydrothermal conditions: I. from isolated cluster to 1-D chain. *Chem. Commun.* **2007**, *18*, 1858–1860. [CrossRef] [PubMed]
13. Tan, H.Q.; Li, Y.G.; Zhang, Z.M.; Qin, C.; Wang, X.L.; Wang, E.B.; Su, Z.M. Chiral Polyoxometalate-Induced Enantiomerically 3D Architectures: A New Route for Synthesis of High-Dimensional Chiral Compounds. *J. Am. Chem. Soc.* **2007**, *129*, 10066–10067. [CrossRef] [PubMed]
14. Ritchie, C.I.; Streb, C.; Thiel, J.; Mitchell, S.G.; Miras, H.N.; Long, D.L.; Boyd, T.; Peacock, R.D.; McGlone, T.; Cronin, L. Reversible Redox Reactions in an Extended Polyoxometalate Framework Solid. *Angew. Chem. Int. Ed.* **2008**, *47*, 6881–6884. [CrossRef] [PubMed]
15. Streb, C.; Ritchie, C.; Long, D.L.; Kögerler, P.; Cronin, L. Modular Assembly of a Functional Polyoxometalate-Based Open Framework Constructed from Unsupported AgI···AgI Interactions. *Angew. Chem. Int. Ed.* **2007**, *46*, 7579–7582. [CrossRef] [PubMed]
16. Long, D.L.; Streb, C.; Song, Y.F.; Mitchell, S.; Cronin, L. Unravelling the Complexities of Polyoxometalates in Solution Using Mass Spectrometry: Protonation versus Heteroatom Inclusion. *J. Am. Chem. Soc.* **2008**, *130*, 1830–1832. [CrossRef]
17. Yoshida, A.; Nakagawa, Y.; Uehara, K.; Hikichi, S.; Mizuno, N. Inorganic Cryptand: Size-Selective Strong Metallic Cation Encapsulation by a Disilicoicosatungstate (Si_2W_{20}) Polyoxometalate. *Angew. Chem. Int. Ed.* **2009**, *48*, 7055–7058. [CrossRef]
18. Micoine, K.; Hasenknopf, B.; Thorimbert, S.; Lacote, E.; Malacria, M. Chiral Recognition of Hybrid Metal Oxide by Peptides. *Angew. Chem. Int. Ed.* **2009**, *48*, 3466–3468. [CrossRef]
19. Müller, A.; Fedin, V.P.; Kuhlmann, C.; Fenske, H.D.; Baum, G.; Bögge, H.; Hauptfleisch, B. 'Adding' stable functional complementary, nucleophilic and electrophilic clusters: A synthetic route to $[\{(SiW_{11}O_{39})Mo_3S_4(H_2O)_3(\mu\text{-}OH)\}_2]^{10-}$ and $[\{(P_2W_{17}O_{61})Mo_3S_4(H_2O)_3(\mu\text{-}OH)\}_2]^{14-}$ as examples. *Chem. Commun.* **1999**, *35*, 1189–1190. [CrossRef]
20. Knaust, J.M.; Inman, C.; Keller, S.W. A host–guest complex between a metal–organic cyclotriveratrylene analog and a polyoxometalate: $[Cu_6(4,7\text{-phenanthroline})_8(MeCN)_4]_2PM_{12}O_{40}$ (M = Mo or W). *Chem. Commun.* **2004**, *40*, 492–493. [CrossRef]
21. Ishii, Y.; Takenaka, Y.; Konishi, K. Porous Organic–Inorganic Assemblies Constructed from Keggin Polyoxometalate Anions and Calix[4]arene–Na^+ Complexes: Structures and Guest-Sorption Profiles. *Angew. Chem. Int. Ed.* **2004**, *43*, 2702–2705. [CrossRef]
22. Wei, M.L.; He, C.; Sun, Q.Z.; Meng, Q.J.; Duan, C.Y. Zeolite Ionic Crystals Assembled through Direct Incorporation of Polyoxometalate Clusters within 3D Metal–Organic Frameworks. *Inorg. Chem.* **2007**, *46*, 5957–5966. [CrossRef] [PubMed]
23. Niu, J.Y.; Wu, Q.; Wang, J.P. 1D and 2D polyoxometalate-based composite compounds. Synthesis and crystal structure of $[\{Ba(DMSO)_5(H_2O)\}_2(SiMo_{12}O_{40})]$ and $[\{Ba(DMSO)_3(H_2O)_3\}\{Ba(DMSO)_5(H_2O)\}(GeMo_{12}O_{40})]$. *J. Chem. Soc. Dalton Trans.* **2002**, *31*, 2512–2516. [CrossRef]
24. Sha, J.Q.; Peng, J.; Tian, A.X.; Liu, H.S.; Chen, J.; Zhang, P.P.; Su, Z.M. Assembly of Multitrack Cu–N Coordination Polymeric Chain-Modified Polyoxometalates Influenced by Polyoxoanion Cluster and Ligand. *Cryst. Growth Des.* **2007**, *7*, 2535–2541. [CrossRef]
25. Cui, X.B.; Sun, Y.Q.; Yang, G.Y. Hydrothermal synthesis and characterization of a novel two-dimensional framework materials constructed from the polyoxometalates and coordination groups: $[As_8^{III}V_{14}^{IV}O_{42}(CO_3)][Cu(en)_2]_3 \cdot 10H_2O$. *Inorg. Chem. Commun.* **2003**, *6*, 259–261. [CrossRef]
26. Cui, X.B.; Xu, J.Q.; Li, Y.; Sun, Y.H.; Yang, G.Y. Hydrothermal Synthesis and Characterization of a Novel Sinusoidal Layer Structure Constructed from Polyoxometalates and Coordination Complex Fragments. *Eur. J. Inorg. Chem.* **2004**, *7*, 1051–1055. [CrossRef]
27. Cui, X.B.; Xu, J.Q.; Sun, Y.H.; Li, Y.; Ye, L.; Yang, G.Y. Hydrothermal synthesis and crystal structure of a novel 1-D chain structure constructed from polyoxometalates and coordination complex fragments. *Inorg. Chem. Commun.* **2004**, *7*, 58–61. [CrossRef]
28. Guo, H.Y.; Li, Z.F.; Zhao, D.C.; Hu, Y.Y.; Xiao, L.N.; Cui, X.B.; Guan, J.Q.; Xu, J.Q. The synthesis and characterization of three organic–inorganic hybrids based on different transition metal complexes and {$As_8V_{14}O_{42}(H_2O)$} clusters. *CrystEngComm* **2014**, *16*, 2251–2259. [CrossRef]
29. Guo, H.Y.; Li, Z.F.; Fu, L.W.; Hu, Y.Y.; Cui, X.B.; Liu, B.J.; Xu, J.N.; Huo, Q.S.; Xu, J.Q. Structural influences of arsenic–vanadium clusters and transition metal complexes on final structures of arsenic–vanadium-based hybrids. *Inorg. Chim. Acta* **2016**, *443*, 118–125. [CrossRef]
30. Guo, H.Y.; Li, Z.F.; Zhang, X.; Fu, L.W.; Hu, Y.Y.; Guo, L.L.; Cui, X.B.; Huo, Q.S.; Xu, J.Q. New self-assembly hybrid compounds based on arsenic–vanadium clusters and transition metal mixed-organic-ligand complexes. *CrystEngComm* **2016**, *18*, 566–579. [CrossRef]
31. Zheng, S.T.; Zhang, J.; Xu, J.Q.; Yang, G.Y. Hybrid inorganic–organic 1-D and 2-D frameworks with {$As_8V_{14}O_{42}$} clusters as building blocks. *J. Solid State Chem.* **2005**, *178*, 3740–3746. [CrossRef]

32. Zheng, S.T.; Zhang, J.; Li, B.; Yang, G.Y. The first solid composed of {As$_4$V$_{16}$O$_{42}$(H$_2$O)} clusters. *Dalton Trans.* **2008**, *41*, 5584–5587. [CrossRef] [PubMed]
33. Zheng, S.T.; Chen, Y.M.; Zhang, J.; Xu, J.Q.; Yang, G.Y. Hybrid Inorganic–Organic 1D and 2D Frameworks with [As$_6$V$_{15}$O$_{42}$]$^{6-}$ Polyoxoanions as Building Blocks. *Eur. J. Inorg. Chem.* **2006**, *9*, 397–406. [CrossRef]
34. Zheng, S.T.; Zhang, M.; Yang, G.Y. [{Zn(enMe)$_2$}$_2$(enMe)$_2${Zn$_2$As$_8$V$_{12}$O$_{40}$(H$_2$O)}]·4H$_2$O: A Hybrid Molecular Material Based on Covalently Linked Inorganic Zn–As–V Clusters and Transition Metal Complexes via enMe Ligands. *Eur. J. Inorg. Chem.* **2004**, *2004–2007*. [CrossRef]
35. Zhou, J.; Zheng, S.T.; Fang, W.H.; Yang, G.Y. A New 2-D Network Containing {As$_4$V$_{16}$O$_{42}$(H$_2$O)} Cluster Units. *Eur. J. Inorg. Chem.* **2009**, *34*, 5057.
36. Qi, Y.F.; Li, Y.G.; Wang, E.B.; Jin, H.; Zhang, Z.M.; Wang, X.L.; Chang, S. Syntheses and structures of four organic–inorganic hybrids based on the surface restricted building unit and various zinc coordination groups. *Inorg. Chim. Acta* **2007**, *360*, 1841–1853. [CrossRef]
37. Qi, Y.F.; Li, Y.G.; Wang, E.B.; Jin, H.; Zhang, Z.M.; Wang, X.L.; Chang, S. Two unprecedented inorganic–organic boxlike and chainlike hybrids based on arsenic–vanadium clusters linked by nickel complexes. *J. Solid State Chem.* **2007**, *180*, 382–389. [CrossRef]
38. Qi, Y.F.; Li, Y.G.; Qin, C.; Wang, E.B.; Jin, H.; Xiao, D.R.; Wang, X.L.; Chang, S. From Chain to Network: Design and Analysis of Novel Organic–Inorganic Assemblies from Organically Functionalized Zinc-Substituted Polyoxovanadates and Zinc Organoamine Subunits. *Inorg. Chem.* **2007**, *46*, 3217–3230. [CrossRef]
39. Zhang, L.J.; Zhao, X.L.; Xu, J.Q.; Wang, T.G. A novel two-dimensional structure containing the first antimony-substituted polyoxovandium clusters: [(Co(en)$_2$)$_2$Sb$^{III}_8$V$^{IV}_{14}$O$_{42}$(H$_2$O)]·6H$_2$O. *J. Chem. Soc. Dalton Trans.* **2002**, *31*, 3275–3276. [CrossRef]
40. Hu, X.X.; Xu, J.Q.; Cui, X.B.; Song, J.F.; Wang, T.G. A novel one-dimensional framework material constructed from antimony-substituted polyoxovanadium clusters: [(C$_2$N$_2$H$_{10}$)$_2$β-{Sb$^{III}_8$V$^{IV}_{14}$O$_{42}$(H$_2$O)}](C$_2$N$_2$H8)·4H$_2$O. *Inorg. Chem. Commun.* **2004**, *7*, 264. [CrossRef]
41. Guo, H.Y.; Zhang, X.; Cui, X.B.; Huo, Q.S.; Xu, J.Q. Vanadoantimonates: From discrete clusters to high dimensional aggregates. *CrystEngComm* **2016**, *18*, 5130–5139. [CrossRef]
42. Antonova, E.; Näther, C.; Kögerler, P.; Bensch, W. Organic Functionalization of Polyoxovanadates: Sb-N Bonds and Charge Control. *Angew. Chem. Int. Ed.* **2011**, *50*, 764–767. [CrossRef] [PubMed]
43. Antonova, E.; Näther, C.; Kögerler, P.; Bensch, W. A C$_2$-symmetric antimonato polyoxovanadate cluster [V$_{16}$Sb$_4$O$_{42}$(H$_2$O)]$^{8-}$ derived from the {V$_{18}$O$_{42}$} archetype. *Dalton Trans.* **2012**, *41*, 6957–6962. [CrossRef]
44. Antonova, E.; Näther, C.; Bensch, W. Antimonato polyoxovanadates with structure directing transition metal complexes: Pseudopolymorphic {Ni(dien)$_2$}$_3$[V$_{15}$Sb$_6$O$_{42}$(H$_2$O)]·nH$_2$O compounds and {Ni(dien)$_2$}$_4$[V$_{16}$Sb$_4$O$_{42}$(H$_2$O)]. *Dalton Trans.* **2012**, *41*, 1338–1344. [CrossRef] [PubMed]
45. Antonova, E.; Näther, C.; Bensch, W. Assembly of [V$_{15}$Sb$_6$O$_{42}$(H$_2$O)]$^{6-}$ cluster shells into higher dimensional aggregates via weak Sb···N/Sb···O intercluster interactions and a new polyoxovanadate with a discrete [V$_{16}$Sb$_4$O$_{42}$(H$_2$O)]$^{8-}$ cluster shell. *CrystEngComm* **2012**, *14*, 6853–6859. [CrossRef]
46. Wutkowski, A.; Näther, C.; Kögerler, P.; Bensch, W. [V$_{16}$Sb$_4$O$_{42}$(H$_2$O){VO(C$_6$H$_{14}$N$_2$)$_2$}$_4$]: A Terminal Expansion to a Polyoxovanadate Archetype. *Inorg. Chem.* **2008**, *47*, 1916–1918. [CrossRef] [PubMed]
47. Wutkowski, A.; Näther, C.; Kögerler, P.; Bensch, W. Antimonato Polyoxovanadate Based Three-Dimensional Framework Exhibiting Ferromagnetic Exchange Interactions: Synthesis, Structural Characterization, and Magnetic Investigation of {[Fe(C$_6$H$_{14}$N$_2$)$_2$]$_3$[V$_{15}$Sb$_6$O$_{42}$(H$_2$O)]}·8H$_2$O. *Inorg. Chem.* **2013**, *52*, 3280–3284. [CrossRef]
48. Kiebach, R.; Näther, C.; Bensch, W. [C$_6$H$_{17}$N$_3$]$_4$[Sb$_4$V$_{16}$O$_{42}$]·2H$_2$O and [NH$_4$]$_4$[Sb$_8$V$_{14}$O$_{42}$]·2H$_2$O—The first isolated Sb derivates of the [V$_{18}$O$_{42}$] family. *Solid. State. Sci.* **2006**, *8*, 964–970. [CrossRef]
49. Kiebach, R.; Näther, C.; Kögerler, P.; Bensch, W. [V$^{IV}_{15}$Sb$^{III}_6$O$_{42}$]$^{6-}$: An antimony analogue of the molecular magnet [V$_{15}$As$_6$O$_{42}$(H$_2$O)]$^{6-}$. *Dalton Trans.* **2007**, *36*, 3221–3223. [CrossRef]
50. Lühmann, H.; Näther, C.; Kögerler, P.; Bensch, W. Solvothermal synthesis and crystal structure of a heterometal-bridged {V$_{15}$Sb$_6$} dimer: [Ni$_2$(tren)$_3$(V$_{15}$Sb$_6$O$_{42}$(H$_2$O)$_{0.5}$)]$_2$[Ni(trenH)$_2$]·H$_2$O. *Inorg. Chim. Acta* **2014**, *421*, 549–552. [CrossRef]
51. Li, Y.; Liu, J.P.; Wang, J.P.; Niu, J.Y. Hydrothermal Synthesis and Crystal Structure of a New 2D Layer Compound [{Ni(en)$_2$}$_2$Sb$_8$V$_{14}$O$_{42}$]·5.5H$_2$O. *Chem. Res. Chin. Univ.* **2009**, *25*, 426–429.
52. Gao, Y.Z.; Han, Z.G.; Xu, Y.Q.; Hu, C.W. pH-Dependent Assembly of Two Novel Organic–inorganic Hybrids Based on Vanadoantimonate Clusters. *J. Cluster. Sci.* **2010**, *21*, 163–171. [CrossRef]
53. Monakhov, K.Y.; Bensch, W.; Kögerler, P. Semimetal-functionalised polyoxovanadates. *Chem. Soc. Rev.* **2015**, *44*, 8443–8483. [CrossRef]
54. Cui, X.B.; Xu, J.Q.; Meng, H.; Zheng, S.T.; Yang, G.Y. A Novel Chainlike As–V–O Polymer Based on a Transition Metal Complex and a Dimeric Polyoxoanion. *Inorg. Chem.* **2004**, *43*, 8005–8009. [CrossRef] [PubMed]
55. You, L.S.; Zhu, Q.Y.; Zhang, X.; Pu, Y.Y.; Bian, G.Q.; Dai, J. A new type of germanium–vanadate cluster, [Ge$_5$V$_6$O$_{21}$(heda)$_6$] (Hheda = N-(2-hydroxyethyl)ethylenediamine). *CrystEngComm* **2013**, *15*, 2411–2415. [CrossRef]
56. Chen, Y.M.; Wang, E.B.; Lin, B.Z.; Wang, S.T. The first polyoxoalkoxovanadium germanate anion with a novel cage-like structure: Solvothermal synthesis and characterization. *Dalton Trans.* **2003**, *32*, 519–520. [CrossRef]
57. Whitfield, T.; Wang, X.; Jacobson, A.J. Vanadogermanate Cluster Anions. *Inorg. Chem.* **2003**, *42*, 3728–3733. [CrossRef]

58. Bi, L.H.; Kortz, U.; Dickman, M.H.; Nellutla, S.; Dalal, N.S.; Keita, B.; Nadjo, L.; Prinz, M.; Neumann, M. Polyoxoanion with Octahedral Germanium(IV) Hetero Atom: Synthesis, Structure, Magnetism, EPR, Electrochemistry and XPS Studies on the Mixed-Valence 14-Vanadogermanate [GeV$^V_{12}$V$^{IV}_2$O$_{40}$]$^{8-}$. *J. Cluster Sci.* **2006**, *17*, 143–165. [CrossRef]
59. Wang, J.; Näther, C.; Kögerler, P.; Bensch, W. Synthesis, crystal structure and magnetism of a new mixed germanium–polyoxovanadate cluster. *Inorg. Chim. Acta* **2010**, *363*, 4399. [CrossRef]
60. Tripathi, A.; Hughbanks, T.; Clearfield, A. The First Framework Solid Composed of Vanadosilicate Clusters. *J. Am. Chem. Soc.* **2003**, *125*, 10528–10529. [CrossRef] [PubMed]
61. Pitzschke, D.; Wang, J.; Hoffmann, R.D.; Pöttgen, R.; Bensch, W. Vanadoantimonates: From discrete clusters to high dimensional aggregates Two Compounds Containing the Mixed Germanium–Vanadium Polyoxothioanion [V$_{14}$Ge$_8$O$_{42}$S$_8$]$^{12-}$. *Angew. Chem. Int. Ed.* **2006**, *45*, 1305–1308. [CrossRef]
62. Gao, Y.Z.; Xu, Y.Q.; Li, S.; Han, Z.G.; Cao, Y.; Cui, F.Y.; Hu, C.W. Syntheses, structures, and magnetism of {V$_{15}$M$_6$O$_{42}$(OH)$_6$(Cl)} (M=Si, Ge). *J. Coord. Chem.* **2010**, *63*, 3373–3383. [CrossRef]
63. Gao, Y.Z.; Xu, Y.Q.; Huang, K.L.; Han, Z.G.; Hu, C.W. Two three-dimensional {V$_{16}$Ge$_4$}$^-$ based open frameworks stabilized by diverse types of CoII-amine bridges and magnetic properties. *Dalton Trans.* **2012**, *41*, 6122–6129. [CrossRef] [PubMed]
64. Wang, J.; Näther, C.; Speldrich, M.; Kögerler, P.; Bensch, W. Chain and layer networks of germanato-polyoxovanadates. *CrystEngComm* **2013**, *15*, 10283. [CrossRef]
65. Wang, J.; Näther, C.; Kögerler, P.; Bensch, W. [V$_{15}$Ge$_6$O$_{42}$S$_6$(H$_2$O)]$^{12-}$, a Thiogermanatopolyoxovanadate Cluster Featuring the Spin Topology of the Molecular Magnet [V$_{15}$As$_6$O$_{42}$(H$_2$O)]$^{6-}$. *Eur. J. Inorg. Chem.* **2012**, 1237–1242. [CrossRef]
66. Zhou, J.; Zhao, J.W.; Wei, Q.; Zhang, J.; Yang, G.Y. Two Tetra-Cd^{II}-Substituted Vanadogermanate Frameworks. *J. Am. Chem. Soc.* **2014**, *136*, 5065–5071. [CrossRef]
67. Gao, G.G.; Li, F.Y.; Xu, L.; Liu, X.Z.; Yang, Y.Y. CO$_2$ Coordination by Inorganic Polyoxoanion in Water. *J. Am. Chem. Soc.* **2008**, *130*, 10838–10839. [CrossRef]
68. Xiao, F.P.; Hao, J.; Zhang, J.; Lv, C.L.; Yin, P.C.; Wang, L.S.; Wei, Y.G. Polyoxometalatocyclophanes: Controlled Assembly of Polyoxometalate-Based Chiral Metallamacrocycles from Achiral Building Blocks. *J. Am. Chem. Soc.* **2010**, *132*, 5956–5957. [CrossRef]
69. Miras, H.N.; Wilson, E.F.; Cronin, L. Unravelling the complexities of inorganic and supramolecular self-assembly in solution with electrospray and cryospray mass spectrometry. *Chem. Commun.* **2009**, 1297–1311. [CrossRef]
70. Sha, J.Q.; Peng, J.; Liu, H.S.; Chen, J.; Tian, A.X.; Zhang, P.P. Asymmetrical Polar Modification of a Bivanadium-Capped Keggin POM by Multiple Cu−N Coordination Polymeric Chains. *Inorg. Chem.* **2007**, *46*, 11183–11189. [CrossRef]
71. Schoedel, A.; Li, M.; Li, D.; O'Keeffe, M.; Yaghi, O.M. Structures of Metal–Organic Frameworks with Rod Secondary Building Units. *Chem. Rev.* **2016**, *116*, 12466. [CrossRef]
72. Singh, C.; Mukhopadhyay, S.; Das, S.K. Polyoxometalate-Supported Bis(2,2'-bipyridine)mono(aqua)nickel(II) Coordination Complex: An Efficient Electrocatalyst for Water Oxidation. *Inorg. Chem.* **2018**, *57*, 6479–6490. [CrossRef] [PubMed]
73. Sheldrick, G.M. A short history of SHELX. *Acta Crystallogr. Sec. A* **2008**, *64*, 112. [CrossRef] [PubMed]
74. Hagrman, P.J.; Zubieta, J. Structural Influences of Organonitrogen Ligands on Vanadium Oxide Solids. Hydrothermal Syntheses and Structures of the Terpyridine Vanadates [V$_2$O$_4$(Terpy)$_2$]$_3$[V$_{10}$O$_{28}$], [VO$_2$(Terpy)][V$_4$O$_{10}$], and [V$_9$O$_{22}$(Terpy)$_3$]. *Inorg. Chem.* **2000**, *39*, 3252–3260. [CrossRef] [PubMed]
75. Brown, I.D.; Altermatt, D. The automatic searching for chemical bonds in inorganic crystal structures. *Acta Cryst.* **1985**, *B41*, 244. [CrossRef]
76. Cui, X.B.; Yang, G.Y. [{(H$_2$O)Ni(enMe)$_2$Mo$^V_4$Mo$^{VI}_4$V$^{IV}_8$(V^{V}O$_4$)O$_{40}$}$_2$[Ni(enMe)$_2$}][Ni(enMe)$_2$]$_4$·8H$_2$O: The First Dimer of Polyoxometalates Linked through Coordination Fragment. *Chem. Lett.* **2002**, *31*, 1238. [CrossRef]
77. Guo, H.Y.; Zhang, T.T.; Lin, P.H.; Zhang, X.; Cui, X.B.; Huo, Q.S.; Xu, J.Q. Preparation, structure and characterization of a series of vanadates. *CrystEngComm* **2017**, *19*, 265. [CrossRef]
78. Keene, T.D.; D'Alessandro, D.M.; Krämer, K.W.; Price, J.R.; Price, D.J.; Decurtins, S.; Kepert, C.J. [V$_{16}$O$_{38}$(CN)]$^{9-}$: A Soluble Mixed-Valence Redox-Active Building Block with Strong Antiferromagnetic Coupling. *Inorg. Chem.* **2012**, *51*, 9192–9199. [CrossRef] [PubMed]
79. Ivanchikova, I.D.; Maksimchuk, N.V.; Skobelev, I.Y.; Kaichev, V.V.; Kholdeeva, O.A. Mesoporous niobium-silicates prepared by evaporation-induced self-assembly as catalysts for selective oxidations with aqueous H$_2$O$_2$. *J. Catal.* **2015**, *332*, 138. [CrossRef]
80. Zalomaeva, O.V.; Evtushok, V.Y.; Maksimov, G.M.; Kholdeeva, O.A. Selective oxidation of pseudocumene and 2-methylnaphthalene with aqueous hydrogen peroxide catalyzed by γ-Keggin divanadium-substituted polyoxotungstate. *J. Organomet. Chem.* **2015**, *793*, 210–216. [CrossRef]
81. Nyman, M.; Burns, P.C. A comprehensive comparison of transition-metal and actinyl polyoxometalates. *Chem. Soc. Rev.* **2012**, *41*, 7354–7367. [CrossRef]
82. Wang, R.; Zhang, G.; Zhao, H. Polyoxometalate as effective catalyst for the deep desulfurization of diesel oil. *Catal. Today* **2010**, *149*, 117. [CrossRef]
83. Gao, H.C.; Yan, Y.; Xu, X.H.; Yu, J.H.; Niu, H.L.; Gao, W.X.; Zhang, W.X.; Jia, M.J. Kinetics and mechanism of thymine degradation by TiO$_2$ photocatalysis. *Chin. J. Catal.* **2015**, *36*, 1811–1824. [CrossRef]

84. Song, X.J.; Yan, Y.; Wang, Y.N.; Hu, D.W.; Xiao, L.N.; Yu, J.H.; Zhang, W.X.; Jia, M.J. Hybrid compounds assembled from copper-triazole complexes and phosphomolybdic acid as advanced catalysts for the oxidation of olefins with oxygen. *Dalton Trans.* **2017**, *46*, 16655–16662. [CrossRef]
85. Ozeki, T.; Yamase, T.; Naruke, H.; Sasaki, Y. X-Ray Structural Characterization of the Protonation Sites in the Dihydrogenhexaniobate Anion. *Bull. Chem. Soc. Jpn.* **1994**, *67*, 3249–3253. [CrossRef]
86. Huang, P.; Qin, C.; Su, Z.M.; Xing, Y.; Wang, X.L.; Shao, K.Z.; Lan, Y.Q.; Wang, E.B. Self-Assembly and Photocatalytic Properties of Polyoxoniobates: $Nb_{24}O_{72}$, $Nb_{32}O_{96}$, and $K_{12}Nb_{96}O_{288}$ Clusters. *J. Am. Chem. Soc.* **2012**, *134*, 14004–14010. [CrossRef]

Review

Recent Advances of Ti/Zr-Substituted Polyoxometalates: From Structural Diversity to Functional Applications

Zhihui Ni [1,*], Hongjin Lv [2,*] and Guoyu Yang [2]

1. Center for Advanced Materials Research, Zhongyuan University of Technology, Zhengzhou 450007, China
2. MOE Key Laboratory of Cluster Science, School of Chemistry and Chemical Engineering, Beijing Institute of Technology, Beijing 102488, China
* Correspondence: nizhihui@zut.edu.cn (Z.N.); hlv@bit.edu.cn (H.L.)

Abstract: Polyoxometalates (POMs), a large family of anionic polynuclear metal–oxo clusters, have received considerable research attention due to their structural versatility and diverse physicochemical properties. Lacunary POMs are key building blocks for the syntheses of functional POMs due to their highly active multidentate O-donor sites. In this review, we have addressed the structural diversities of Ti/Zr-substituted POMs based on the polymerization number of POM building blocks and the number of Ti and Zr centers. The synthetic strategies and relevant catalytic applications of some representative Ti/Zr-substituted POMs have been discussed in detail. Finally, the outlook on the future development of this area is also prospected.

Keywords: polyoxometalates; titanium; zirconium; transition metal substitution

1. Introduction

Polyoxometalates (POMs), as anionic metal-oxide clusters with diverse nuclearities, elemental compositions and physicochemical properties, have usually been constructed through the self-assembly of reactive oxometallate precursors in aqueous or organic reaction systems. [1–4] POMs can serve as crucial intermediates in the reaction pathway from water-soluble metal ions to insoluble metal oxides, and isolation of these molecular intermediates enable insightful elucidation on the formation mechanism and control over reaction pathways. POMs exhibit special characteristics of high negative charges, rich redox properties, good thermal stability, and readily available organic grafting [5,6], leading to wide applications in catalysis [7], magnetism [8], material science [9], electrochemistry [10], luminescence [11], etc.

As an important derivative of plenary POMs, lacunary POMs can be easily formed by removing one to several [MO_6] (M = Mo, W) building blocks from prototypal architectures such as the Keggin or Wells–Dawson type POMs [12]. These lacunary POMs usually show high coordination reactivity and oxidative and thermal stability. Their high negative charge and nucleophilic oxygen-enriched surfaces render them suitable inorganic, diamagnetic, multidentate nucleophilic ligands toward the electrophilic center. Transition metals (TM) or lanthanide (Ln) cations can be easily incorporated into the defect sites of lacunary POM ligands to construct metal-substituted POMs, which can exhibit unique physicochemical properties depending on the types of incorporated metal ions [13–20]. Metal-substituted POMs (MSPs) typically possess a higher negative charge density than that of the plenary POMs due to the substitution of a high oxidation state M^{6+} ion (e.g., W^{6+}, Mo^{6+}) with a low oxidation state M^{n+} ion (usually n = 1–3) [21]. To date, a wide variety of MSPs have been prepared, especially by the transition metals like manganese, iron, cobalt, nickel, copper and zinc in the fourth period and the lanthanides in the sixth period of the periodic table [22,23]. In contrast, the research on the syntheses of titanium- and zirconium-substituted POMs is still in a very early stage, which could be mainly attributed to the following two reasons: (a) the easy hydrolysis of Ti^{4+}/Zr^{4+} salts in

aqueous synthesis, and (b) the high tendency to formation of isolate oligomeric structures through intermolecular dehydration of terminal hydroxyls.

In this review, we have mainly focused on the structural diversities of Ti/Zr-substituted POMs according to the polymerization number of POM building blocks and the number of titanium or zirconium atoms. The representative catalytic application of Ti/Zr-substituted POMs has also been discussed. Finally, a perspective of this research area is also proposed. It is expected that this review could provide research insights into the controllable design and syntheses of Ti/Zr-substituted POMs derivatives with interesting catalytic properties.

2. The Syntheses and Structures of Ti/Zr-substituted POMs
2.1. Ti/Zr-Substituted Monomeric POMs

It is well known that titanium/zirconium-based compounds (e.g., TiO_2, ZrO_2) have been widely used in the fields of energy conversion, catalysis, and environmental treatment. As interesting molecular models of TiO_2 and ZrO_2 structures, the syntheses of titanium-/zirconium-substituted POMs could be dated back to the 1980s. In 1983, Knoth et al. reported the first case of Ti-substituted Keggin-type monomeric $[TiW_{11}PO_{40}]^{5-}$ polyoxoanion cluster (Figure 1a), which was prepared by the reaction of monovacant $(Bu_4N)_4H_3W_{11}PO_{39}$ with titanium tetrachloride in dichloroethane solution [24]. In 2000, Kholdeeva and co-workers also reported two similar cases of $[PTiW_{11}O_{40}]^{5-}$ and $[PTiW_{11}O_{41}]^{7-}$ polyoxoanions [25]. Qu et al. reported Ti-substituted Dawson-type monomeric $\alpha_2\text{-}[P_2W_{17}(TiO_2)O_{61}]^{8-}$ polyoxoanion (Figure 1b), which was synthesized from vacant heteropolytungstate precursors $\alpha_2\text{-}[P_2W_{17}O_{61}]^{10-}$ and $Ti(SO_4)_2$ using an aqueous solution-based synthetic approach [26]. Successively, Ti_3-substituted monomeric POM has also been reported with multi-lacunary POMs $\alpha\text{-}1,2,3\text{-}[P_2W_{15}O_{56}]^{12-}$ as precursors. For instance, Nomiya et. al. successfully reported a tris-[peroxotitanium(IV)]-substituted α-Dawson monomeric $[\alpha\text{-}1,2,3\text{-}P_2W_{15}(TiO_2)_3O_{56}(OH)_3]^{9-}$ polyoxoanion (**1**, Figure 1c), which are derived from $\{[\alpha\text{-}1,2,3\text{-}P_2W_{15}Ti_3O_{59}(OH)_3]_4[\mu_3\text{-}Ti(OH)_3]_4Cl\}^{33-}$ (**2a**) in 30% aqueous hydrogen peroxide solution. Thereinto, the four bridging Ti octahedral groups of **2a** were considered as the crucial roles for the synthesis of polyoxoanion **1** [27]. Subsequently, they prepared $[[\{Ti(OH)(ox)\}_2(\mu\text{-}O)](\alpha\text{-}PW_{11}O_{39})]^{7-}$ (Figure 1d) by using the trilacunary species of $[A\text{-}PW_9O_{34}]^{9-}$ and the anionic titanium(IV) complex as precursors with the molar ratio of 1:2 under acidic conditions. The molecular structure can be recognized as a hybrid containing one mono-lacunary POM ligand and two octahedral Ti-oxo moieties [28]. Then, a tetra-Ti-substituted di-lacunary α-Keggin monomeric $[[\{Ti(ox)(H_2O)\}_4(\mu\text{-}O)_3](\alpha\text{-}PW_{10}O_{37})]^{7-}$ polyoxoanion (Figure 1e) was also prepared, which was constructed by using the dimeric dititanium(IV)-substituted POM $[(\alpha\text{-}1,2\text{-}PW_{10}Ti_2O_{39})_2]^{10-}$ as precursor under acidic conditions [29]. Additionally, $[[\{Ti(H_2O)_3\}_2\{Ti(H_2O)_2\}_2(\mu\text{-}O)_3(SO_4)](PW_{10}O_{37})]^-$ polyoxoanion (Figure 1f) was synthesized through the reaction of $Ti(SO_4)_2$ with $[(\alpha\text{-}1,2\text{-}PW_{10}Ti_2O_{38})_2O_2]^{10-}$ and $[(\alpha\text{-}1,2,3\text{-}PW_9Ti_3O_{37})_2O_3]^{12-}$ under strongly acidic conditions, where the tetra-titanium(IV) oxide cluster was anchored onto the binding sites of lacunary Keggin POM [30]. In 2018, An et al. reported two organic-inorganic hybrid POMs, $[(H_2O)_4(3\text{-}Hpic)_2Ln][(H_2O)_5(3\text{-}Hpic)_2Ln][PW_{10}Ti_2O_{40}]^-$ (Ln= Ce, Nd, Sm), where binuclear Ti-substituted Keggin-type $[PW_{10}Ti_2O_{40}]^{7-}$ polyoxoanion was further modified by four Ln-3-Hpic coordinating groups [31]. Recently, Poblet et al. successfully prepared the other organic-inorganic hybrid POM, $[B\text{-}\alpha\text{-}SbW_9O_{33}(^tBuSiO)_3Ti(O^iPr)]^{3-}$, by anchoring $Ti(O^iPr)$ moiety on the silanol-functionalized an antimony-containing trilacunary POM ligand. The resulting complex has been utilized as a catalyst for the catalytic epoxidation of alkenes [32].

Compared to Ti-substituted POMs, the exploration of Zr-containing POMs has seldom been studied. In 1985, Chauveau et al. reported a compound of Zr-containing POM, $[ZrW_5O_{19}H_2]^{2-}$, which was considered as the Lindqvist-type structure. However, it is still doubted about the exact structure given the presented low signal-to-noise ratio and incorrect intensity ratio of ^{183}W NMR data [33]. Subsequently, Villanneau et al. has been unambiguously determined the structure as $[W_5O_{18}Zr(H_2O)_3]^{2-}$ by using EXFAS data [34,35]. Meanwhile, the same group also reported a similar compound with the structural formula of $[\{W_5O_{18}Zr(\mu\text{-}OH)\}_2]^{6-}$ (Figure 2a). In 2009, Sokolov et al. reported two mono-Zr-substituted Dawson-type monomeric polyoxoanion clusters $[\{(H_2O)_2ZrP_2W_{17}O_{61}\}]^{6-}$ and $[Zr(L\text{-}OOCCH(OH)CH_2COO)P_2W_{17}O_{61}]^{8-}$ (Figure 2b) through the reaction of monovacant α_2-$[P_2W_{17}O_{61}]^{10-}$ ligand and Zr salt under aqueous solution conditions [36]. The latter complex exhibited chirality due to the presence of chiral L-malic acid ligand. Later on, other organic ligand-modified Zr-POM, (tpp)-Zr-$(PW_{11}O_{39})$[TBA]$_5$ (tpp referring to ternary porphyrin) (Figure 2c) [37] and (Pc)-Zr-$(PW_{11}O_{39})$[TBA]$_5$ (Pc referring to Phthalocyanine) [38], has been reported by Drain et al. in 2009 and 2013, respectively.

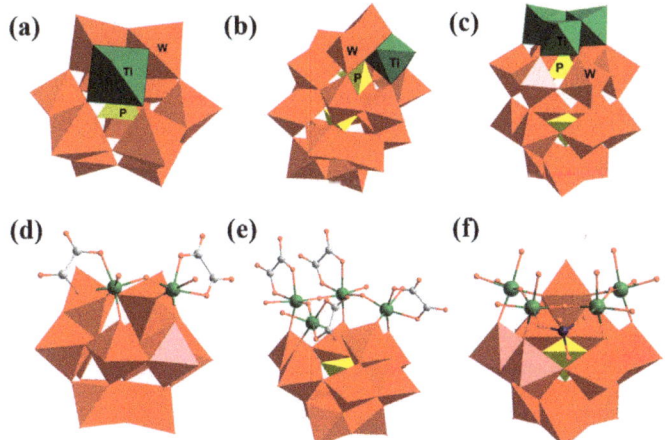

Figure 1. The ball-and-stick and polyhedral illustration of (**a**) $[TiW_{11}PO_{40}]^{5-}$ polyoxoanion, (**b**) α_2-$[P_2W_{17}(TiO_2)O_{61}]^{8-}$ polyoxoanion, (**c**) $[\alpha\text{-}1,2,3\text{-}P_2W_{15}(TiO_2)_3O_{56}(OH)_3]^{9-}$ polyoxoanion, (**d**) $[[\{Ti(OH)(ox)\}_2(\mu\text{-}O)](\alpha\text{-}PW_{11}O_{39})]^{7-}$ polyoxoanion; (**e**) $[[\{Ti(ox)(H_2O)\}_4(\mu\text{-}O)_3](\alpha\text{-}PW_{10}O_{37})]^{7-}$ polyoxoanion, and (**f**) $[[\{Ti(H_2O)_3\}_2\{Ti(H_2O)_2\}_2(\mu\text{-}O)_3(SO_4)](PW_{10}O_{37})]^-$ polyoxoanion. Color codes: WO$_6$, red; PO$_4$, yellow; Ti, green; S, dark blue; C, gray; O, red.

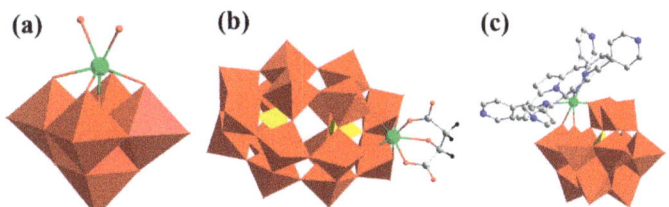

Figure 2. The ball-and-stick and polyhedral illustration of (**a**) $[W_5O_{18}Zr(\mu\text{-}OH)_2]^{6-}$ polyanion, (**b**) $[Zr(L\text{-}OOCCH(OH)CH_2COO)P_2W_{17}O_{61}]^{8-}$ polyanion and (**c**) (tpp)Zr$(PW_{11}O_{39})$[TBA]$_5$. Color codes: WO$_6$, red; PO$_4$, yellow; Zr, bright green; N, blue; C, gray; O, red.

2.2. Ti/Zr-Substituted Dimeric POMs

In addition to the monomeric POMs, some presentative Ti/Zr-substituted dimeric POMs have also been reviewed herein in detail. In 1993, Finke et al. reported the first hexa-Ti-substituted sandwich-type dimeric silicotungstate, $[A\text{-}\beta\text{-}Si_2W_{18}Ti_6O_{77}]^{14-}$ (Figure 3a), which was prepared by the reaction of $[A\text{-}\beta\text{-}HSiW_9O_{34}]^{9-}$ with $Ti(O)(C_2O_4)_2^{2-}$ or Ti(O)SO$_4$ in a regulated pH environment. The structural formulation has been lately corrected as $[A\text{-}\beta\text{-}(SiW_9O_{37})_2(Ti\text{-}O\text{-}Ti)_3]^{14-}$, implying the dimerization of two hypothetical

"[A-β-SiW$_9$(TiOH)$_3$O$_{37}$]$^{7-}$" Keggin units through the linkage of Ti-O-Ti bridges [39]. In 2000, Kholdeeva and co-workers reported a Ti$_2$-substituted dimeric POMs [(PTiW$_{11}$O$_{39}$)$_2$OH]$^{7-}$ which was prepared using [PTiW$_{11}$O$_{40}$]$^{5-}$ subunit as reaction materials [25]. Then, Nomiya and co-workers prepared a similar hexa-Ti-substituted sandwich-type dimeric POMs except for the replacement of {A-SiW$_9$O$_{34}$} with {A-PW$_9$O$_{34}$} [40]. Subsequently, Cronin and co-workers reported a hexa-Ti-substituted tungstoarsenate, K$_6$[Ti$_4$(H$_2$O)$_{10}$(AsTiW$_8$O$_{33}$)$_2$]·30H$_2$O, where two {AsTiW$_8$O$_{33}$}fragments were used to encapsulate a {Ti$_4$(H$_2$O)$_{10}$}$^{16+}$ moiety [41]. Additionally, they also reported the first mono-Ti-substituted tungstoantimonate [TiO(SbW$_9$O$_{33}$)$_2$]$^{16-}$, which was two B-α-{Sb^{III}W$_9$O$_{33}$} fragments linked by five sodium cations and an unprecedented square pyramidal Ti(O)O$_4$ group with a terminal Ti = O bond. In 2013, Kortz et al. reported two Ti-substituted phosphotungstates, [Ti$_8$(C$_2$O$_4$)$_8$P$_2$W$_{18}$O$_{76}$(H$_2$O)$_4$]$^{18-}$ (Figure 3b) and [Ti$_6$(C$_2$O$_4$)$_4$P$_4$W$_{32}$O$_{124}$]$^{20-}$ (Figure 3c). The former is the Ti$_8$-substituted Keggin-type phosphotungstates, consisting of two {PW$_9$} units encapsulating eight titanium centers bridged with two Ti–O–Ti bonds. The latter represents the first Ti$_6$-substituted Dawson-type phosphotungstates, which are constructed by two di-Ti-substituted {P$_2$W$_{16}$} units connected via two Ti(C$_2$O$_4$) moieties [42]. In 2015, Nomiya reported the first hexa-Ti-substituted Well–Dawson phosphotungstates [{α-P$_2$W$_{15}$Ti$_3$O$_{60}$(OH)$_2$}$_2$(Cp*Rh)$_2$]$^{16-}$ (Figure 3d). The polyoxoanion was constructed with two tri-Ti-substituted protonated Wells–Dawson subunits "[P$_2$W$_{15}$Ti$_3$O$_{60}$(OH)$_2$]$^{10-}$" bridged by the two organometallic Cp*Rh^{2+} groups [43]. Additionally, they also reported the first tetra-Ti^{IV}-1,2-substituted α-Keggin polyoxotungstate in aqueous solution, [α,α-P$_2$W$_{20}$Ti$_4$O$_{78}$]$^{10-}$ (Figure 3e). The polyoxoanion consisted of a dimeric anhydride form of two [α-1,2-PW$_{10}$Ti$_2$O$_{40}$]$^{7-}$ Keggin units linked with two Ti–O–Ti bonds [44]. Similar structures have also been reported by Mizuno and Wang's groups, respectively [45,46]. With continuous research, more di-Ti-substituted POMs have been further developed. In 2007, Kortz et al. reported a special di-Ti-substituted tungstodiarsenate (III) [Ti$_2$(OH)$_2$As$_2$W$_{19}$O$_{67}$(H$_2$O)]$^{8-}$ (Figure 3f), prepared by the reaction of TiOSO$_4$ and K$_{14}$[As$_2$W$_{19}$O$_{67}$(H$_2$O)] in a 2:1 molar ratio in acidic media (pH 2). The polyoxoanion was a sandwich-type structure with nominal C$_{2v}$ symmetry, which was constructed with two (B-α-As^{III}W$_9$O$_{33}$) Keggin moieties linked by an octahedral {WO$_5$(H$_2$O)} fragment and two unprecedented square-pyramidal {TiO$_4$(OH)} groups [47]. Nomiya et al. further reported the synthesis of a novel molecular solid Brønsted acid based on the Dawson-type sandwich POM [Ti$_2${P$_2$W$_{15}$O$_{54}$(OH$_2$)$_2$}$_2$]$^{8-}$ (Figure 3g) [48]. Subsequently, they synthesized a similar structure using mono-lacunary Dawson precursor K$_{10}$[α$_2$-P$_2$W$_{17}$O$_{61}$]·23H$_2$O [49]. In 2015, Li's group synthesized two isomorphic di-Ti-substituted Keggin-type phosphotungstate ([(Ti$_2$O)(PW$_{11}$O$_{39}$)$_2$]$^{8-}$, Figure 3h) containing dissimilar copper under hydrothermal condition. The resulting organic–inorganic hybrid assemblies contained a rare corner-sharing double-Keggin type POM architecture in the Ti-POM species, which was further connected with the butterfly-type [Cu^{II}Lo] units to form a 1-D chain and a square plane, respectively [50].

In contrast to the diverse structures of Ti-substituted POMs, Zr-substituted dimeric POMs have been far less reported. Some presentative examples are summarized below. In 2003, May et al. reported an example of a dimeric structure of mono-Zr-substituted Kegging-type POM, [Zr(PMo$_{12}$O$_{40}$)(PMo$_{11}$O$_{39}$)]$^{6-}$ (Figure 4a), which also represented the first crystallographic determination of the [PMo$_{11}$O$_{39}$]$^{7-}$ anion [51]. The similar Keggin-type chiral phosphotungstate and borotungstate were also reported by Liu and Xue's groups in 2009, respectively [52,53]. In 2006, Nomiya et al. also reported the first Zr-substituted Well-Dawson phosphotungstate [Zr(α$_2$-P$_2$W$_{17}$O$_{61}$)$_2$]$^{16-}$ (Figure 4b), deriving from mono-lacunary precursor [α$_2$-P$_2$W$_{17}$O$_{61}$]$^{10-}$ [54]. Similar mono-Zr-substituted Well-Dawson phosphotungstate POMs ([{P$_2$W$_{15}$O$_{54}$(H$_2$O)$_2$}$_2$Zr]$^{12-}$, and [{P$_2$W$_{15}$O$_{54}$(H$_2$O)$_2$}Zr{P$_2$W$_{17}$O$_{61}$}]$^{14-}$) have been further reported by Hill et al. in 2007 [55]. To increase the nuclearity of Zr centers, Kholdeeva et al. prepared three Zr$_2$-substituted Keggin-type phosphotungstate ([{PW$_{11}$O$_{39}$Zr(μ-OH)}$_2$]$^{8-}$, [{PW$_{11}$O$_{39}$Zr(μ-OH)}$_2$]$^{8-}$, and [{PW$_{11}$O$_{39}$Zr$_2$}$_2$(μ-OH)(μ-O)]$^{9-}$) [56]. Then, Mizuno et al. synthesized a di-Zr-substituted Keggin-type silicotungstate [(γ-SiW$_{10}$O$_{36}$)$_2$Zr$_2$(μ-OH)$_2$]$^{10-}$ (Figure 4c) [57], and Sokolov et al. reported di-Zr-substituted Dawson-type phosphotungstate [{(H$_2$O)Zr(μ_2-OH)(P$_2$W$_{17}$O$_{61}$)}$_2$]$^{14-}$ [58]. In 2011, Villanneau et al. reported two Zr$_2$-containing POMs derivatives [{PW$_9$O$_{34}${PO(R)}$_2$}$_2${Zr(H$_2$O)(μ-OH)}$_2$]$^{4-}$ and [{PW$_9$O$_{34}${PO(R)}$_2$}$_2${Zr(DMF)(μ-OH)}$_2$]$^{4-}$ (R = Ph, ^{t}Bu), which were obtained by using [(^{n}Bu$_4$N)$_3$Na$_2$[PW$_9$O$_{34}${PO(R)}$_2$] and ZrOCl$_2$·8H$_2$O [59]. Subsequently, the same group also reported a similar structure in 2013 [60]. In 2005, Hill et al. reported the first chiral tri-Zr-substituted Dawson-type phosphotungstate {[α-P$_2$W$_{15}$O$_{55}$(H$_2$O)]Zr$_3$(μ_3-O)(H$_2$O)(L-tartH)[α-P$_2$W$_{16}$O$_{59}$]}$^{15-}$ (Figure 4d) and {[α-P$_2$W$_{15}$O$_{55}$(H$_2$O)]Zr$_3$(μ_3-O)(H$_2$O)(D-tartH)[α-P$_2$W$_{16}$O$_{59}$]}$^{15-}$ [61]. After that, Cadot et. al. reported a tri-Zr(IV)-substituted sandwich-type Keggin POM [Zr$_3$O(OH)$_2$(SiW$_9$O$_{34}$)$_2$]$^{12-}$, which consists of a [Zr$_3$O(OH)$_2$] triangular central cluster closely embedded between two A-α-[SiW$_9$O$_{34}$]$^{10-}$ subunits [62]. Subsequently, three cases of isomorphic compounds were also reported by Xue, Nomiya and Yang's groups, respectively [63–65]. Among these tri-Zr-substituted POMs, it is worth mentioning that Yang's group reported the first tri-Zr^{IV}-substituted POM [Zr$_3$(μ_2-OH)$_2$(μ_2-O)(A-α-GeW$_9$O$_{34}$)(1,4,9-α-P$_2$W$_{15}$O$_{56}$)]$^{14-}$ (Figure 4e), where the tri-Zr centers were stabilized by mixed types of tri-lacunary POM ligands including Keggin-type [A-α-GeW$_9$O$_{34}$]$^{10-}$ and Dawson-type [1,4,9-α-P$_2$W$_{15}$O$_{56}$]$^{12-}$ units [66]. In addition, a number of tetra-Zr substituted POMs have also been prepared. For instance, Pope et al. reported a Zr$_4$-substituted phosphotungstate, [Zr$_4$(μ_3-O)$_2$(μ_2-OH)$_2$(H$_2$O)$_4$(P$_2$W$_{16}$O$_{59}$)$_2$]$^{14-}$ (Figure 4f). Therein, the divacant lacunary {P$_2$W$_{16}$O$_{59}$} ligands were derived from the plenary Wells–Dawson (α-P$_2$W$_{18}$O$_{62}$) polyoxoanion [67]. Then, similar structures were reported by Hill and Li's groups in 2005 and 2013, respectively [68,69]. Subsequently, tetra-Zr substituted Keggin-type silicotungstates [(γ-SiW$_{10}$O$_{36}$)$_2$Zr$_4$(μ_4-O)(μ-OH)$_6$]$^{8-}$ (Figure 4g) and five other similar structures were successively reported [70–77]. The nuclearity of Zr substitution has been further improved to six by Kortz and co-workers. they reported the first hexa-Zr-substituted dimeric tungstoarsenates [Zr$_6$O$_4$(OH)$_4$(H$_2$O)$_2$(CH$_3$COO)$_5$(AsW$_9$O$_{33}$)$_2$]$^{11-}$ (Figure 4h). In the polyoxoanion, the unprecedented hexa-Zr unit is perfectly located at the cavity formed by two (B-α-AsW$_9$O$_{33}$) fragments lying at an angle of about 74° with respect to each other [78].

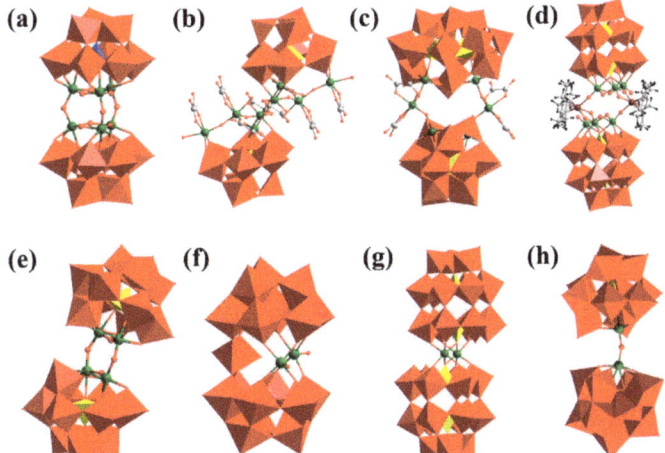

Figure 3. The ball-and-stick and polyhedral illustration of (**a**) [A-β-Si$_2$W$_{18}$Ti$_6$O$_{77}$]$^{14-}$ polyoxoanion, (**b**) [Ti$_8$(C$_2$O$_4$)$_8$P$_2$W$_{18}$O$_{76}$(H$_2$O)$_4$]$^{18-}$ polyoxoanion, (**c**) [Ti$_6$(C$_2$O$_4$)$_4$P$_4$W$_{32}$O$_{124}$]$^{20-}$ polyoxoanion, (**d**) P$_2$W$_{15}$Ti$_3$O$_{60}$(OH)$_2$}$_2$(Cp*Rh)$_2$]$^{16-}$ polyoxoanion; (**e**) [α,α-P$_2$W$_{20}$Ti$_4$O$_{78}$]$^{10-}$ polyoxoanion; (**f**) [Ti$_2$(OH)$_2$As$_2$W$_{19}$O$_{67}$(H$_2$O)]$^{8-}$ polyoxoanion; (**g**) [Ti$_2${P$_2$W$_{15}$O$_{54}$(OH$_2$)$_2$}$_2$]$^{8-}$ polyoxoanion and (**h**) [(Ti$_2$O)(PW$_{11}$O$_{39}$)$_2$]$^{8-}$ polyoxoanion. Color codes: WO$_6$, red; PO$_4$, yellow; SiO$_4$, light blue; Ti, green; As, lavender; C, gray; O, red; H, black.

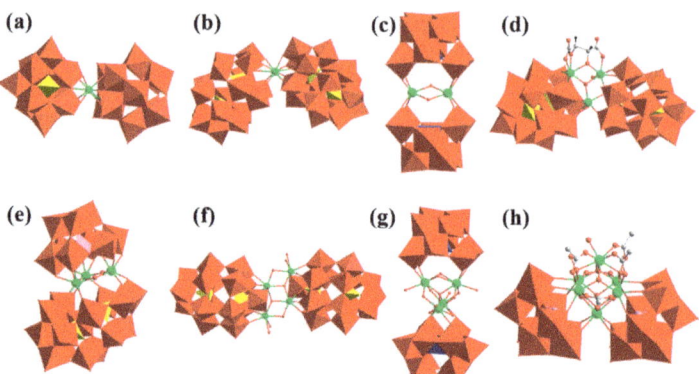

Figure 4. The ball-and-stick and polyhedral illustration of (**a**) $[Zr(PMo_{12}O_{40})(PMo_{11}O_{39})]^{6-}$ polyoxoanion, (**b**) $[Zr(\alpha_2\text{-}P_2W_{17}O_{61})_2]^{16-}$ polyoxoanion, (**c**) $[(\gamma\text{-}SiW_{10}O_{36})_2Zr_2(\mu\text{-}OH)_2]^{10-}$ polyoxoanion, (**d**) $\{[\alpha\text{-}P_2W_{15}O_{55}(H_2O)]Zr_3(\mu_3\text{-}O)(H_2O)(L\text{-tartH})[\alpha\text{-}P_2W_{16}O_{59}]\}^{15-}$ polyoxoanion; (**e**) $[Zr_3(\mu_2\text{-}OH)_2(\mu_2\text{-}O)(A\text{-}\alpha\text{-}GeW_9O_{34})(1,4,9\text{-}\alpha\text{-}P_2W_{15}O_{56})]^{14-}$ polyoxoanion; (**f**) $[Zr_4(\mu_3\text{-}O)_2(\mu_2\text{-}OH)_2(H_2O)_4(P_2W_{16}O_{59})_2]^{14-}$ polyoxoanion; (**g**) $[(\gamma\text{-}SiW_{10}O_{36})_2Zr_4(\mu_4\text{-}O)(\mu\text{-}OH)_6]^{8-}$ polyoxoanion and (**h**) $[Zr_6O_4(OH)_4(H_2O)_2(CH_3COO)_5(AsW_9O_{33})_2]^{11-}$ polyoxoanion. Color codes: W/MoO_6, red; PO_4, yellow; SiO_4, light blue; GeO_4, turquoise; Zr, bright green; As, lavender; C, gray; O, red; H, black.

2.3. Ti/Zr-Substituted Trimeric POMs

In contrast to the dimeric POM structures, the syntheses of Ti/Zr-substituted trimeric POMs have rarely been reported. The early reported Zr-substituted trimeric POM is $Zr_6O_2(OH)_4(H_2O)_3(\beta\text{-}SiW_{10}O_{37})_3]^{14-}$ (Figure 5a), which was synthesized using an aqueous solution-based method by Kortz et al. in 2006. This polyoxoanion consists of three β_{23}-$SiW_{10}O_{37}$ units linked by an unprecedented $Zr_6O_2(OH)_4(H_2O)_3$ cluster with C_1 point group symmetry [79]. The similar polyoxoanion $[Zr_6(\mu_3\text{-}O)_3(OH)_3(OAc)(H_2O)(\beta\text{-}GeW_{10}O_{37})_3]^{16-}$ was also synthesized via hydrothermal method by Yang's group in 2019, except that $\{\beta\text{-}SiW_{10}O_{37}\}$ was replaced by $\{\beta\text{-}GeW_{10}O_{37}\}$ [80]. Late on, Kortz et al. reported another Zr$_6$-substituted silicotungstate $[Zr_6(O_2)_6(OH)_6(\gamma\text{-}SiW_{10}O_{36})_3]^{18-}$ (Figure 5b) in 2008, where 6-Peroxo-6-Zr Crown embedded in a triangular polyoxoanion. This polyoxoanion is composed of three $[\gamma\text{-}SiW_{10}O_{36}]^{8-}$ units encapsulating the unprecedented $[Zr_6(O_2)_6(OH)_6]^{6+}$ wheel, while it can also be considered as a cyclic assembly of three fused $\{Zr_2(O_2)_2(OH)_2(\gamma\text{-}SiW_{10}O_{36})\}$ monomers. This work belongs to the first structurally characterized Zr-peroxo POM with side-on, bridging peroxo units [81]. Additionally, Kortz et al. reported the first examples of Ti-containing trimeric polytungstates, two cyclic Ti$_9$-containing trimeric POMs, $[(\alpha\text{-}Ti_3PW_9O_{38})_3(PO_4)]^{18-}$ (Figure 5c) and $[(\alpha\text{-}Ti_3SiW_9O_{37}OH)_3(TiO_3(OH_2)_3)]^{17-}$ (Figure 5d) using the solution-based synthetic method. Both compounds were composed of three $(Ti_3XW_9O_{37})$ units (X = P or Si) bridged with three Ti-O-Ti bonds and a capping group (tetrahedral PO_4 or octahedral TiO_6) [82]. Liu's group reported two similar trimeric nine-Ti^{IV} contained tungstogermanates $\{K\subset[(Ge(OH)O_3)(GeW_9Ti_3O_{38}H_2)_3]\}^{14-}$ (Figure 5e) and $\{K\subset[(SO_4)(GeW_9Ti_3O_{38}H_3)_3]\}^{10-}$. The two compounds were obtained from the reactions between $K_8[\gamma\text{-}GeW_{10}O_{36}]$ [83] and $TiO(SO_4)$ under different pH conditions. The former polyoxoanion consisted of three tri-Ti^{IV}-substituted Keggin fragments $[GeW_9Ti_3O_{38}]$ and a GeO_4 tetrahedral linker bridged with both Ti–O–Ti and Ti–O–Ge bonds, and the structure of the latter one is similar except for the replacement of SO_4 with GeO_4 [84]. Recently, a novel trimer compound $[\{Ca_6(CO_3)(\mu_3\text{-}OH)(OH_2)_{18}\}(P_2W_{15}Ti_3O_{61})_3Ca(OH_2)_3]^{19-}$ (Figure 5f) was reported that contains a hexacalcium cluster cation, one carbonate anion, and one calcium cation assembled on a trimeric tri-Ti-substituted Wells–Dawson polyoxometalates. This complex was obtained through the reaction of calcium chloride with the monomeric trititanium(IV)-substituted Wells–Dawson POM species "$[P_2W_{15}Ti_3O_{59}(OH)_3]^{9-}$". During the synthesis, the $[Ca_6(CO_3)(\mu_3\text{-}OH)(OH_2)_{18}]^{9+}$ cluster cation, composed of six calcium cations linked by one μ_6-carbonato anion and one μ_3-OH⁻ anion, assembled with one calcium ion, a trimeric "$[P_2W_{15}Ti_3O_{59}(OH)_3]^{9-}$" species to form the target product. The compound is an unprecedented POM species containing an alkaline-earth-metal cluster cation, and it is also the first example of alkaline-earth-metal ions clustered around a Ti-substituted POM [85].

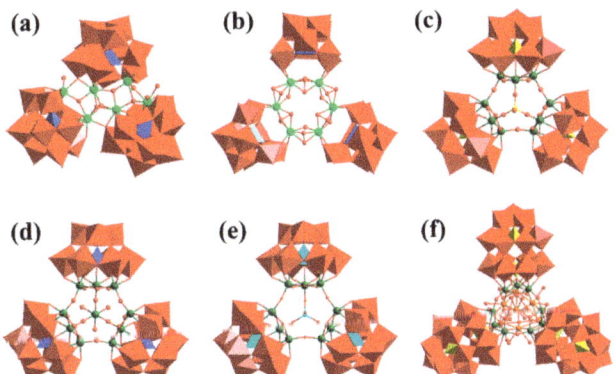

Figure 5. The ball-and-stick and polyhedral illustration of (**a**) $Zr_6O_2(OH)_4(H_2O)_3(\beta\text{-}SiW_{10}O_{37})_3]^{14-}$ polyoxoanion, (**b**) $[Zr_6(O_2)_6(OH)_6(\gamma\text{-}SiW_{10}O_{36})_3]^{18-}$ polyoxoanion, (**c**) $[(\alpha\text{-}Ti_3PW_9O_{38})_3(PO_4)]^{18-}$ polyoxoanion, (**d**) $[(\alpha\text{-}Ti_3SiW_9O_{37}OH)_3(TiO_3(OH_2)_3)]^{17-}$ polyoxoanion, (**e**) $\{K \subset [(Ge(OH)O_3)(GeW_9Ti_3O_{38}H_2)_3]\}^{14-}$ polyoxoanion and (**f**) $[\{Ca_6(CO_3)(\mu_3\text{-}OH)(OH_2)_{18}\}(P_2W_{15}Ti_3O_{61})_3Ca(OH_2)_3]^{19-}$ polyoxoanion. Color codes: WO_6, red; PO_4, yellow; SiO_4, light blue; GeO_4, turquoise; Ti, green; Zr, bright green; Ca, orange; C, gray; O, red.

2.4. Ti/Zr-Substituted Tetrameric POMs

In this section, a number of Ti/Zr-substituted tetrameric POMs will be briefly introduced. The early example is a dodeca-Ti-substituted Dawson-type tetrameric, $[\{Ti_3P_2W_{15}O_{57.5}(OH)_3\}_4]^{24-}$ (Figure 6a), representing a supramolecular phosphotungstate reported by Kortz's group in 2003 [86]. The polyoxoanion was composed of four lacunary $[P_2W_{15}O_{56}]^{12-}$ Well–Dawson building blocks linked with terminal Ti-O bonds, resulting in a structure with T_d symmetry. The $\{Ti_{12}O_{46}\}$ core of the polyoxoanion is composed of four groups of three edge-shared, corner-linked TiO_6 octahedra. Such a rare arrangement resembles one set of the four corner-shared faces of an octahedron, described as a "reversed Keggin structure", which is very similar to the $[As_4Mo_{12}O_{50}]^{8-}$ geometry reported by Sasaki and Nishikawa [87]. Apart from this Ti_{12} cluster, they also discovered another deca-Ti-substituted tetrameric species $[\{Ti_3P_2W_{15}O_{57.5}(OH)_3\}_2\{Ti_2P_2W_{16}O_{60}(OH)\}_2]^{26-}$ containing two $\{Ti_3P_2W_{15}\}$ and two $\{Ti_2P_2W_{16}\}$ fragments, therefore resulting in a structure with C_{2v} symmetry. Meanwhile, Nomiya et al. also reported two multi-Ti-substituted tetrameric POMs. The first one is a giant "tetrapod"-shaped dodeca-Ti-substituted Dawson-type tetrameric phosphotungstate, $[(\alpha\text{-}1,2,3\text{-}P_2W_{15}Ti_3O_{60.5})_4Cl]^{37-}$ (Figure 6b), which contains the four Wells-Dawson units fused together through Ti–O–Ti bonds. The structure exhibits an approximately T_d symmetry, where the four Ti_3O_6 facets of "$P_2W_{15}Ti_3$" occupied four alternate facets of an octahedron, and the one Cl⁻ ion was encapsulated in the central octahedral cavity [88]. It is noted that a similar structure $[(P_2W_{15}Ti_3O_{60.5})_4(NH_4)]^{35-}$ was also reported in 2011, except that Cl⁻ was replaced by NH_4^+ [89]. The other example belongs to a "tetrapod"-shaped Ti_{16}-substituted Dawson-type tetrameric phosphotungstate $[(\alpha\text{-}1,2,3\text{-}P_2W_{15}Ti_3O_{62})_4\{\mu_3\text{-}Ti(OH)_3\}_4Cl]^{45-}$ (Figure 6c). The polyoxoanion, prepared by the reaction of $[P_2W_{15}O_{56}]^{12-}$ with an excess amount of $TiCl_4$ in aqueous solution, was composed of four tri-Ti^{IV}-1,2,3-substituted α-Dawson substructures, four $Ti(OH)_3$ bridging groups, and one encapsulated chloride ion [90]. Subsequently, Nomiya et al. also reported three similar Ti_{16}-substituted POMs, $[(\alpha\text{-}P_2W_{15}Ti_3O_{59}(OH)_3)_4\{\mu_3\text{-}Ti(H_2O)_3\}_4X]^{21-}$

(X = Br$^-$, I$^-$, and NO$_3{}^-$), except that halogen atoms and some oxygen atoms are protonated [91]. In 2004, Kortz et al. synthesized a unique cyclic octa-Ti- substituted tetrameric tungstosilicate [{β-Ti$_2$SiW$_{10}$O$_{39}$}$_4$]$^{24-}$ (Figure 6d) assembly under mild, one-pot reaction conditions. The polyoxoanion is composed of four {β-Ti$_2$SiW$_{10}$O$_{39}$} Keggin fragments bridged with Ti–O–Ti bonds, leading to a cyclic assembly. The successful preparation of this compound provides future possibilities for preparing even larger wheel-shaped polyoxotungstates and other discrete nanomolecular objects of similar size, structure, and function as those made with polyoxometalates [92]. Then, Kortz's group also successfully prepared a hepta-Ti substituted arsenotungstate [Ti$_6$(TiO$_6$)(AsW$_9$O$_{33}$)$_4$]$^{20-}$ (Figure 6e) in 2014 using a simple one-pot procedure. The polyoxoanion contains a novel Ti$_7$-core consisting of a central TiO$_6$ octahedron surrounded by six TiO$_5$ square pyramids, which was further capped by four trilacunary {As^{III}W$_9$} fragments, leading to an assembly with T_d point-group symmetry [93]. In 2019, Yang's group also reported a similar structure [Ti$_7$O$_6$(SbW$_9$O$_{33}$)$_4$]$^{20-}$ by replacing {AsW$_9$O$_{33}$} with {SbW$_9$O$_{33}$} [94]. In 2022, Yang's group further synthesized a ring-shaped 12-Ti-substituted poly(polyoxometalate) [{K$_2$Na(H$_2$O)$_3$}@{(Ti$_2$O)$_2$(Ti$_4$O$_4$)$_2$(A-α-1,3,5-GeW$_9$O$_{36}$)$_2$(A-α-2,3,4-GeW$_9$O$_{36}$)$_2$}]$^{25-}$ (Figure 6f) under hydrothermal conditions, which represents the highest number of Ti centers in Keggin-type poly(POM) family to date. In this structure, two types of novel chiral trivacant [GeW$_9$O$_{36}$]$^{14-}$ (A-α-1,3,5-GeW$_9$O$_{36}$ and A-α-2,3,4-GeW$_9$O$_{36}$) fragments have been first discovered in POM chemistry, and four [GeW$_9$O$_{36}$]$^{14-}$ fragments are alternately connected by two Ti$_2$O and two Ti$_4$O$_4$ cores to form a ring-shaped poly(POM) [95].

In addition to these Ti-substituted POMs, some Zr-substituted tetrameric POMs have also been actively investigated. For instance, a Zr$_4$-substituted tungstoselenites, [(α-SeW$_9$O$_{34}$){Zr(H$_2$O)}{WO(H$_2$O)}(WO$_2$)(SeO$_3$){α-SeW$_8$O$_{31}$Zr(H$_2$O)}]$_2{}^{12-}$ (Figure 7a), has been firstly constructed using {α-SeW$_9$} building blocks. Such a tetrameric structure can be divided into two same subunits, each containing a dimer sandwich-type structure that consists of a well-known trivacant Keggin-type {α-SeW$_9$O$_{34}$} building blocks [96]. Then, Yang's group continuously reported eight tetrameric Zr-substituted POMs. The first example represents a new tetra-Zr-substituted tungstophosphate, {Zr$_2$[SbP$_2$W$_4$(OH)$_2$O$_{21}$][α$_2$-PW$_{10}$O$_{38}$]}$_2{}^{20-}$ (Figure 7b) through the hydrothermal reaction of the [B-α-SbW$_9$O$_{33}$]$^{9-}$ building block with Zr^{4+} cations and PO$_4{}^{3-}$ anions in the presence of dimethylamine hydrochloride in NaOAc-HOAc buffer solution. The compound exhibits a toroidal structure formed by two divacant [α$_2$-PW$_{10}$O$_{38}$]$^{11-}$ units and two [SbP$_2$W$_4$(OH)$_2$O$_{21}$]$^{7-}$ fragments linked by four Zr^{4+} cations. It is noted that the triangular pyramidal SbO$_3$ in the [B-α-SbW$_9$O$_{33}$]$^{9-}$ precursor was replaced by tetrahedral PO$_4$ unit in the final compound, and the pendant SbO$_3$ derives from the dissociation of the [B-α-SbW$_9$O$_{33}$]$^{9-}$ precursor [97]. Then, the same group also reported a Zr(IV)-substituted tetramer polyoxotungstate, [Zr$_4$(β-GeW$_{10}$O$_{38}$)$_2$(A-α-PW$_9$O$_{34}$)$_2$]$^{26-}$ (Figure 7c) [98], which was obtained in a one-pot reaction of the hexalacunary polyanion [P$_2$W$_{12}$O$_{48}$]$^{12-}$ and trilacunary [GeW$_9$O$_{34}$]$^{10-}$ polyanion precursors with Zr^{4+} in a slightly alkali aqueous solution in the presence of borates. Subsequently, a Zr$_9$-substituted tetrameric germanotungstate, [{Zr$_5$(μ$_3$-OH)$_4$(OH)$_2$}@{Zr$_2$(OAc)$_2$(α-GeW$_{10}$O$_{38}$)$_2$}$_2$]$^{22-}$ (Figure 7d) was constructed by two novel sandwich-type dimers [Zr$_2$(OAc)$_2$(α$_2$-GeW$_{10}$O$_{38}$)$_2$]$^{18-}$ and one unique [Zr$_5$(μ$_3$-OH)$_4$(OH)$_2$]$^{14+}$ core in an approximately orthogonal fashion, showing a staggering tetrahedral polyoxoanion [80]. In addition, Yang et al. also reported a series of ring-shaped Zr$_8$-substituted silicotungstates, [{Zr$_2$(OH)$_2$(α-SiW$_{10}$O$_{38}$)}$_2${Zr$_2$(OH)$_2$(β-SiW$_{10}$O$_{38}$)}$_2$]$^{24-}$ (Figure 7e) (dap = 1,3-diamino-propane), [(Zr$_2$(OH)$_2$)$_2$(Zr$_2$BO(OH)$_4$)$_2$(β-SiW$_{10}$O$_{38}$)$_4$]$^{26-}$ (Figure 7f) and [(Zr$_2$BO(OH)$_4$)$_2$(Zr$_2$B$_2$O$_2$(OH)$_5$)$_2$(β-SiW$_{10}$O$_{38}$)$_4$]$^{28-}$ [99]. The latter two compounds first provided the possibility of introducing Zr–B–O linkage to the lacunary sites. Very recently, they also reported a di-Zr-substituted polyoxotungstate [ZrSb$_4$(OH)O$_2$(A-α-PW$_8$O$_{32}$)(A-α-PW$_9$O$_{34}$)]$_2{}^{18-}$ (Figure 7g) [100] and hepta-Zr-incorporated polyoxometalate [SbZr$_7$O$_6$(OH)$_4$(B-α-GeW$_9$O$_{34}$)$_2$(B-α-GeW$_{11}$O$_{39}$)$_2$]$^{21-}$ (Figure 7h) [101], these works greatly enriched the family of Zr-substituted POMs.

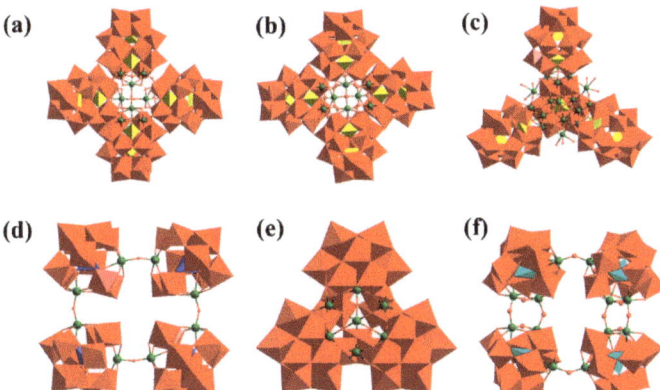

Figure 6. The ball-and-stick and polyhedral illustration of (**a**) $[\{Ti_3P_2W_{15}O_{57.5}(OH)_3\}_4]^{24-}$ polyoxoanion, (**b**) $[(\alpha\text{-}1,2,3\text{-}P_2W_{15}Ti_3O_{60.5})_4Cl]^{37-}$ polyoxoanion, (**c**) $[(\alpha\text{-}1,2,3\text{-}P_2W_{15}Ti_3O_{62})_4\{\mu_3\text{-}Ti(OH)_3\}_4Cl]^{45-}$ polyoxoanion, (**d**) $[\{\beta\text{-}Ti_2SiW_{10}O_{39}\}_4]^{24-}$ polyoxoanion, (**e**) $[Ti_6(TiO_6)(AsW_9O_{33})_4]^{20-}$ polyoxoanion and (**f**) $[\{K_2Na(H_2O)_3\}@\{(Ti_2O)_2\text{-}(Ti_4O_4)_2(A\text{-}\alpha\text{-}1,3,5\text{-}GeW_9O_{36})_2(A\text{-}\alpha\text{-}2,3,4\text{-}GeW_9O_{36})_2\}]^{25-}$ polyoxoanion. Color codes: WO_6, red; PO_4, yellow; SiO_4, light blue; GeO_4, turquoise; Ti, green; As, lavender; O, red.

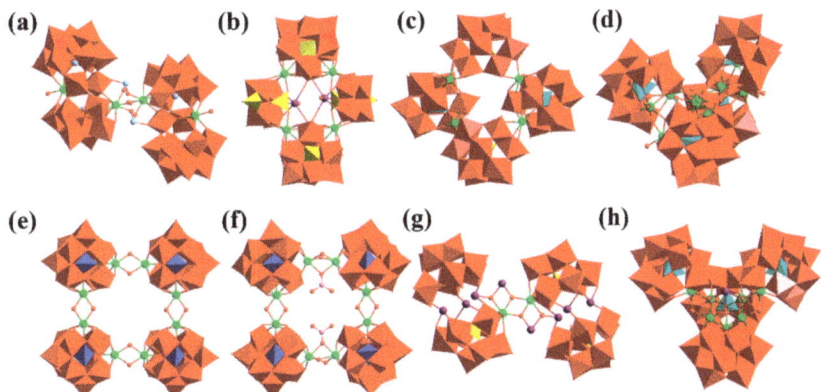

Figure 7. The ball-and-stick and polyhedral illustration of (**a**) $[(\alpha\text{-}SeW_9O_{34})\{Zr(H_2O)\}\{WO(H_2O)\}(WO_2)(SeO_3)\{\alpha\text{-}SeW_8O_{31}Zr(H_2O)\}]_2^{12-}$ polyoxoanion, (**b**) $\{Zr_2[SbP_2W_4(OH)_2O_{21}][\alpha_2\text{-}PW_{10}O_{38}]\}_2^{20-}$ polyoxoanion, (**c**) $[Zr_4(\beta\text{-}GeW_{10}O_{38})_2(A\text{-}\alpha\text{-}PW_9O_{34})_2]^{26-}$ polyoxoanion; (**d**) $[\{Zr_5(\mu_3\text{-}OH)_4(OH)_2\}@\{Zr_2(OAc)_2(\alpha\text{-}GeW_{10}O_{38})_2\}_2]^{22-}$ polyoxoanion, (**e**) $[\{Zr_2(OH)_2(\alpha\text{-}SiW_{10}O_{38})\}_2\{Zr_2(OH)_2(\beta\text{-}SiW_{10}O_{38})\}_2]^{24-}$ polyoxoanion, (**f**) $[(Zr_2(OH)_2)_2(Zr_2BO(OH)_4)_2(\beta\text{-}SiW_{10}O_{38})_4]^{26-}$ polyoxoanion, (**g**) $[ZrSb_4(OH)O_2(A\text{-}\alpha\text{-}PW_8O_{32})(A\text{-}\alpha\text{-}PW_9O_{34})]_2^{18-}$ polyanion and (**h**) $[SbZr_7O_6(OH)_4(B\text{-}\alpha\text{-}GeW_9O_{34})_2(B\text{-}\alpha\text{-}GeW_{11}O_{39})_2]^{21-}$ polyoxoanion. Color codes: WO_6, red; PO_4, yellow; SiO_4, light blue; GeO_4, turquoise; Se, pale blue; Sb, violet; Zr, bright green; B, rose; O, red.

2.5. Ti/Zr-Substituted Multimeric POMs

Compared to those mono-, di-, tri- and tetrameric POM structures, there are very few reports on the preparation of multimeric Ti/Zr-substituted POMs. To date, only two related compounds have been reported. The first example belongs to an octa-Ti-substituted Dawson-type supramolecular polyoxoanion reported by Kortz's group in 2003 [86]. However, the polyoxoanion was described by the preliminary and incomplete formula "$Ti_8P_{12}W_{84}$" or "$(Ti_2P_2W_{15})_2(Ti_2P_2W_{16})_2(P_2W_{11})_2$" (Figure 8a) due to the poor quality of the crystallographic data. The second representative example is the gigantic Zr_{24}-cluster-substituted Keggin-type germanotungstates $[Zr_{24}O_{22}(OH)_{10}(H_2O)_2(W_2O_{10}H)_2(GeW_9O_{34})_4(GeW_8O_{31})_2]^{32-}$ (Figure 8b) reported by Yang's group in 2014 [102]. The polyoxoanion was successfully synthesized under hydrothermal conditions, which contains the largest $[Zr_{24}O_{22}(OH)_{10}(H_2O)_2]$ cluster among all reported Zr-based poly(polyoxometalate)s to date. Detailed structural analyses of this complex showed that the centrosymmetric Zr_{24}-cluster-based hexamer contained two symmetry-related $[Zr_{12}O_{11}(OH)_5(H_2O)(W_2O_{10}H)(GeW_9O_{34})_2(GeW_8O_{31})]^{16-}$ trimers linked via six μ_3-oxo bridges, which were further encapsulated by different POM fragments including B-α-GeW_9O_{34}, B-α-GeW_8O_{31}, and W_2O_{10}. Catalytic experiments also showed that this compound worked as a good catalyst for the oxygenation of thioethers to sulfoxides/sulfones in the presence of H_2O_2, which could be attributed to the unique redox property of oxygen-enriched polyoxotungstate fragments as well as the Lewis acidity of the Zr_{24} cluster. Although the synthesis of multimeric POMs is rather difficult, these pioneering works provide some insights and future direction for the exploration of this specific research area.

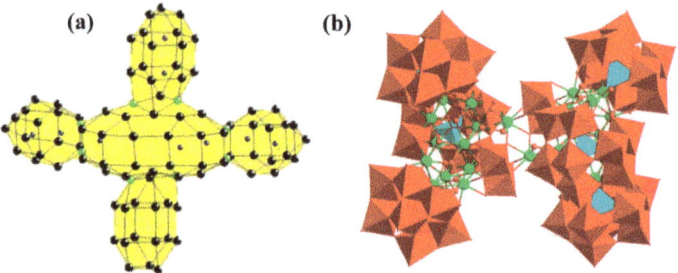

Figure 8. The ball-and-stick and polyhedral illustration of (**a**) $(Ti_2P_2W_{15})_2(Ti_2P_2W_{16})_2(P_2W_{11})_2$ polyoxoanion, (**b**) $[Zr_{24}O_{22}(OH)_{10}(H_2O)_2(W_2O_{10}H)_2(GeW_9O_{34})_4(GeW_8O_{31})_2]^{32-}$ polyoxoanion. Color codes: WO_6, red; GeO_4, turquoise; Zr, bright green; O, red. Adapted with permission from ref. [86], copyright 2003 John Wiley and Sons.

3. The Applications of Representative Ti/Zr-Substituted POMs

It is well known that transition-metal clusters are a unique area in inorganic chemistry, considering their vital contribution to the blossom of modern chemistry as well as their potential application as structural models for various industrial and biological catalytic processes [103–106]. To date, the Ti/Zr-substituted POMs have been widely investigated as catalysts for the oxidation of organic substrates. For example, Poblet et al. investigated the oxidation of alkenes by H_2O_2 catalyzed using Ti(IV)-containing POMs, which were models of Ti single-site catalysts at the DFT computational level. The catalytic mechanism of the C_2H_4 epoxidation with H_2O_2 mediated by $[PTi(OH)W_{11}O_{39}]^{4-}$ and $[Ti_2(OH)_2As_2W_{19}O_{67}(H_2O)]^{8-}$ can be processed by following two main steps (Scheme 1): (i) H_2O_2 was activated to form the titanium-peroxo or -hydroperxo intermediate, and (ii) the reactive intermediate further attack alkene to form the epoxide and water [107]. Kholdeeva and co-workers also reported several cases of various organic catalytic studies with Ti-containing POMs. For example, they investigated the mechanism of thioether oxidation of $(Bu_4N)_7\{[PW_{11}O_{39}Ti]_2OH\}$ dimeric heteropolytungstate in 2000 [108]. In 2004, they also reported a protonated titanium peroxo complex $[Bu_4N]_4[HPTi(O_2)W_{11}O_{39}]$ and found that protonated titanium peroxo complex has a higher redox potential, which can improve the catalytic performance [109]. In the same year, Ti(IV)-monosubstituted Keggin-type POMs were reported to exhibit excellent catalytic oxidation properties with H_2O_2 [110–112]. Subsequently, Kholdeeva et al. also reported the epoxidation of a range of alkenes easily proceeds with aqueous H_2O_2 as oxidant and the dititanium-containing 19-tungstodiarsenate (III) as catalyst [113]. In 2012, the same group also published two works on alkene oxidation by Ti-containing POMs. They claimed that the energy barrier for the heterolytic oxygen transfer from the reactive Ti hydroperoxo intermediate was significantly reduced by the protonated Ti-containing POM, as revealed by the kinetic and DFT studies, thereby greatly enhancing the activity and selectivity of alkene oxidation [114,115]. Subsequently, Poblet and Guillemot reported the alkene epoxidation catalyzed by the hybrid [B-α-SbW$_9$O$_{33}$(^{t}BuSiO)$_3$Ti(O^{i}Pr)]$^{3-}$, [PW$_9$O$_{34}$(^{t}BuSiO)$_3$Ti(O^{i}Pr)]$^{3-}$, and Ti-complexe of silanol functionalized POMs [SbW$_9$O$_{33}$(RSiO)$_3$Ti(O^{i}Pr)]$^{3-}$, respectively [32,116,117]. Based on the research of alkene epoxidation catalysis, Kholdeeva et al. also revealed the mechanism of thioether oxidation of Ti-substituted POMs by kinetic modeling and DFT calculations. Two possible models regarding the active group were proposed: (1) the active group is the terminal Ti–OH group for the mononuclear, and (2) the active group is the bridging Ti$_2$(μ-OH) moiety for the multinuclear [118]. Subsequently, Yang's group also reported two cases of the catalytic oxidation of thioethers using Ti$_7$- and Ti$_{12}$-substituted POMs, respectively, which both exhibited good catalytic properties [94,95]. In addition, Li's group investigated the photocatalytic degradation of MB with Ti$_2$-substituted POM units under UV irradiation, which proved that Ti-substituted Keggin-type POMs showed better photocatalytic activities than that of typical Keggin-type POMs [50]. Some Ti-substituted POMs have also been reported with good electrocatalytic properties [119].

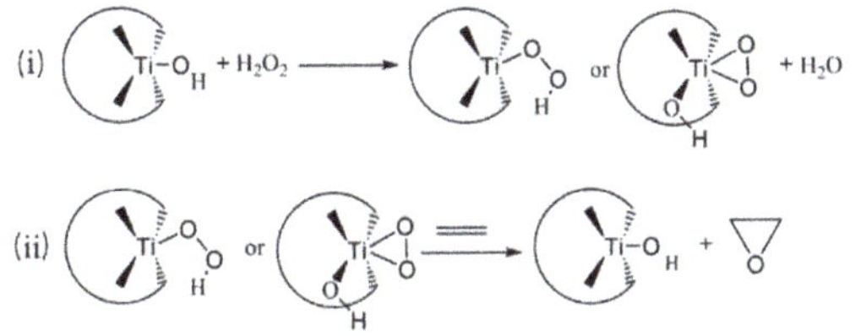

Scheme 1. Schematic diagram of the proposed mechanistic processes.

Compared with the applications of Ti-substituted POMs, a part of Zr-substituted POMs also exhibited good catalytic oxidation of thioether, electrocatalytic and nonlinear optical properties, etc. For example, Yang's groups reported a few works on the oxidation of sulfide using Zr$_2$-, Zr$_4$-, Zr$_7$- and Zr$_8$-substituted POMs, respectively [76,99–101]. These compounds showed remarkable heterogeneous catalysts for the catalytic oxidation of sulfides into the corresponding sulfones with H$_2$O$_2$. The same groups also reported several Chiral Zr-substituted POMs which have excellent nonlinear optical properties [75,77]. In addition, transition-metal-substituted POMs are air- and water-stable Lewis acids that were often used in organic reactions [120], and in the last two decades, most Zr-substituted POMs also were reported to be used as optimal Lewis acid catalysts to hydrolyze the O = C-NH- bonds in proteins or peptides (Scheme 2), leading to the formation of amino acids [121–126]. These works implied the potential applications of Zr-substituted POMs in biological systems.

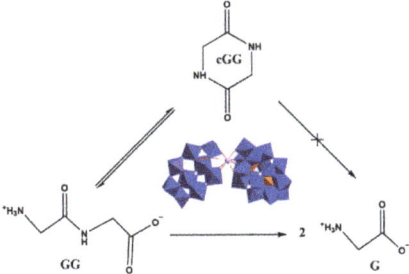

Scheme 2. Schematic diagram of hydrolysis of GG in the Presence of POMs.

4. Conclusions and Perspectives

In summary, this review has mainly addressed the development of Ti/Zr-substituted POMs with an emphasis on structural diversity, synthetic approaches, and potential catalytic applications. According to the overview of these reported Ti/Zr-substituted POMs, we can conclude that (1) the solution-based synthetic approach is an effective method for the syntheses of Ti/Zr-substituted POMs; (2) interesting and attracting POM structures could be often obtained via the hydrothermal synthetic strategy. These successful synthetic strategies and the persistent and dedicated efforts of chemists over the past decades have greatly contributed to the vast and beautiful array of Ti/Zr-substituted POMs. However, there are still bottlenecks in Ti/Zr-substituted POM chemistry with respect to the controllable assembly of target POM structures, the exploration of novel synthetic approaches, as well as the insightful understanding of catalytic mechanisms using Ti/Zr-substituted POM catalysts. Therefore, we believe that the development of other synthetic methods, for instance, the mixed solvent diffusion method, ionothermal approach, templated modular assembly method or the combination with existing solution/hydrothermal approaches, would provide new blood to the synthetic chemistry of Ti/Zr-substituted POMs. Moreover, the ground-breaking exploration of new catalytic functionalities of these Ti/Zr-substituted POMs should also be strengthened in the future. Finally, we hope this critical review could provide research insights into the controllable design and syntheses of Ti/Zr-substituted POMs derivatives and, in the meantime, attract more researchers to join the research community of POM Chemistry or the related interdisciplinary research areas.

Author Contributions: Literature research, Z.N.; writing—original draft preparation, Z.N.; writing—review and editing, H.L. and G.Y.; funding acquisition, Z.N., H.L., and G.Y. All authors have read and agreed to the published version of the manuscript.

Funding: Project supported by Key Projects of Science and Technology of Henan Province (No. 212102210442), the National Natural Science Foundation of China (Nos. 21871025 and 21831001).

Institutional Review Board Statement: Not applicable.

Informed Consent Statement: Not applicable.

Acknowledgments: The authors would like to express their gratitude to Cambridge Crystallographic Data Centre (CCDC) for providing access to the CSD Enterprise.

Conflicts of Interest: The authors declare no conflict of interest.

References

1. Cronin, L.; Müller, A. Special POM themed issue. *Chem. Soc. Rev.* **2012**, *41*, 7325–7648.
2. Misra, A.; Kozma, K.; Streb, C.; Nyman, M. Beyond charge balance: Counter-cations in polyoxometalate chemistry. *Angew. Chem. Int. Ed.* **2020**, *59*, 596–612. [CrossRef] [PubMed]
3. Liu, J.-X.; Zhang, X.-B.; Li, Y.-L.; Huang, S.-L.; Yang, G.-Y. Polyoxometalate functionalized architectures. *Coord. Chem. Rev.* **2020**, *414*, 213260–213275. [CrossRef]
4. Pope, M.T.; Müller, A. *Polyoxometalates: From Platonic Solids to Anti-Retroviral Activity*; Kluwer Academic Publishers: Dordrecht, The Netherlands, 1994.
5. Coronado, E.; Giménez-Saiz, C.; Gómez-García, C.J. Recent advances in polyoxometalate-containing molecular conductors. *Coord. Chem. Rev.* **2005**, *249*, 1776–1796. [CrossRef]
6. Zhang, J.; Xiao, F.P.; Hao, J.; Wei, Y.G. The chemistry of organoimido derivatives of polyoxometalates. *Dalton Trans.* **2012**, *41*, 3599–3615. [CrossRef]
7. Wang, S.S.; Yang, G.Y. Recent advances in polyoxometalate-catalyzed reactions. *Chem. Rev.* **2015**, *115*, 4893–4962. [CrossRef]
8. Clemente-Juan, J.M.; Coronado, E.; Gaita-Ariño, A. Magnetic polyoxometalates: From molecular magnetism to molecular spintronics and quantum computing. *Chem. Soc. Rev.* **2012**, *41*, 7464–7478. [CrossRef]
9. Long, D.L.; Burkholder, E.; Cronin, L. Polyoxometalate clusters, nanostructures and materials: From self-assembly to designer materials and devices. *Chem. Soc. Rev.* **2007**, *36*, 105–121. [CrossRef]
10. Gupta, R.; Khan, I.; Hussain, F.; Bossoch, A.M.; Mbomekalle, I.M.; Oliveira, P.D.; Sadakane, M.; Kato, C.; Ichihashi, K.; Inoue, K.; et al. Two new sandwich-type manganese {Mn_5}-substituted polyoxotungstates: Syntheses, crystal structures, electrochemistry, and magnetic properties. *Inorg. Chem.* **2017**, *56*, 8759–8767. [CrossRef]

11. Xu, B.B.; Lu, M.; Kang, J.; Wang, D.G.; Brown, J.; Peng, Z.H. Synthesis and optical properties of conjugated polymers containing polyoxometalate clusters as side-chain pendants. *Chem. Mater.* **2005**, *17*, 2841–2851. [CrossRef]
12. Rong, C.; Pope, M.T. Lacunary polyoxometalate anions are. Pi.-acceptor ligands. Characteriza-tion of some tungstoruthenate (II, III, IV, V) heteropolyanions and their atom-transfer reactivity. *J. Am. Chem. Soc.* **1992**, *114*, 2932–2938. [CrossRef]
13. Li, C.; Jimbo, A.; Yamaguchi, K.; Suzuki, K. A protecting group strategy to access stable lacunary polyoxomolybdates for introducing multinuclear metal clusters. *Chem. Sci.* **2021**, *12*, 1240–1244. [CrossRef] [PubMed]
14. Wang, Z.-W.; Zhao, Q.; Chen, C.-A.; Sun, J.-J.; Lv, H.J.; Yang, G.-Y. Chiral {Ni$_6$PW$_9$} cluste-organic framework: Synthesis, structure, and properties. *Inorg. Chem.* **2022**, *61*, 7477–7483. [CrossRef] [PubMed]
15. Qin, L.; Wang, R.J.; Xin, X.; Zhang, M.; Liu, T.F.; Lv, H.J.; Yang, G.-Y. Synergy of nitrogen vacancies and intercalation of carbon species for enhancing sunlight photocatalytic hydrogen production of carbon nitride. *Appl. Catal. B Environ.* **2022**, *314*, 121497–121509.
16. Yonesato, K.; Yamazoe, S.; Yokogawa, D.; Yamaguchi, K.; Suzuki, K. A molecular hybrid of an atomically precise silver nanocluster and polyoxometalates for H$_2$ cleavage into protons and electrons. *Angew. Chem. Int. Ed.* **2021**, *60*, 16994–16998. [CrossRef]
17. Chen, Y.; Guo, Z.-W.; Li, X.-X.; Zheng, S.-T.; Yang, G.-Y. Multicomponent cooperative assembly of nanoscale boron-rich polyoxotungstates {B$_{30}$Si$_6$Ni$_{12}$Ln$_6$W$_{27}$(OH)$_{26}$O$_{168}$}, {B$_{30}$Si$_5$Ni$_{12}$Ln$_7$W$_{27}$(OH)$_{26}$O$_{166}$(H$_2$O)}, and {B$_{22}$Si$_4$Ni$_{12}$Ln$_4$W$_{36}$(OH)$_{12}$O$_{178}$}. *CCS Chem.* **2021**, *3*, 1232–1241.
18. Xie, S.S.; Jiang, J.J.; Wang, D.; Tang, Z.G.; Mi, R.F.; Chen, L.J.; Zhao, J.W. Tricarboxylic-ligand-decorated lanthanoid-inserted heteropolyoxometalates built by mixed-heteroatom-directing polyoxotungstate units: Syntheses, structures, and electrochemical. *Inorg. Chem.* **2021**, *60*, 7536–7544. [CrossRef]
19. Cai, J.; Ye, R.; Jia, K.; Qiao, X.; Zhao, L.; Liu, J.; Sun, W. pH-controlled construction of lanthanide clusters from lacunary polyoxometalate with single-molecule magnet behavior. *Inorg. Chem. Commun.* **2020**, *112*, 107694. [CrossRef]
20. Li, C.; Mizuno, N.; Yamaguchi, K.; Suzuki, K. Self-Assembly of Anionic Polyoxometalate–Organic Architectures Based on Lacunary Phosphomolybdates and Pyridyl Ligands. *J. Am. Chem. Soc.* **2019**, *141*, 7687–7692. [CrossRef]
21. Khajavian, R.; Jodaian, V.; Taghipour, F.; Mague, J.T.; Mirzaei, M. Roles of organic fragments in redirecting crystal/molecular structures of inorganic-organic hybrids based on lacunary Keggin-type polyoxometalates. *Molecules* **2021**, *26*, 5994. [CrossRef]
22. Zheng, S.-T.; Yang, G.-Y. Recent advances in paramagnetic-TM-substituted polyoxometalates (TM = Mn, Fe, Co, Ni, Cu). *Chem. Soc. Rev.* **2012**, *41*, 7623–7646. [CrossRef] [PubMed]
23. Zhao, J.W.; Li, Y.Z.; Chen, L.J.; Yang, G.Y. Research progress on polyoxometalate-based transition-metal–rare-earth heterometallic derived materials: Synthetic strategies, structural overview and functional applications. *Chem. Commun.* **2016**, *52*, 4418–4445. [CrossRef]
24. Knoth, W.H.; Domaille, P.J.; Roe, D.C. Halometal derivatives of W$_{12}$PO$_{40}{}^{3-}$ and related ^{183}W NMR studies. *Inorg. Chem.* **1983**, *22*, 198–201. [CrossRef]
25. Kholdeeva, O.A.; Maksimov, G.M.; Maksimovskaya, R.I.; Kovaleva, L.A.; Fedotov, M.A.; Grigoriev, V.A.; Hill, C.L. A dimeric titanium-containing polyoxometalate. Synthesis, characterization, and catalysis of H$_2$O$_2$-based thioether oxidation. *Inorg. Chem.* **2000**, *39*, 3828–3837.26. [CrossRef] [PubMed]
26. Qu, L.-Y.; Shan, Q.-J.; Gong, J.; Lu, R.-Q.; Wang, D.-R. Synthesis, properties and characterization of Dawson-type tungstophosphate heteropoly complexes substituted by titanium and peroxotitanium. *J. Chem. Soc. Dalton Trans.* **1997**, *23*, 4525–4528. [CrossRef]
27. Sakai, Y.; Kitakoga, Y.; Hayashi, K.; Yoza, K.; Nomiya, K. Isolation and molecular structure of a monomeric, tris [peroxotitanium (IV)]-substituted α-Dawson polyoxometalate derived from the tetrameric anhydride form composed of four tris [titanium (IV)]-substituted α-Dawson substructures and four bridging titanium (IV) octahedral groups. *Eur. J. Inorg. Chem.* **2004**, *23*, 4646–4652.
28. Hayashi, K.; Takahashi, M.; Nomiya, K. Novel Ti−O−Ti bonding species constructed in a metal-oxide cluster. *Dalton Trans.* **2005**, *23*, 3751–3756. [CrossRef]
29. Hayashi, K.; Murakami, H.; Nomiya, K. Novel Ti−O−Ti bonding species constructed in a metal-oxide cluster: Reaction products of bis (oxalato) oxotitanate (IV) with the dimeric, 1,2-dititanium (IV)-substituted Keggin polyoxotungstate. *Inorg. Chem.* **2006**, *45*, 8078–8085. [CrossRef]
30. Nomiya, K.; Mouri, Y.; Sakai, Y.; Matsunaga, S. Reaction products of titanium (IV) sulfate with the two, dimeric precursors, 1,2,3-tri-titanium (IV)- and 1,2-di-titanium (IV)-substituted α-Keggin polyoxometalates (POMs), under acidic conditions. A tetra-titanium (IV) oxide cluster and one coordinated sulfate ion grafted on a di-lacunary Keggin POM. *Inorg. Chem. Commun.* **2012**, *19*, 10–14.
31. An, H.Y.; Zhang, Y.M.; Hou, Y.J.; Hu, T.; Yang, W.; Chang, S.Z.; Zhang, J.J. Hybrid dimers based on metal-substituted Keggin polyoxometalates (metal = Ti, Ln) for cyanosilylation catalysis. *Dalton Trans.* **2018**, *47*, 9079–9089. [CrossRef]
32. Solé-Daura, A.; Zhang, T.; Fouilloux, H.; Robert, C.; Thomas, C.M.; Chamoreau, L.-M.; Carbó, J.J.; Proust, A.; Guillemot, G.; Poblet, J.M. Catalyst design for alkene epoxidation by molecular analogues of heterogeneous titanium-silicalite catalysts. *ACS Catal.* **2020**, *10*, 4737–4750. [CrossRef]
33. Chauveau, F.; Eberle, J.; Lefebvre, J. Synthèse de molécules élaborées à partir de métaux de transition: Un polyoxométallate mixte de tungstène VI et de zirconium IV. *Nouv. J. Chim.* **1985**, *9*, 315.
34. Villanneau, R.; Carabineiro, H.; Carrier, X.; Thouvenot, R.; Herson, P.; Lemos, F.; Ribeiro, F.R.; Che, M. Synthesis and characterization of Zr (IV) polyoxotungstates as molecular analogues of zirconia-supported tungsten catalysts. *J. Phys. Chem. B* **2004**, *108*, 12465–12471. [CrossRef]

35. Carabineiro, H.; Villanneau, R.; Carrier, X.; Herson, P.; Lemos, F.; Ribeiro, F.R.; Proust, A.; Che, M. Zirconium-substituted isopolytungstates: Structural models for zirconia-supported tungsten catalysts. *Inorg. Chem.* **2006**, *45*, 1915–1923. [CrossRef]
36. Sokolov, M.N.; Izarova, N.V.; Peresypkina, E.V.; Mainichev, D.A.; Fedin, V.P. Zirconium and hafnium aqua complexes $[(H_2O)_3M(P_2W_{17}O_{61})]^{6-}$: Synthesis, characterization and substitution of water by chiral ligand. *Inorg. Chim. Acta* **2009**, *362*, 3756–3762. [CrossRef]
37. Falber, A.; Burton-Pye, B.P.; Radivojevic, I.; Todaro, L.; Saleh, R.; Francesconi, L.C.; Drain, C.M. Ternary porphyrinato Hf^{IV} and Zr^{IV} polyoxometalate complexes. *Eur. J. Inorg. Chem.* **2009**, *2009*, 2459–2466. [CrossRef]
38. Radivojevic, I.; Ithisuphalap, K.; Burton-Pye, B.P.; Saleh, R.; Francesconi, L.C.; Drain, C.M. Ternary phthalocyanato Hf (IV) and Zr (IV) polyoxometalate complexes. *RSC Adv.* **2013**, *3*, 2174–2177. [CrossRef]
39. Lin, Y.; Weakley, T.J.R.; Rapko, B.; Finke, R.G. Polyoxoanions derived from A-β-$SiW_9O_{34}^{10-}$: Synthesis, single-crystal structural determination, and solution structural characterization by ^{183}W NMR and IR of A-β-$Si_2W_{18}Ti_6O_{77}^{14-}$. *Inorg. Chem.* **1993**, *32*, 5095–5101. [CrossRef]
40. Nomiya, K.; Takahashi, M.; Ohsawa, K.; Widegren, J.A. Synthesis and characterization of tri-titanium (IV)-1,2,3-substituted α-Keggin polyoxotungstates with heteroatoms P and Si. Crystal structure of the dimeric, Ti–O–Ti bridged anhydride form $K_{10}H_2[\alpha,\alpha-P_2W_{18}Ti_6O_{77}]\cdot17H_2O$ and confirmation of dimeric forms in aqueous solution by ultracentrifugation molecular weight measurements. *J. Chem. Soc. Dalton Trans.* **2001**, *19*, 2872–2878. [CrossRef]
41. McGlone, T.; Vila-Nadal, L.; Miras, H.N.; Long, D.L.; Poblet, J.M.; Cronin, L. Assembly of titanium embedded polyoxometalates with unprecedented structural features. *Dalton Trans.* **2010**, *39*, 11599–11604. [CrossRef]
42. Al-Kadamany, G.; Bassil, B.S.; Raad, F.; Kortz, U. The oxalato-titanium-containing tungstophosphate (V) dimers, $[Ti_8(C_2O_4)_8P_2W_{18}O_{76}(H_2O)_4]^{18-}$ and $[Ti_6(C_2O_4)_4P_4W_{32}O_{124}]^{20-}$. *J Clust. Sci.* **2014**, *25*, 867–878. [CrossRef]
43. Matsuki, Y.; Hoshino, T.; Takaku, S.; Matsunaga, S.; Nomiya, K. Synthesis and molecular structure of a water-soluble, dimeric tri-titanium (IV)-substituted Wells–Dawson polyoxometalate containing two bridging $(C_5Me_5)Rh^{2+}$ groups. *Inorg. Chem.* **2015**, *54*, 11105–11113. [CrossRef] [PubMed]
44. Nomiya, K.; Takahashi, M.; Widegren, J.A.; Aizawa, T.; Sakai, Y.; Kasuga, N.C. Synthesis and pH-variable ultracentrifugation molecular weight measurements of the dimeric, Ti–O–Ti bridged anhydride form of a novel di-Ti^{IV}-1,2-substituted α-Keggin polyoxotungstate. Molecular structure of the $[(\alpha-1,2-PW_{10}Ti_2O_{39})_2]^{10-}$ polyoxoanion. *J. Chem. Soc. Dalton Trans.* **2002**, *19*, 3679–3685. [CrossRef]
45. Goto, Y.; Kamata, K.; Yamaguchi, K.; Uehara, K.; Hikichi, S.; Mizuno, N. Synthesis, structural characterization, and catalytic performance of dititanium-substituted γ-Keggin silicotungstate. *Inorg. Chem.* **2006**, *45*, 2347–2356. [CrossRef] [PubMed]
46. Tan, R.X.; Li, D.L.; Wu, H.B.; Zhang, C.L.; Wang, X.H. Synthesis and structure of dititanium-containing 10-tungstogermanate $[\{\gamma-GeTi_2W_{10}O_{36}(OH)_2\}_2(\mu-O)_2]^{8-}$. *Inorg. Chem. Commun.* **2008**, *11*, 835–836. [CrossRef]
47. Hussain, F.; Bassil, B.S.; Kortz, U.; Kholdeeva, O.A.; Timofeeva, M.N.; de Oliveira, P.; Keita, B.; Nadjo, L. Dititanium-containing 19-tungstodiarsenate (III) $[Ti_2(OH)_2As_2W_{19}O_{67}(H_2O)]^{8-}$: Synthesis, structure, electrochemistry, and oxidation catalysis. *Chem. Eur. J.* **2007**, *13*, 4733–4742. [CrossRef]
48. Murakami, H.; Hayashi, T.; Tsukada, I.; Hasegawa, T.; Yoshida, S.; Miyano, R.; Kato, C.N.; Nomiya, K. Novel solid-state $8H^+$-heteropolyacid. Synthesis and molecular structure of a free-acid form of a Dawson-Type sandwich complex, $[Ti_2\{P_2W_{15}O_{54}(OH_2)_2\}_2]^{8-}$. *Bull. Chem. Soc. Jpn.* **2007**, *80*, 2161–2169. [CrossRef]
49. Yoshida, S.; Murakami, H.; Sakai, Y.; Nomiya, K. Syntheses, molecular structures and pH-dependent monomer-dimer equilibria of Dawson α_2-monotitanium (IV)-substituted polyoxometalates. *Dalton Trans.* **2008**, *34*, 4630–4638. [CrossRef]
50. Xu, L.J.; Zhou, W.Z.; Zhang, L.Y.; Li, B.; Zang, H.Y.; Wang, Y.H.; Li, Y.G. Organic-inorganic hybrid assemblies based on Ti-substituted polyoxometalates for photocatalytic dye degradation. *CrystEngComm* **2015**, *17*, 3708–3714. [CrossRef]
51. Gaunt, A.J.; May, J.; Collison, D.; Fox, O.D. A novel zirconium polyoxometalate complex that contains both a coordinated saturated anion, $[PMo_{12}O_{40}]^{3-}$, and a coordinated unsaturated anion, $[PMo_{11}O_{39}]^{7-}$. *Inorg. Chem.* **2003**, *42*, 5049–5051. [CrossRef]
52. Cai, L.L.; Li, Y.X.; Yu, C.J.; Ji, H.M.; Liu, Y.; Liu, S.X. Spontaneous resolution of a chiral polyoxometalate: Synthesis, crystal structures and properties. *Inorg. Chim. Acta* **2009**, *362*, 2895–2899. [CrossRef]
53. Niu, Y.J.; Liu, B.; Xue, G.L.; Hu, H.M.; Fu, F.; Wang, J.W. A new sandwich polyoxometalate based on Keggin-type monolacunary polyoxotungstoborate anion, $[Zr(\alpha-BW_{11}O_{39})_2]^{14-}$. *Inorg. Chem. Commun.* **2009**, *12*, 853–855. [CrossRef]
54. Kato, C.N.; Shinohara, A.; Hayashi, K.; Nomiya, K. Syntheses and X-ray crystal structures of zirconium (IV) and hafnium (IV) complexes containing monovacant Wells-Dawson and Keggin polyoxotungstates. *Inorg. Chem.* **2006**, *45*, 8108–8119. [CrossRef] [PubMed]
55. Fang, X.; Hill, C.L. Multiple reversible protonation of polyoxoanion surfaces: Direct observation of dynamic structural effects from proton transfer. *Angew. Chem. Int. Ed.* **2007**, *46*, 3877–3880. [CrossRef]
56. Kholdeeva, O.A.; Maksimov, G.M.; Maksimovskaya, R.I.; Vanina, M.P.; Trubitsina, T.A.; Naumov, D.Y.; Kolesov, B.A.; Antonova, N.S.; Carbó, J.J.; Poblet, J.M. Zr^{IV}-monosubstituted Keggin-type dimeric polyoxometalates: Synthesis, characterization, catalysis of H_2O_2-based oxidations, and theoretical study. *Inorg. Chem.* **2006**, *45*, 7224–7234. [CrossRef]
57. Yamaguchi, S.; Kikukawa, Y.; Tsuchida, K.; Nakagawa, Y.; Uehara, K.; Yamaguchi, K.; Mizuno, N. Synthesis and structural characterization of a γ-Keggin-type dimeric silicotungstate with a Bis (μ-hydroxo) dizirconium core $[(\gamma-SiW_{10}O_{36})_2Zr_2(\mu-OH)_2]^{10-}$. *Inorg. Chem.* **2007**, *46*, 8502–8504. [CrossRef]

58. Sokolov, M.N.; Izarova, N.V.; Peresypkina, E.V.; Virovets, A.V.; Fedin, V.P. Synthesis and structures of dinuclear Zr^{IV} and Hf^{IV} hydroxo complexes with the monolacunar Keggin and Dawson anions. *Russ. Chem. Bull.* **2009**, *58*, 507–512. [CrossRef]
59. Villanneau, R.; Racimor, D.; Messner-Henning, E.; Rousselière, H.; Picart, S.; Thouvenot, R.; Proust, A. Insights into the coordination chemistry of phosphonate derivatives of heteropolyoxotungstates. *Inorg. Chem.* **2011**, *50*, 1164–1166. [CrossRef]
60. Villanneau, R.; Djamâa, A.B.; Chamoreau, L.-M.; Gontard, G.; Proust, A. Bisorganophosphonyl and -organoarsenyl derivatives of heteropolytungstates as hard ligands for early-transition-metal and lanthanide cations. *Eur. J. Inorg. Chem.* **2013**, *2013*, 1815–1820. [CrossRef]
61. Fang, X.; Anderson, T.M.; Hill, C.L. Enantiomerically pure polytungstates: Chirality transfer through zirconium coordination centers to nanosized inorganic clusters. *Angew. Chem. Int. Ed.* **2005**, *44*, 3540–3544. [CrossRef]
62. Leclerc-Laronze, N.; Marrot, J.; Haouas, M.; Taulelle, F.; Cadot, E. Slow-proton dynamics within a zirconium-containing sandwich-like complex based on the trivacant anion α-$[SiW_9O_{34}]^{10-}$ synthesis, structure and NMR spectroscopy. *Eur. J. Inorg. Chem.* **2008**, *31*, 4920–4926. [CrossRef]
63. Chen, L.L.; Li, L.L.; Liu, B.; Xue, G.L.; Hu, H.M.; Fu, F.; Wang, J.W. A zirconium-containing sandwich-type dimer based on trivacant α- and β-$[GeW_9O_{34}]^{10-}$ units, $[Zr_3O(OH)_2(\alpha\text{-}GeW_9O_{34})(\beta\text{-}GeW_9O_{34})]^{12-}$. *Inorg. Chem. Commun.* **2009**, *12*, 1035–1037. [CrossRef]
64. Saku, Y.; Sakai, Y.; Shinohara, A.; Hayashi, K.; Yoshida, S.; Kato, C.N.; Yoza, K.; Nomiya, K. Sandwich-type Hf^{IV} and Zr^{IV} complexes composed of tri-lacunary Keggin polyoxometalates: Structure of $[M_3(\mu\text{-}OH)_3(A\text{-}\alpha\text{-}PW_9O_{34})_2]^{9-}$ (M = Hf and Zr). *Dalton Trans.* **2009**, *5*, 805–813. [CrossRef]
65. Wei, K.Y.; Yang, T.; Qin, S.J.; Ma, X.; Li, X.X.; Yang, G.Y. Hydrothermal synthesis, structural characterization and proton-conducting property of a 3-D framework based on Zr_3Na_3-substituted polyoxometalate building blocks. *Chin. J. Struct. Chem.* **2016**, *35*, 1461–1468.
66. Zhang, Z.; Zhao, J.-W.; Yang, G.-Y. Tri-Zr^{IV} substituted sandwiched polyoxometalate with mixed trilacunary Keggin-/Dawson-type polyoxotungstate units. *Eur. J. Inorg. Chem.* **2017**, *2017*, 3244–3247. [CrossRef]
67. Gaunt, A.J.; May, I.; Collison, D.; Travis Holman, K.; Pope, M.T. Polyoxometal cations within polyoxometalate anions. Seven-coordinate uranium and zirconium heteroatom groups in $[(UO_2)_{12}(\mu_3\text{-}O)_4(\mu_2\text{-}H_2O)_{12}(P_2W_{15}O_{56})_4]^{32-}$ and $[Zr_4(\mu_3\text{-}O)_2(\mu_2\text{-}OH)_2(H_2O)_4(P_2W_{16}O_{59})_2]^{14-}$. *J. Mol. Struct.* **2003**, *656*, 101–106. [CrossRef]
68. Fang, X.; Anderson, T.M.; Hou, Y.; Hill, C.L. Stereoisomerism in polyoxometalates: Structural and spectroscopic studies of bis(malate)-functionalized cluster systems. *Chem. Commun.* **2005**, *40*, 5044–5046. [CrossRef] [PubMed]
69. Li, D.; Han, H.Y.; Wang, Y.G.; Wang, X.; Li, Y.G.; Wang, E.B. Modification of tetranuclear zirconium-substituted polyoxometalates-syntheses, structures, and peroxidase-like catalytic activities. *Eur. J. Inorg. Chem.* **2013**, *2013*, 1926–1934. [CrossRef]
70. Kikukawa, Y.; Yamaguchi, S.; Tsuchida, K.; Nakagawa, Y.; Uehara, K.; Yamaguchi, K.; Mizuno, N. Synthesis and catalysis of di- and tetranuclear metal sandwich-type silicotungstates $[(\gamma\text{-}SiW_{10}O_{36})_2M_2(\mu\text{-}OH)_2]^{10-}$ and $[(\gamma\text{-}SiW_{10}O_{36})_2M_4(\mu_4\text{-}O)(\mu\text{-}OH)_6]^{8-}$ (M = Zr or Hf). *J. Am. Chem. Soc.* **2008**, *130*, 5472–5478. [CrossRef]
71. Chen, L.L.; Liu, Y.; Chen, S.H.; Hu, H.M.; Fu, F.; Wang, J.W.; Xue, G.L. Acetate-functionalized zirconium-substituted tungstogermanate, $[Zr_4O_2(OH)_2(CH_3COO)_2(\alpha\text{-}GeW_{10}O_{37})_2]^{12-}$. *J. Clust. Sci.* **2009**, *20*, 331–340. [CrossRef]
72. Zhang, W.; Liu, S.-X.; Zhang, C.-D.; Tan, R.-K.; Ma, F.-J.; Li, S.-J.; Zhang, Y.-Y. An acetate-functionalized tetranuclear zirconium sandwiching polyoxometalate complex. *Eur. J. Inorg. Chem.* **2010**, *2010*, 3473–3477. [CrossRef]
73. Zhang, Z.; Wang, Y.-L.; Yang, G.-Y. Two inorganic-organic hybrid polyoxotungstates constructed from tetra-Zr^{IV}-substituted sandwich-type germanotungstates functionalized by tris ligand. *Inorg. Chem. Commun.* **2017**, *85*, 32–36. [CrossRef]
74. Ni, Z.-H.; Zhang, Z.; Yang, G.-Y. Two new tetra-Zr (IV)-substituted sandwich-type polyoxometalates functionalized by different organic amine ligands. *J. Clust. Sci.* **2018**, *29*, 1185–1191. [CrossRef]
75. Ni, Z.-H.; Li, H.-L.; Li, X.-Y.; Yang, G.-Y. Zr_4-substituted polyoxometalate dimers decorated by D-tartaric acid/glycolic acid: Syntheses, structures and optical/electrochemical properties. *CrystEngComm* **2019**, *21*, 876–883. [CrossRef]
76. Wang, Y.-L.; Zhang, Z.; Li, H.-L.; Li, X.-Y.; Yang, G.-Y. A new oxalate-functionalized tetra-Zr^{IV}-substituted sandwich-type silicotungstate: Hydrothermal synthesis, structural characterization and catalytic oxidation of thioethers. *Eur. J. Inorg. Chem.* **2019**, *2019*, 417–422. [CrossRef]
77. Wang, Y.L.; Zhao, J.W.; Zhang, Z.; Sun, J.J.; Li, X.Y.; Yang, B.F.; Yang, G.Y. Enantiomeric polyoxometalates based on malate chirality-inducing tetra-Zr^{IV}-substituted Keggin dimeric clusters. *Inorg. Chem.* **2019**, *58*, 4657–4664. [CrossRef]
78. Al-Kadamany, G.; Mal, S.S.; Milev, B.; Donoeva, B.G.; Maksimovskaya, R.I.; Kholdeeva, O.A.; Kortz, U. Hexazirconium- and hexahafnium-containing tungstoarsenates (III) and their oxidation catalysis properties. *Chem. Eur. J.* **2010**, *16*, 11797–11800. [CrossRef]
79. Bassil, S.B.; Dickman, M.H.; Kortz, U. Synthesis and structure of asymmetric zirconium-substituted silicotungstates, $[Zr_6O_2(OH)_4(H_2O)_3(\beta\text{-}SiW_{10}O_{37})_3]^{14-}$ and $[Zr_4O_2(OH)_2(H_2O)_4(\beta\text{-}SiW_{10}O_{37})_2]^{10-}$. *Inorg. Chem.* **2006**, *45*, 2394–2396. [CrossRef]
80. Zhang, Z.; Li, H.L.; Wang, Y.L.; Yang, G.Y. Syntheses, structures, and electrochemical properties of three new acetate-functionalized zirconium-substituted germanotungstates: From dimer to tetramer. *Inorg. Chem.* **2019**, *58*, 2372–2378. [CrossRef]
81. Bassil, B.S.; Mal, S.S.; Dickman, M.H.; Kortz, U.; Oelrich, H.; Walder, L. 6-peroxo-6-zirconium crown and its hafnium analogue embedded in a triangular polyanion: $[M_6(O_2)_6(OH)_6(\gamma\text{-}SiW_{10}O_{36})_3]^{18-}$ (M = Zr, Hf). *J. Am. Chem. Soc.* **2008**, *130*, 6696–6697. [CrossRef]

82. Al-Kadamany, G.A.; Hussain, F.; Mal, S.S.; Dickman, M.H.; Leclerc-Laronze, N.; Marrot, J.; Cadot, E.; Kortz, U. Cyclic Ti$_9$ Keggin trimers with tetrahedral (PO$_4$) or octahedral (TiO$_6$) capping groups. *Inorg. Chem.* **2008**, *47*, 8574–8576. [CrossRef] [PubMed]
83. Nsouli, N.H.; Bassil, B.S.; Dickman, M.H.; Kortz, U.; Keita, B.; Nadjo, L. Synthesis and structure of dilacunary decatungstogermanate, [γ-GeW$_{10}$O$_{36}$]$^{8-}$. *Inorg. Chem.* **2006**, *45*, 3858–3860. [CrossRef]
84. Ren, Y.-H.; Liu, S.-X.; Cao, R.-G.; Zhao, X.-Y. Cao, J.-F.; Gao, C.-Y. Two trimeric tri-Ti^{IV}-substituted Keggin tungstogermanates based on tetrahedral linkers. *Inorg. Chem. Commun.* **2008**, *11*, 1320–1322. [CrossRef]
85. Hoshino, T.; Isobe, R.; Kaneko, T.; Matsuki, Y.; Nomiya, K. Synthesis and molecular structure of a novel compound containing a carbonate-bridged hexacalcium cluster cation assembled on a trimeric trititanium (IV)-substituted Wells-Dawson polyoxometalate. *Inorg. Chem.* **2017**, *56*, 9585–9593. [CrossRef]
86. Kortz, U.; Hamzeh, S.S.; Nasser, N.A. Supramolecular structures of titanium (IV)-substituted Wells-Dawson polyoxotungstates. *Chem. Eur. J.* **2003**, *9*, 2945–2952. [CrossRef]
87. Nishikawa, T.; Sasaki, Y. The crystal structure of monium dodecamolybdotetraarsenate (V) tetrahydrate, (NH$_4$)$_4$H$_4$As$_4$Mo$_{12}$O$_{50}$·4H$_2$O. *Chem. Lett.* **1975**, *4*, 1185–1186. [CrossRef]
88. Sakai, Y.; Yoza, K.; Katoa, C.N.; Nomiya, K. A first example of polyoxotungstate-based giant molecule. Synthesis and molecular structure of a tetrapod-shaped Ti–O–Ti bridged anhydride form of Dawson tri-titanium (IV)-substituted polyoxotungstate. *Dalton Trans.* **2003**, *18*, 3581–3586. [CrossRef]
89. Sakai, Y.; Ohta, S.; Shintoyo, Y.; Yoshida, S.; Taguchi, Y.; Matsuki, Y.; Matsunaga, S.; Nomiya, K. Encapsulation of anion/cation in the central cavity of tetrameric polyoxometalate, composed of four trititanium (IV)-substituted α-Dawson subunits, initiated by protonation/deprotonation of the bridging oxygen atoms on the intramolecular surface. *Inorg. Chem.* **2011**, *50*, 6575–6583. [CrossRef]
90. Sakai, Y.; Yoza, K.; Kato, C.N.; Nomiya, K. Tetrameric, trititanium (IV)-substituted polyoxo-tungstates with an α-Dawson substructure as soluble metal-oxide analogues: Molecular structure of the Giant "Tetrapod" [(α-1,2,3-P$_2$W$_{15}$Ti$_3$O$_{62}$)$_4${μ$_3$-Ti(OH)$_3$}$_4$Cl]$^{45-}$. *Chem. Eur. J.* **2003**, *9*, 4077–4083. [CrossRef]
91. Sakai, Y.; Yoshida, S.; Hasegawa, T.; Murakami, H.; Nomiya, K. Tetrameric, tri-titanium (IV)-substituted polyoxometalates with an α-Dawson substructure as soluble metal analogues. synthesis and molecular structure of three giant "tetrapods" encapsulating different anions (Br$^-$, I$^-$, and NO$_3^-$). *Bull. Chem. Soc.* **2007**, *80*, 1965–1974. [CrossRef]
92. Hussain, F.; Bassil, B.S.; Bi, L.H.; Reicke, M.; Kortz, U. Structural control on the nanomolecular scale: Self-assembly of the polyoxotungstate wheel [{β-Ti$_2$SiW$_{10}$O$_{39}$}$_4$]$^{24-}$. *Angew. Chem. Int. Ed.* **2004**, *43*, 3485–3488. [CrossRef] [PubMed]
93. Wang, K.-Y.; Bassil, B.S.; Lin, Z.-G.; Haider, A.; Cao, J.; Stephan, H.; Viehweger, K.; Kortz, U. Ti$_7$-containing, tetrahedral 36-tungsto-4-arsenate (III) [Ti$_6$(TiO$_6$)(AsW$_9$O$_{33}$)$_4$]$^{20-}$. *Dalton Trans.* **2014**, *43*, 16143–16146. [CrossRef] [PubMed]
94. Li, H.-L.; Lian, C.; Yin, D.-P.; Jia, Z.-Y.; Yang, G.-Y. A new hepta-nuclear Ti-oxo-cluster-substituted tungstoantimonate and its catalytic oxidation of thioethers. *Cryst. Growth Des.* **2019**, *19*, 376–380. [CrossRef]
95. Li, H.-L.; Lian, C.; Yang, G.-Y. A ring-shaped 12-Ti-substituted poly (polyoxometalate): Synthesis, structure, and catalytic properties. *Sci. China Chem.* **2022**, *65*, 892–897. [CrossRef]
96. Chen, W.-C.; Yan, L.-K.; Wu, C.-X.; Wang, X.-L.; Shao, K.-Z.; Su, Z.-M.; Wang, E.-B. Assembly of Keggin-/Dawson-type polyoxotungstate clusters with different metal units and SeO$_3^{2-}$ heteroanion templates. *Cryst. Growth Des.* **2014**, *14*, 5099–5110. [CrossRef]
97. Li, H.L.; Wang, Y.L.; Zhang, Z.; Yang, B.F.; Yang, G.Y. A new tetra-Zr (IV)-substituted polyoxotungstate aggregate. *Dalton Trans.* **2018**, *47*, 14017–14024. [CrossRef]
98. Zhang, Z.; Wang, Y.L.; Yang, G.Y. An unprecedented Zr containing polyoxometalate tetramer with mixed trilacunary/dilacunary Keggin-type polyoxotungstate units. *Acta Crystallogr. Sect. C Struct. Chem.* **2018**, *74*, 1284–1288. [CrossRef]
99. Zhang, Z.; Wang, Y.L.; Liu, Y.; Hang, S.-L.; Yang, G.Y. Three ring-shaped Zr (IV)-substituted silicotungstates: Syntheses, structures and their properties. *Nanoscale* **2020**, *12*, 18333–18341. [CrossRef]
100. Li, H.L.; Lian, C.; Yin, D.P.; Yang, G.Y. Three Zr (IV)-substituted polyoxotungstate aggregates: Structural transformation from tungstoantimonate to tungstophosphate induced by pH. *Inorg. Chem.* **2020**, *59*, 12842–12849. [CrossRef]
101. Zhang, P.Y.; Wang, Y.; Yao, L.Y.; Yang, G.Y. Hepta-Zr-incorporated polyoxometalate assembly. *Inorg. Chem.* **2022**, *61*, 10410–10416. [CrossRef]
102. Huang, L.; Wang, S.S.; Zhao, J.W.; Cheng, L.; Yang, G.Y. Synergistic combination of multi-Zr^{IV} cations and lacunary Keggin germanotungstates leading to a gigantic Zr$_{24}$-cluster-substituted polyoxometalate. *J. Am. Chem. Soc.* **2014**, *136*, 7637–7642. [CrossRef] [PubMed]
103. Chou, P.-T.; Chi, Y.; Chung, M.-W.; Lin, C.-C. Harvesting luminescence via harnessing the photophysical properties of transition metal complexes. *Coord. Chem. Rev.* **2011**, *255*, 2653–2665. [CrossRef]
104. Ma, D.-L.; Ma, V.P.-Y.; Chan, D.S.-H.; Leung, K.-H.; He, H.-Z.; Leung, C.-H. Recent advances in luminescent heavy metal complexes for sensing. *Coord. Chem. Rev.* **2012**, *256*, 3087–3113. [CrossRef]
105. Daniel, C. Photochemistry and photophysics of transition metal complexes: Quantum chemistry. *Coord. Chem. Rev.* **2015**, *282–283*, 19–32. [CrossRef]
106. Gómez-Coca, S.; Aravena, D.; Morales, R.; Ruiz, E. Large magnetic anisotropy in mononuclear metal complexes. *Coord. Chem. Rev.* **2015**, *289–290*, 379–392. [CrossRef]

107. Antonova, N.S.; Carbo, J.J.; Kortz, U.; Kholdeeva, O.A.; Poblet, J.M. Mechanistic insights into alkene epoxidation with H_2O_2 by Ti-and other TM-containing polyoxometalates: Role of the metal nature and coordination environment. *J. Am. Chem. Soc.* **2010**, *132*, 7488–7497. [CrossRef]
108. Kholdeeva, O.A.; Kovaleva, L.A.; Maksimovskaya, R.I.; Maksimov, G.M. Kinetics and mechanism of thioether oxidation with H_2O_2 in the presence of Ti (IV)-substituted heteropolytungstates. *J. Mol. Catal. A Chem.* **2000**, *158*, 223–229. [CrossRef]
109. Kholdeeva, O.A.; Trubitsina, T.A.; Maksimovskaya, R.I.; Golovin, A.V.; Neiwert, W.A.; Kolesov, B.A.; López, X.; Poblet, J.M. First isolated active titanium peroxo complex: Characterization and theoretical study. *Inorg. Chem.* **2004**, *43*, 2284–2292. [CrossRef]
110. Kholdeeva, O.A.; Trubitsina, T.A.; Maksimov, G.M.; Golovin, A.V.; Maksimovskaya, R.I. Synthesis, characterization, and reactivity of Ti (IV)-monosubstituted Keggin polyoxometalates. *Inorg. Chem.* **2005**, *44*, 1635–1642. [CrossRef]
111. Kholdeeva, O.A.; Trubitsina, T.A.; Timofeeva, M.N.; Maksimov, G.M.; Maksimovskaya, R.I.; Rogov, V.A. The role of protons in cyclohexene oxidation with H_2O_2 catalysed by Ti (IV)-monosubstituted Keggin polyoxometalate. *J. Mol. Catal. A Chem.* **2005**, *232*, 173–178. [CrossRef]
112. Kholdeeva, O.A.; Maksimovskaya, R.I. Titanium- and zirconium-monosubstituted polyoxometalates as molecular models for studying mechanisms of oxidation catalysis. *J. Mol. Catal. A Chem.* **2007**, *262*, 7–24. [CrossRef]
113. Donoeva, B.G.; Trubitsina, T.A.; Antonova, N.S.; Carbó, J.J.; Poblet, J.M.; Al-Kadamany, G.; Kortz, U.; Kholdeeva, O.A. Epoxidation of alkenes with H_2O_2 catalyzed by dititanium-containing 19-tungstodiarsenate (III): Experimental and theoretical studies. *Eur. J. Inorg. Chem.* **2010**, *2010*, 5312–5317. [CrossRef]
114. Jimenez-Lozano, P.; Ivanchikova, I.D.; Kholdeeva, O.A.; Poblet, J.M.; Carbo, J.J. Alkene oxidation by Ti-containing polyoxometalates. unambiguous characterization of the role of the protonation state. *Chem. Commun.* **2012**, *48*, 9266–9268. [CrossRef]
115. Jimenez-Lozano, P.; Skobelev, I.Y.; Kholdeeva, O.A.; Poblet, J.M.; Carbo, J.J. Alkene epoxidation catalyzed by Ti-containing polyoxometalates: Unprecedented beta-oxygen transfer mechanism. *Inorg. Chem.* **2016**, *55*, 6080–6084. [CrossRef] [PubMed]
116. Zhang, T.; Mazaud, L.; Chamoreau, L.-M.; Paris, C.; Proust, A.; Guillemot, G. Unveiling the active surface sites in heterogeneous titanium-based silicalite epoxidation catalysts: Input of silanol-functionalized polyoxotungstates as soluble analogues. *ACS Catal.* **2018**, *8*, 2330–2342. [CrossRef]
117. Zhang, T.; Solé-Daura, A.; Fouilloux, H.; Poblet, J.M.; Proust, A.; Carbó, J.J.; Guillemot, G. Reaction pathway discrimination in alkene oxidation reactions by designed Ti-siloxy-polyoxometalates. *ChemCatChem* **2021**, *13*, 1220–1229. [CrossRef]
118. Skobelev, I.Y.; Zalomaeva, O.V.; Kholdeeva, O.A.; Poblet, J.M.; Carbo, J.J. Mechanism of thioether oxidation over di- and tetrameric Ti centres: Kinetic and DFT studies based on model Ti-containing polyoxometalates. *Chem. Eur. J.* **2015**, *21*, 14496–14506. [CrossRef]
119. Wang, K.-Y.; Lin, Z.G.; Bassil, B.S.; Xing, X.L.; Haider, A.; Keita, B.; Zhang, G.J.; Silvestru, C.; Kortz, U. Ti_2–containing 18-tungsto-2-arsenate (III) monolacunary host and the incorporation of a phenylantimony (III) guest. *Inorg. Chem.* **2015**, *54*, 10530–10532. [CrossRef]
120. Dupré, N.; Rémy, P.; Micoine, K.; Boglio, C.; Thorimbert, S.; Lacôte, E.; Hasenknopf, B.; Malacria, M. Chemoselective catalysis with organosoluble Lewis acidic polyoxotungstates. *Chem. Eur. J.* **2010**, *16*, 7256–7264. [CrossRef]
121. Absillis, G.; Parac-Vogt, T.N. Peptide bond hydrolysis catalyzed by the Wells–Dawson $Zr(\alpha_2-P_2W_{17}O_{61})_2$ polyoxometalate. *Inorg. Chem.* **2012**, *51*, 9902–9910. [CrossRef]
122. Stroobants, K.; Absillis, G.; Moelants, E.; Proost, P.; Parac-Vogt, T.N. Regioselective hydrolysis of human serum albumin by Zr (IV)-substituted polyoxotungstates at the interface of positively charged protein surface patches and negatively charged amino acid residues. *Chem. Eur. J.* **2014**, *20*, 3894–3897. [CrossRef] [PubMed]
123. Stroobants, K.; Goovaerts, V.; Absillis, G.; Bruylants, G.; Moelants, E.; Proost, P.; Parac-Vogt, T.N. Molecular origin of the hydrolytic activity and fixed regioselectivity of a Zr (IV)-substituted polyoxotungstate as artificial protease. *Chem. Eur. J.* **2014**, *20*, 9567–9577. [CrossRef] [PubMed]
124. Ly, H.G.; Absillis, G.; Janssens, R.; Proost, P.; Parac-Vogt, T.N. Highly amino acid selective hydrolysis of myoglobin at aspartate residues as promoted by zirconium(IV)-substituted polyoxometalates. *Angew. Chem. Int. Ed.* **2015**, *54*, 7391–7394. [CrossRef] [PubMed]
125. Ly, H.G.; Mihaylov, T.; Absillis, G.; Pierloot, K.; Parac-Vogt, T.N. Reactivity of dimeric tetrazirconium (IV) Wells-Dawson polyoxometalate toward dipeptide hydrolysis studied by a combined experimental and density functional theory approach. *Inorg. Chem.* **2015**, *54*, 11477–11492. [CrossRef] [PubMed]
126. Sap, A.; De Zitter, E.; Van Meervelt, L.; Parac-Vogt, T.N. Structural characterization of the complex between hen egg-white lysozyme and Zr(IV) -substituted Keggin polyoxometalate as artificial protease. *Chem. Eur. J.* **2015**, *21*, 11692–11695. [CrossRef]

Review

Recent Advances of Anderson-Type Polyoxometalates as Catalysts Largely for Oxidative Transformations of Organic Molecules

Zheyu Wei [1], Jingjing Wang [2], Han Yu [1,2,*], Sheng Han [2,*] and Yongge Wei [1,*]

[1] Department of Chemistry, Tsinghua University, Beijing 100062, China
[2] School of Chemistry and Environmental Engineering, Shanghai Institute of Technology, Shanghai 201418, China
* Correspondence: hanyu0220@tsinghua.edu.cn (H.Y.); hansheng654321@sina.com (S.H.); yonggewei@tsinghua.edu.cn (Y.W.)

Abstract: Anderson-type ($[XM_6O_{24}]^{n-}$) polyoxometalates (POMs) are a class of polymetallic-oxygen cluster inorganic compounds with special structures and properties. They have been paid extensive attention by researchers now, due to their chemical modification and designability, which have been widely applied in the fields of materials, catalysis and medicine. In contemporary years, the application of Anderson-type POMs in catalytic organic oxidation reaction has gradually shown great significance for the research of green catalytic process. In this paper, we investigate the application of Anderson-type POMs in organic synthesis reaction, and these works are summarized according to the different structure of POMs. This will provide a new strategy for further investigation of the catalytic application of Anderson-type POMs and the study of green catalysis.

Keywords: polyoxometalate; Anderson structure; organic synthesis; molecular catalysis; green chemistry

Citation: Wei, Z.; Wang, J.; Yu, H.; Han, S.; Wei, Y. Recent Advances of Anderson-Type Polyoxometalates as Catalysts Largely for Oxidative Transformations of Organic Molecules. *Molecules* **2022**, *27*, 5212. https://doi.org/10.3390/molecules27165212

Academic Editor: Xiaobing Cui

Received: 13 July 2022
Accepted: 11 August 2022
Published: 16 August 2022

Publisher's Note: MDPI stays neutral with regard to jurisdictional claims in published maps and institutional affiliations.

Copyright: © 2022 by the authors. Licensee MDPI, Basel, Switzerland. This article is an open access article distributed under the terms and conditions of the Creative Commons Attribution (CC BY) license (https://creativecommons.org/licenses/by/4.0/).

1. Introduction

The Anderson type of polyoxometalates (abbreviated as POMs) are an important class of structures in oxygen-bridged polymetallic cluster compounds, which can be expressed as $[XM_6O_{24}]^{n-}$ or $[H_x(XO_6)M_6O_{18}]^{n-}$ (M=Mo or W); its central heteroatom X can be replaced with the majority of transition metal atoms like Fe, Co, Cu, Ni and Mn, etc. The stable configuration of Anderson-type POMs is a flat and circular structure (Figure 1a,b). There are three different types of oxygen atoms on the surface of Anderson-type POMs, with different coordination modes, so the reactivity of different sites on the surface is fairly different. Additionally, the class of protonation of the surface oxygen atoms of the POMs are also significantly varied, which can be divided into the following two types (Figure 1c): one non-protonation μ_3-O on the surface of Anderson-type POMs called type A; the other possesses protonated μ_3-O called type B. Provided that the valence state of the central atom is high (oxidation state > 4), Anderson-type POMs are also found to exist in a fold isomer configuration. This is similar to ammonium heptamolybdate (Figure 1d,e). However, for the central metal atoms in the low valence states, this structure requires organic ligand protection to keep steady (Figure 1f) [1–5].

The oxygen atoms of Anderson-type POMs are surrounded by six octahedrons around the central hetero-atoms by sharing an edge to form a common planar ring structure, otherwise a folded-twisted structure. Moreover, the central hetero-atoms are varied and modified easily, so the structure has a high coordination activity. Their structure and properties can be more multifarious by further modification.

Classical structures of POMs are mostly uniform particles with nano, otherwise sub-nano, size. Due to the octahedral connection and easily modified characteristics of Anderson-type POMs, this leads them to become a good sub-nano building unit. They can be used to design and synthesize a variety of special sizes and properties of organic–inorganic POMs compounds which show exceedingly important application value in materials, medicine, catalysis and other fields.

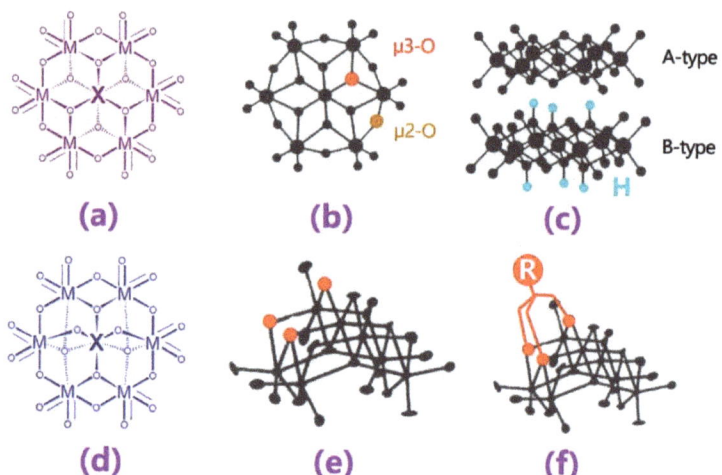

Figure 1. (a) Structure diagram of α-Anderson-type POMs; (b,c) Ball and stick model of α-Anderson-type POMs; (d) Structure diagram of β-Anderson-type POMs; (e) Ball and stick model of β-Anderson-type POMs; (f) Organic modified β-Anderson-type POMs.

The previous research about Anderson-type POMs mainly attach great importance to the synthesis and preparation of new compounds and characterization, etc.; the Wang group, Wei group, Zhou group, Niu group and Kreb group have made an army of contributions in this field.

In 2014, Wei groups developed a liquid-phase diffusion single-crystal growth sample addition method and a special and sample addition tube [6]. The invention can significantly improve the working efficiency of the single-crystal cultivation, and the quality of the obtained single crystal by this method is also significantly improved. The Wei group designed a large number of different Anderson-type POMs and crystal of their organic derivatives via this approach (Figure 2). They afterward defined a series of structure of new Anderson-type POMs and the relevant derivatives [7–13] through single-crystal X-ray diffraction, infrared spectrum, NMR spectrum and liquid chromatography, which greatly promoted the development of synthesis Anderson-type POMs and structural chemistry.

With the structural development of Anderson-type POMs, a mounting number of researchers began to study the modification of POMs which controlled the application in the catalytic field [14–19]. These studies illustrate that Anderson-type POMs not only possess acid-based properties, but also excellent redox performance for the contributor that is present in central atoms. Owing to their structural stability, they can also be used as catalysts and applied in homogeneous and heterogeneous reactions, and even through phase-transfer catalyst or ionic liquid catalyst after appropriate chemical modification. Therefore, Anderson-type POMs are regarded as new sub-nano molecules with high catalytic activity. Nowadays, the research about Anderson-type POMs materials and their functions has gradually become a hot field in polyacid chemistry [20–25]. Nevertheless, the early work in the catalytic application of Anderson-type POMs mainly focused on desulfurization of oxidative and the treatment of industrial wastewater [26–30]. The applications in organic reaction are still the confronting great challenge.

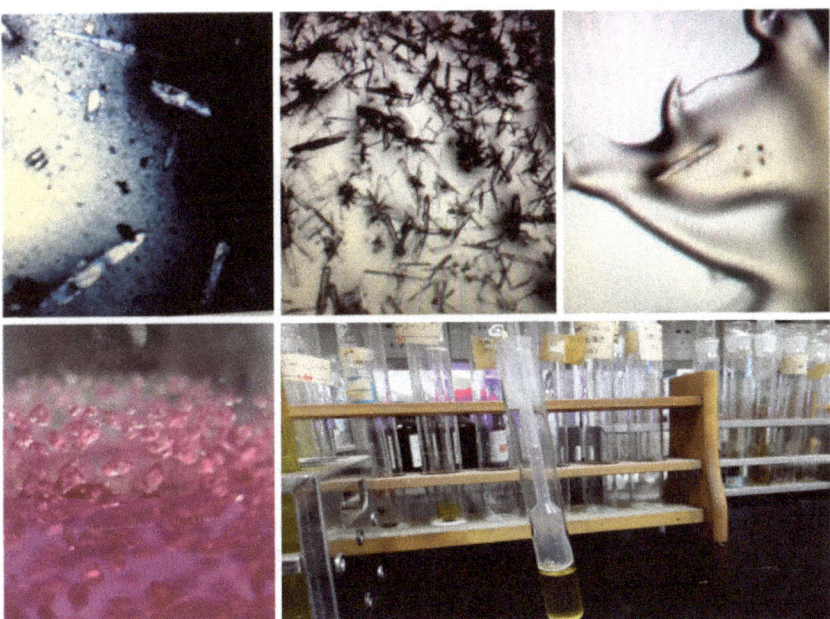

Figure 2. Several Anderson-type POMs crystals prepared by the special sample tube for liquid-phase diffusion technique [6–13].

It is universally acknowledged that progress of society depends on the development of organic synthetic chemistry. Reducing or avoiding producing harmful products is the question of organic synthesis reaction and the focus of "green" chemical research [31–34]. With the development of Anderson-type POMs in the catalytic reaction, the researchers found that the replacement of a traditional metal or non-metal catalyst is feasible with Anderson-type POMs or their derivatives [35–51]. This system not only can improve the catalytic efficiency, but also generate less by-product in the process. This is exactly an environmentally friendly, green and efficient catalytic system.

The variety of Anderson-type POMs catalysts have been reported. This review summarizes previous research studies based on recent literature. These works are roughly divided into two major categories according to the anion structure (Figure 3). For convenience of description, we define the relevant letter for the following categories: (1) The catalytic application of simple Anderson-type POMs (abbreviated as P); (2) Catalytic application of organic modified Anderson-type POMs derivatives (abbreviated as PO). In these two categories, the catalysts are further divided into two minor categories according to the cationic form of Anderson-type POMs: (1) Simple inorganic cations (such as $NH_4^+.Na^+.H^+.K^+$, the following is abbreviated as $[I]^+$); (2) Organic cations ($[O]^+$).

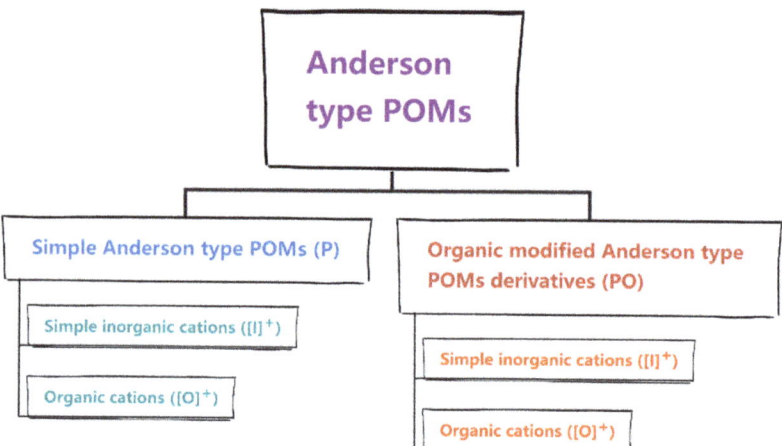

Figure 3. Tree-structure classification of Anderson-type POMs in this paper.

2. The Application of Anderson-Type POMs in Catalytic Synthesis

2.1. Catalyzation of Anderson-Type POMs (P)

[I]$^+$P

In 2016, Kadam and Gokavi investigated the dynamics and catalytic mechanism of the Anderson-type POMs, chromium (III) molybdate catalyst, which originally formed via the reaction between [O=Cr^V(OH)$_6$Mo$_6$O$_{18}$]$^{3-}$ from the oxidation of H[Cr^{III}(OH)$_6$Mo$_6$O$_{18}$]$^{2-}$ and hydrazide without free radicals in the course of the reaction process. This work provides good ideas and inspiration for oxidation hydrazide by Anderson-type POMs [35].

In 2017, the Wei group reported a simple, mild and efficient method of aerobic oxidation amine by inorganic ligand-loaded non-precious metal catalyst (NH$_4$)$_n$[MMo$_6$O$_{18}$(OH)$_6$] (M=Cu^{2+}; Fe^{3+}; Co^{3+}; Ni^{2+}; Zn^{2+}, n = 3 or 4) in water at 100 °C via one step, demonstrating that the catalytic activity and selectivity can be significantly improved by transforming the central metal atoms. In the presence of Anderson-type POMs, with O$_2$ as the unique oxidant, the catalytic oxidation of fundamental amine and secondary amines, and the coupling reaction of alcohol and amines can be achieved to generate the relevant imines. This new catalytic system provides a new method for catalytic oxidation reaction through inorganic ligand-supported metal catalysts [36].

Soon after, Wei and collaborators proposed a Cu-based Anderson-type POM, (NH$_4$)$_4$[Cu(OH)$_6$Mo$_6$O$_{18}$] used for oxidation of carboxylic acid to aldehyde in water (Figure 4). This system used oxygen as the sole oxidant in the mild condition and was also suitable for various aldehyde derivatives with different functional groups [37]. In this catalytic process, the catalyst (NH$_4$)$_4$[Cu(OH)$_6$Mo$_6$O$_{18}$] can be recycled at least six times and maintain high catalytic activity. This method of preparing for carboxylic acid is not only simple to operate, but also avoids the use of expensive and toxic raw materials. The versatility of this catalytic route makes it possible for industrial applications.

In 2018, the Wei Group used inorganic ligand-supported Fe-based Anderson-type POMs (NH$_4$)$_3$[Fe(OH)$_6$Mo$_6$O$_{18}$] to prepare imines by aerobic oxidation of aldehyde/ketone and amine oxidation coupling with oxygen as the sole oxidation [38]. The catalyst is comprised of an inexpensive system (NH$_4$)$_6$Mo$_7$O$_{24}$·4H$_2$O and Fe$_2$(SO$_4$)$_3$. This system can be applied in broad substrates with low loss of catalytic activity, proving that this catalytic system has great potential in catalytic chemical transformation. Additionally, the stably inorganic skeleton for the catalyst provides good stability and is reusable in the course of the reaction process; for this reason, the catalytic oxidation of halide and amine can be easily improved to the gram level and has the potential to apply in industry.

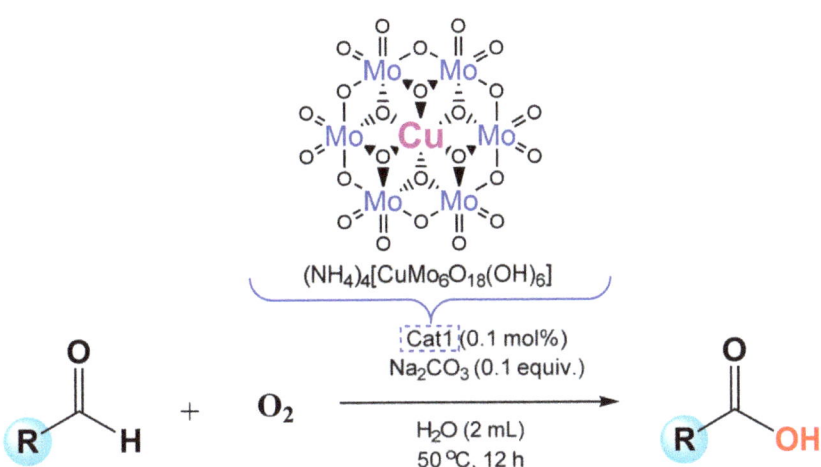

Figure 4. A method for preparing carboxylic acids by oxidizing aldehydes in water using $(NH_4)_4[Cu(OH)_6Mo_6O_{18}]$.

After that, the Wei group published an inorganic ligand-supported iodine catalyst $(NH_4)_5[IMo_6O_{24}]$ using an efficient method of aerobic alcohol oxidation. This system is compatible with multiple functional groups with high selectivity and good stability [39]. Based on experimental results, they proposed a preliminary mechanism of iodine catalyst $(NH_4)_5[IMo_6O_{24}]$ for alcohol oxidation (Figure 5). This process is similar to the enzymatic oxidation reaction, which can be divided into two independent semi-reactions: $(NH_4)_5[I^{VII}Mo_6^{VI}O_{24}]$ mediated alcohol oxidation reactions and dioxygen-coupled oxidation reaction with $[I^{V}Mo_6^{VI}O_{24}]$. For the iodo-molybdic acid catalyst by the inorganic ligands supported, two oxidizing equivalents required for oxidation are stored at the iodine center. Meanwhile, the addition of additives exerts a significant influence on the progress of the reaction, because the presence of Cl^- and Ac^- as an electron transfer medium promotes the electron transfer efficiency of $(NH_4)_5[I^{VII}Mo_6^{VI}O_{24}]$. This proposed mechanism is meaningful for the application of new Anderson-type POMs in organic reactions. This catalyst is a kind of efficient and mild inorganic ligand coordination catalytic system with high valence of iodine. The system has a wide range of substrate tolerance, good selectivity and recyclability, which avoids using the toxic organic ligands and toxic oxidant. Compared with the expensive organic iodine reagent, this system is convenient to apply in medicine, spices and food additives. Besides, the structure of $(NH_4)_5[IMo_6O_{24}]$ also provides more insight on designing the new Anderson-type POMs by replacing the metal ion in the backbone.

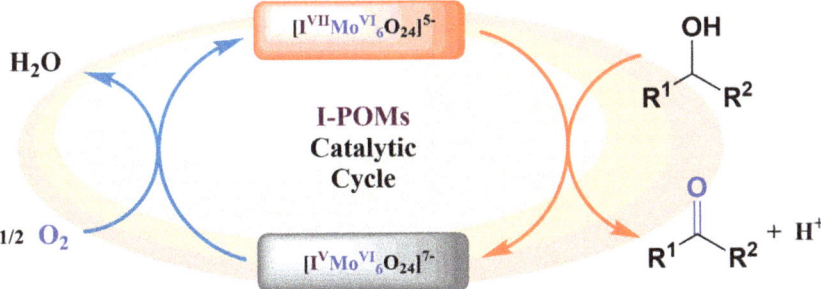

Figure 5. Proposed mechanism for the I-POM-catalyzed oxidation of alcohols.

Subsequently, Wei and co-worker prepared a Zn-based Anderson-type POM, $(NH_4)_4$ $[Zn(OH)_6Mo_6O_{18}]$, which was used as a catalyst for oxide cross-coupling reaction of halide and amine, oxide self-coupling reaction of amine, and the halide oxide reaction [40]. In this catalytic system (Figure 6), they chose benzyl amine and benzyl chloride as model substrates for the oxidative cross-coupling reaction with generating N-benzylidenebenzylamine in 32% yield, 47% selectivity. By screening the solvent, acetonitrile showed the best yield and selectivity with 89% and 90%, respectively. Additionally, shortening and extending the reaction time also affected the yield. In addition, the optimal reaction temperature at 60 °C and 1.0 mol% was the optimal amount of catalyst. Although the reaction uses O_2 as an oxygen source, relevant imines, aldehyde and ketone can still be efficiently prepared. The inorganic ligand-supported zinc-based Anderson-type POMs are not only easy to prepare and synthesize by the hydrothermal method, but also easy to recover due to the heterogeneity of the reaction. Compared with precious noble metal catalysts such as rhodium, ruthenium, and palladium, this catalyst avoids the use of toxic, air and water-sensitive organic ligands. So, Anderson-type POMs still have great potential as a heterogeneous catalyst in organic reactions.

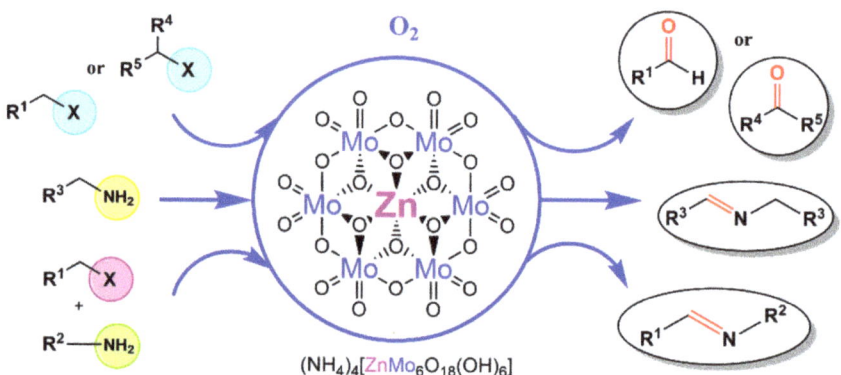

Figure 6. Zn-based Anderson-type POMs catalyze the oxidation of halides to aldehydes or ketones, and amines to sub-units.

In the same year, Sawant and co-worker investigated the application of Co-based Anderson-type POMs, $Co^{III}(OH)_6Mo_6O_{18}]^{3-}$, in catalyzed oxidation of acetaminophen in the aqueous media when the pH values are 1 or 2 [41]. In this reaction, the electron transfer from neutral acetaminophen to anion, and then in the step of rate determination, the free radicals are further oxidized to N-acetylquinone imide as an intermediate, and what is more, hydrolyze to obtain benzoquinone and acetic acid. The experimental results reveal that the formation of a weak complex among the reactants promotes the reaction.

In 2019, the Wei group reported Cu-based Anderson-type POMs, $(NH_4)_4[Cu(OH)_6 Mo_6O_{18}]$, which were used to study the aerobic oxidation of alcohol [42]. The classical transition metal complex catalytic system required complex organic ligands or nitroso radicals as auxiliary catalysts, and even strong oxidants. Nevertheless, when many organic compounds are in contact with these strong oxidants, it can easily result in exploding. Therefore, the traditional oxidation reaction of alcohol possesses complex process, high-price, severe and dangerous reaction conditions. Although, this catalyst is synthesized in water from cheap ammonium heptamolybdate and copper sulfate. The results showed that this catalytic system was suitable for various substrates in the oxidation of alcohol with excellent selectivity and activity. This method is not only safe and efficient, but also environmental. It has the potential to achieve industrial application. The experimental results show that the catalytic mechanism of the catalytic system is as follows (Figure 7) [42]: The two oxidation equivalents required for oxidation are not only stored in the copper center, but can be transferred to six marginalized MoO_6 on the unit for $(NH_4)_4[Cu(OH)_6Mo_6O_{18}]$. The Cu-based Anderson-type POMs react with alcohol to generate active substance A, while A and E, as a pair of isoforms, can interconvert, possibly due to the transfer of electrons to the inorganic ligands MoO_6 via the intramolecular oxygen bridge Cu-O-Mo. Thus, one of the Mo cells goes from positive hexavalent to positive pentavalent. Anderson-type POMs A firstly activate molecular oxygen generation activity species B. Subsequently, species B hybridizes by the O=O bond to obtain the highly reactive metal oxygen species C as an active oxidant. During this catalytic reaction, the detection by MS indicates that water is involved in the reaction, producing hydrogen peroxide, so that the catalytic system is closer to the galactose oxidase reaction. Cu^{II} and the hydroxyl radical act together as a single-electron oxidant to be a two-electron alcohol oxidation reaction. In the reaction of galactose oxidase, a single center which is copper reacts with oxygen, producing hydrogen peroxide as a by-product. High active species C reacts with ethanol to generate intermediate D, and then H atoms are extracting to generate aldehyde and regenerate E. Finally, the intermediate E (Cu^{II}) was re-oxidized by hydrogen peroxide to substance A (Cu^{I}).

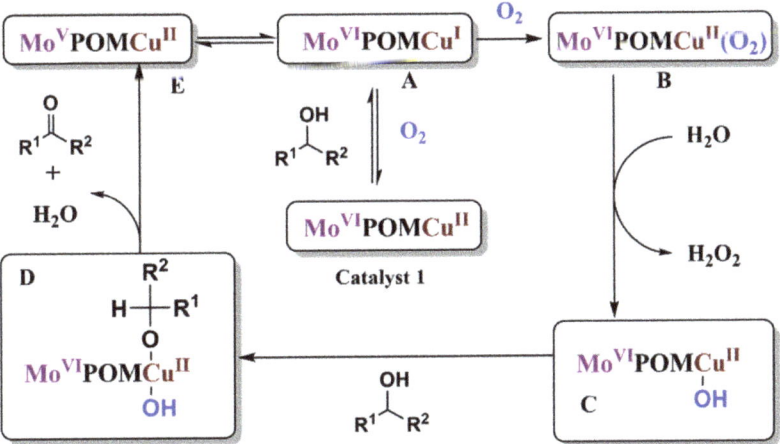

Figure 7. The catalytic mechanism of alcohol oxidation by using $(NH_4)_4[Cu(OH)_6Mo_6O_{18}]$.

At the same time, the Wei group also prepared Fe-based Anderson-type POMs, $(NH_4)_3[Fe(OH)_6Mo_6O_{18}]$ and applied them in olefination epoxidation reaction with 30% H_2O_2 as oxidant. This catalytic system does not require additional reductant or free radicals as initiators, and the experimental operation is simple, completely avoiding the use of expensive, toxic precious metal catalysts. Moreover, it does not need air/water sensitive and commercially unavailable organic ligands or tungsten POMs. $(NH_4)_3[Fe(OH)_6Mo_6O_{18}]$ can be obtained by one-step reaction in 100 °C water. This system has successfully converted the various aromatic and aliphatic alkenes into the corresponding epoxy compounds, with a good yield and selectivity [43]. Subsequently, Wei and collaborators further used this system in oxidative esterification reaction of various aldehydes with alcohol, and the corresponding esters were achieved under mild conditions with high yields, including several drug molecules and natural products [44].

After that, the Wei group and co-worker developed Cr-based Anderson-type POMs, $(NH_4)_3[CrMo_6O_{18}(OH)_6]$. The various primary and secondary amines, and even dual primary amine, were successfully converted into the corresponding formamides and methylated using this catalyst with methanol as a potential formylation reagent. Compared with the high-valence Cr-based catalyst including CrO_3 and $K_2Cr_2O_7$, etc. This one with low-valent chromium catalyst is more effective, safe, and green environmental [45]. Wei then reported another work using $(NH_4)_3[Fe(OH)_6Mo_6O_{18}]$ catalyzed formic acid and amine coupling to generate formamide in mild conditions [46]. In the same year, the

Xu group also developed Fe-based Anderson-type POMs, $Na_3Fe(OH)_6Mo_6O_{18} \cdot 5H_2O$ to prepare for 5-formyl-2-furanformic acid from 5-hydroxyl methylfuranic acid. The related results also further proved that the catalyst has high catalytic activity under an aerobic environment [47].

In 2020, Wei and co-worker proposed Co-based Anderson-type POMs $(NH_4)_3[Co(OH)_6 Mo_6O_{18}]$. In the presence of this catalyst, it is effective in achieving from alcohol to esters when KCl is used as an additive and 30% H_2O_2 as oxidant (Figure 8). As the important complexes in the fields of biology, medicine, and fine chemicals, esters are usually prepared by carboxylic acids or active derivatives (acyl chloride and anhydride) reacted with alcohol. The Co-based Anderson-type POMs catalytic system is cheap, stable, safe and efficient and can successfully make alcohol to esters under mild conditions. They first explored the substrate range of benzyl alcohol and fatty alcohol oxidation coupled with methanol to generate methyl esters. The results demonstrate that various substituted benzyl alcohols as well as fatty alcohols were performed efficiently and highly selectively; the highest yield can reach 99%. Subsequently, on the basis of the above work, they further extended the catalytic system to the esterification reaction of benzyl alcohol and fatty alcohol with other alcohols, the corresponding ester products were also obtained. More importantly, the natural product octranactone can also be prepared by this method with 66% yield. In this catalytic system, the experiment shows that the additive can significantly enhance the selectivity and activity of the catalytic reaction system. The single-crystal X-ray diffraction test showed that the chloride ion in the additive linked with $CoMo_6$ forming a supramolecular dimer 2 ($CoMo_6Cl$) served as the key catalytic intermediate, which then achieves alcohol-to-ester conversion through nucleophilic addition and β-H elimination. The catalyst enables to achieve an oxidation coupling reaction of various alcohols (aromatic and aliphatic) under mild conditions and produce the corresponding esters with high yield [48]. This work shows that Anderson-type POMs, as an important class of single-metal ion inorganic molecule carriers, are an excellent and environmentally friendly catalyst with great potential applications in organic reactions. As a multifunctional catalyst, it changes the structure and properties of the central metal, so that Anderson-type POMs show excellent catalytic activity and extremely high catalytic efficiency in the green catalytic oxidation of alcohol.

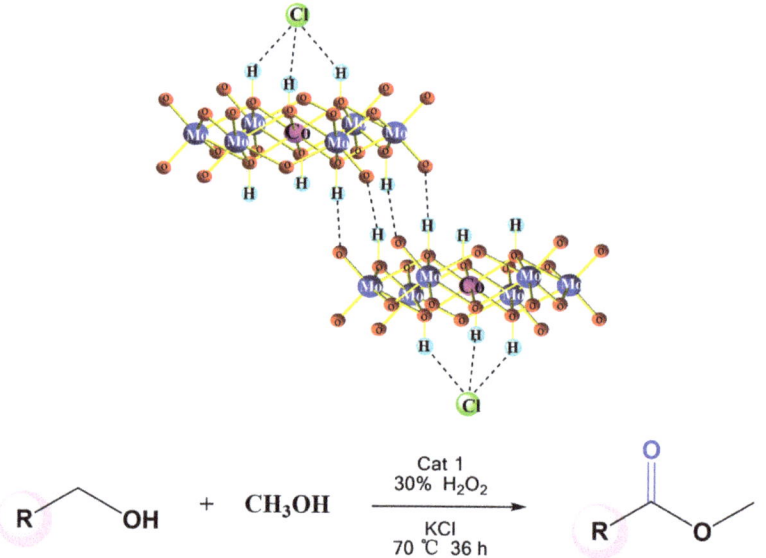

Figure 8. The supramolecular structure of dimer 2($CoMo_6Cl$) and its catalytic route.

2.2. Catalyzation of Organic Modified Anderson-Type POMs (PO)

2.2.1. [I]$^+$PO

In 2017, Wei and co-workers reported an organic ligand-stable β-isomer Anderson-type POMs: [NH$_4$]β-{[H$_3$NC(CH$_2$O)$_3$]$_2$MnMo$_6$O$_{18}$}[49]. The sub-nano-cluster acts as a highly active catalyst applied in selectively catalyzed oxidation of cyclohexanone, cyclohexanol, or a mixture thereof. The results show that the special butterfly topology of Anderson-type POMs significantly improves the catalytic activity. Additionally, due to its high selectivity and high catalytic conversion efficiency, it can enable use for the green industrial synthesis of adipic acid.

2.2.2. [O]$^+$PO

In 2017, the Yu group developed the first organic modified hybrid Fe-based Anderson-type POMs [N (C$_4$H$_9$)$_4$]$_3$[FeMo$_6$O$_{18}$(OH)$_3${(OCH$_2$)$_3$CNH$_2$}] for the aerobic oxidation of aldehydes to acids in water (Figure 9) [50] with O$_2$ (1 atm). They found that the alkalinity of additives played a great impact on yield. Benzaldehyde, which was used as substrate, reacted with different additives at 50 °C. When Na$_2$CO$_3$ (pK$_b$ = 3.67) was used as additive, the yield of benzoic acid was 95%. When NaHCO$_3$ (pK$_b$ = 7.95) was added in the system, the yield of benzoic acid decreased to 63%. However, the additive replaced by Na$_2$SO$_3$ (pK$_b$ = 6.8) the yield reduced to 26%. While Na$_2$SO$_4$ (pK$_b$ = 12.0) was chosen as additive, the lowest yield of benzoic acid was 5%. For organic alkaline additives, such as CH$_3$COONa and Et$_3$N, the yield of benzoic acid was 83% and 89%, respectively. Regarding additives containing neutral salts, such as NaBr, KCl, and NaCl, the desired product was obtained in moderate yield. The acid additive, NH$_4$Cl, avoided the process reaction. These results indicate that the additive to Fe-based Anderson-type POMs, Fe^{III}Mo$_6$, has a great effect on the activity of the catalyst, perhaps because of the high tunability of the acid-based properties of POMs. Subsequently, they also studied the effect of the catalyst dosage and the reaction temperature on catalytic activity. Changing the amount of the catalyst has little effect on the yield of the product. The yield of benzoic acid reached the maximum at 50 °C, but increasing or decreasing the temperature showed the different result. When the reaction temperature increased to 70 °C, the yield decreases to 96%, possibly due to reduced contact surface between the oxygen (gas phase) and the catalyst. When the reaction was performed in a nitrogen sphere, the yield of the obtained products was 5% less. As a control experiment, the experiments conducted without Fe^{III}Mo$_6$, only a small amount of product generation can be detected. This Fe-based Anderson-type POMs system was carried out with oxygen as the sole oxidant under extremely mild aqueous solution conditions, and suitable to a variety of functional group aldehydes. This method is environmentally friendly, has a low cost and high catalytic efficiency, as well as the potential application in industry.

The Wei group and co-worker also found that halogen ions (X$^-$) can be incorporated with sub-nanoscale organoalkoxyl ligands-modified Al-based Anderson-type POMs([[(n-C$_4$H$_9$)$_4$N]$_3${AlMo$_6$O$_{18}$(OH)$_3$[(OCH$_2$)$_3$CCH$_3$]]) in solution, forming the stable supramolecular complex with the binding constant K = 1.53 × 10^3. This system was used in oxidation to aldehydes from alcohol. Crystal structure analysis demonstrated this behavior that binding occurs between the halogen ion X$^-$ and the unmodified three hydroxyl groups on the surface of {AlMo$_6$O$_{18}$(OH)$_3$[(OCH$_2$)$_3$CCH$_3$]}$^{3-}$, forming multiple X···H-O [51]. The interaction in this supramolecular sub-nano-cluster system means that its catalytic activity for oxidation of aldehydes can be adjusted by introduction of halogen–halogen ions and water. Chlorine ions inhibit by blocking the active center of the cluster, and the catalytic activity of the cluster can be reactivated by replacing the chloride with water supra-molecules. The result shows that the presence of multiple synergistic hydrogen bonds is key to overcome the electrostatic repulsion between halogen ions and Al-based Anderson-type POMs. This work not only enriches the supramolecular chemistry of Anderson-type POMs, but also provides a new way of selective catalysts for oxidation reactions.

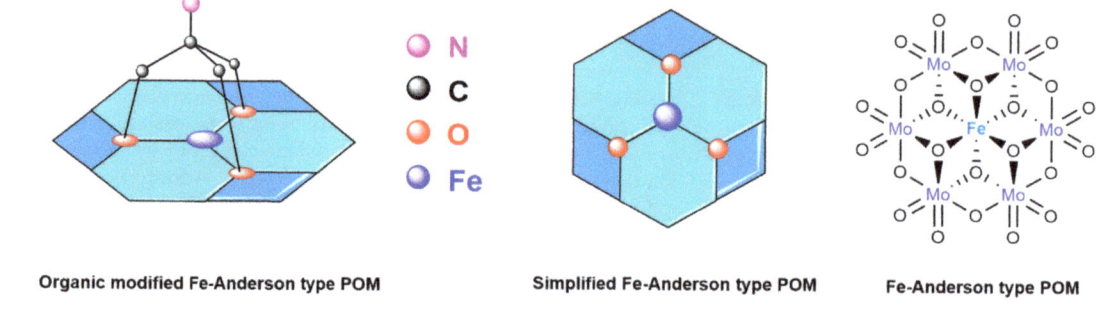

Figure 9. Catalyst anion structure and the catalytic reaction route.

3. Summary and Outlook

In order to solve the environmental crisis brought by the traditional production of chemicals, the research and development of green chemical technology had gradually become one of the hot fields in chemical research since the 1990s. The use of chemical technology and methods to reduce or eliminate the production of substances that are detrimental to human health and the ecological environment is one of the research priorities in the green industry of chemicals.

Advances and innovations in chemical technology are often driven by new catalytic materials and new catalytic technologies. In the research in the catalytic field, as a new type of high-efficiency, green, cheap and safe catalyst for Anderson-type POMs, they possess excellent redox activity compared with other inorganic acids. At the same time, it has measure and controlled acidity, as well as excellent dual-functional properties in the catalytic reactions. Moreover, most POMs possess good solvability. Apart from that, they not only have a definite structure and size, but also can be further modified by organic group, giving them more excellent features. In addition, Anderson-type POMs as a catalyst have more selectivity, few side reactions and retard the corrosion for the equipment. Therefore, the catalytic application of Anderson-type POMs has a great prospect and research value in scientific exploration and green chemical technology (Figure 10).

In combination with relevant literature, due to the instability of Anderson-type POMs, sometimes it is necessary to perform a series of organic–inorganic hybrid modifications for the parent POMs. The simple Anderson-type POMs are widely used in organic synthesis reactions, and the organic group modified POMs derivatives can perform relevant structural modification for specific catalytic reactions. It can be seen that the difference of the molecular structure of POMs has a certain influence on the specific catalytic reaction, especially in the aspects of catalytic efficiency, selectivity and yield. The POMs modified by different inorganic cations or organic ligands have their specific characteristics functional, which enlightens us. We can modify and design POMs catalysts for special organic catalytic reactions, and have achieved the desired effect. Anderson-type POMs displayed remarkable catalytic characteristics in oxidation of alcohol. They also have extremely high efficiency in the formation of C-N, C-O and C-C bonds. By comparison to the traditional precious metal catalyst, Anderson-type POMs are not only easy to synthesize, but also have high catalytic activity and recyclability. It can be expected that this catalytic system will also play the significant role in the selective oxidation of hydrocarbons and will have overwhelmingly wide application in the field of industrial catalysis.

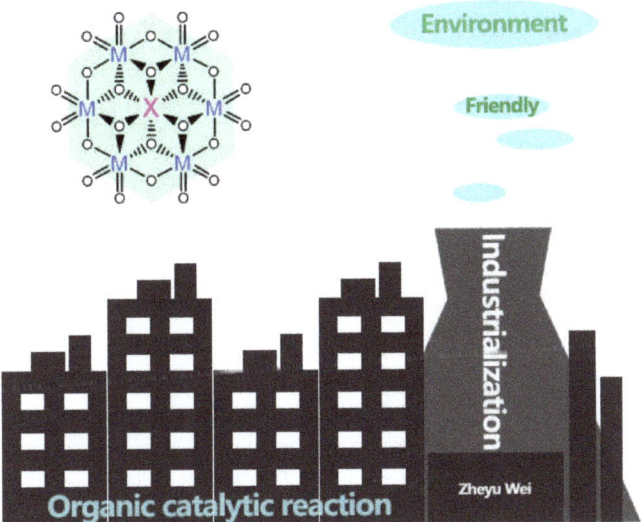

Figure 10. Industrialization expectation of Anderson-type POMs catalyst.

This paper aims at paying much attention to investigate Anderson-type POMs in organic reactions and provide some new strategies and ideas for researchers. It is believed that, in the near future, a mounting number of Anderson-type POMs will be used in organic reactions, which will promote the development of organic synthetic chemistry.

Author Contributions: Literature research, Z.W.; writing—original draft preparation, Z.W.; writing—review and editing, J.W., H.Y., S.H. and Y.W.; visualization, Z.W.; supervision, H.Y., S.H. and Y.W.; funding acquisition, Y.W. All authors have read and agreed to the published version of the manuscript.

Funding: Project supported by the National Natural Science Foundation of China (Nos. 21971134, 21225103) and 2021–2023 Chinese-Serbian bilateral intergovernmental personnel exchange project No1 of Ministry of Science and Technology of the People's Republic of China.

Institutional Review Board Statement: Not applicable.

Informed Consent Statement: Not applicable.

Conflicts of Interest: The authors declare no conflict of interest.

References

1. Chen, W.L.; Wang, E.B. *Multi-Acid Chemistry*; Science Press: Beijing, China, 2013; pp. 1–9.
2. Wang, N. Research progress of Anderson type polyoxometalates and their derivatives. *J. Chang. Norm. Univ.* **2015**, *34*, 58–61.
3. Wang, F. Self Assembly Synthesis, Structure and Properties of Anderson type Polyoxometalates and Biomolecules. Ph.D. Thesis, Northeast Normal University, Shenyang, China, 2007. (In Chinese)
4. Qin, Z.X.; Li, Q.; Huang, Y.C.; Zhang, J.W.; Li, G.; Wei, Y.G. New progress of Anderson type polyoxometalates modified by controlled alkylation. *Chin. Sci. Bull.* **2018**, *63*, 3263–3276. [CrossRef]
5. Zhang, J.W.; Huang, Y.C.; Hao, J.; Wei, Y.G. β-{Cr[RC(CH$_2$O)$_3$]$_2$Mo$_6$O$_{18}$}$^{3-}$: The first organically-functionalized β isomer of Anderson-type polyoxometalates. *Inorg. Chem. Front.* **2017**, *4*, 1215–1218. [CrossRef]
6. Wei, Z.Y.; Wei, Y.G. Liquid Phase Diffusion Single Crystal Growth Sampling Method and Special Sampling Tube. Patent No.:CN104152980A, 2014.
7. Zhang, J.W.; Huang, Y.C.; Zhang, J.; She, S.; Hao, J.; Wei, Y.G. A direct anchoring of Anderson-type polyoxometalates in aqueous media with tripodal ligands especially containing the carboxyl group. *Dalton Trans.* **2014**, *43*, 2722–2725. [CrossRef]

8. Zhang, J.; Li, Q.; Zeng, M.; Huang, Y.; Zhang, J.; Hao, J.; Wei, Y.G. The proton-controlled synthesis of unprecedented diol functionalized Anderson-type POM. *Chem. Commun.* **2016**, *52*, 2378–2381. [CrossRef]
9. Huang, Y.C.; Zhang, J.W.; Hao, J.; Wei, Y.G. A general and highly regioselective synthesis approach to multi-functionalized organoimido derivatives of Polyoxometalates. *Sci. Rep.* **2016**, *6*, 24759. [CrossRef]
10. Zhu, L.; Chen, K.; Hao, J.; Wei, Z.Y.; Zhang, H.C.; Yin, P.C.; Wei, Y.G. Synthesis and Crystallization Behavior of Surfactants with Hexamolybdate as the Polar Headgroup. *Inorg. Chem.* **2015**, *54*, 6075–6077. [CrossRef]
11. Xu, Q.; Yuan, S.; Zhu, L.; Hao, J.; Wei, Y.G. Synthesis of novel bis (Triol)-functionalized Anderson clusters serving as potential synthons for forming organic–inorganic hybrid chains. *Chem. Commun.* **2017**, *53*, 5283–5286. [CrossRef]
12. Zhang, Y.S.; Jia, H.L.; Li, Q.; Huang, Y.C.; Wei, Y.G. Synthesis and characterization of an unprecedented water-soluble tris-functionalized Anderson-type polyoxometalate. *J. Mol. Struct.* **2020**, *1219*, 128555. [CrossRef]
13. Li, Q.; Wei, Y.G. Unprecedented monofunctionalized β-Anderson clusters:$[R_1R_2C(CH_2O)_2Mn^{IV}W_6O_{22}]^{6-}$, a class of potential candidates for new inorganic linkers. *Chem. Commun.* **2021**, *57*, 3865–3868. [CrossRef]
14. Hill, C.L.; Harrup, M.K. Thermal multi-electron transfer catalysis by polyoxometalates. Application to the practical problem of sustained, selective oxidation of hydrogen sulfide to sulfur. *J. Mol. Catal. A Chem.* **1996**, *106*, 57–66.
15. André, D.; Belletti, A.; Cavani, F.; Degrand, H.; Dubois, J.L.; Trifirò, F. Molybdenum oxide-based systems, prepared starting from Anderson-type polyoxometalates precursors, as catalysts for the oxidation of i-C4 hydrocarbons. In *Studies in Surface Science and Catalysis*; Elsevier: Amsterdam, The Netherlands, 2005; Volume 155, pp. 57–66.
16. Liu, Y.; Liu, S.X.; Cao, R.G.; Ji, H.M.; Zhang, S.W.; Ren, Y.H. Hydrothermal assembly and luminescence property of lanthanide-containing Anderson polyoxometalates. *J. Solid State Chem.* **2008**, *181*, 2237–2242. [CrossRef]
17. Sheshmani, S.; Fashapoyeha, M.A.; Mirzaei, M.; Rad, B.A.; Ghortolmesh, S.N.; Yousefi, M. Preparation, characterization and catalytic application of some polyoxometalates with Keggin, Wells-Dawson and Preyssler structures. *Indian J. Chem. A.* **2011**, *50*, 1725–1729.
18. Ye, T.; Wang, J.; Dong, G.; Jiang, Y.; Feng, C.; Yang, Y. Recent Progress in the Application of Polyoxometalates for Dye-Sensitized/Organic Solar Cells. *Chin. J. Chem.* **2016**, *34*, 747–756. [CrossRef]
19. Ong, B.C.; Hong, K.L.; Tay, C.Y.; Lim, T.T.; Dong, Z. Polyoxometalates for bifunctional applications: Catalytic dye degradation and anticancer activity. *Chemosphere.* **2021**, *286*, 131869. [CrossRef]
20. Hu, C.W.; Zhen, H.; Xu, L.; Wang, E.B. Recent progress in the study of oxidation and oxidation catalysis of Polyoxometalates. *J. Mol. Sci.* **1997**, *1*, 45–55. (In Chinese)
21. Wu, P.S.; Zhang, H.Y.; Xu, L.; De, G.J.R.H.; Hu, C.W.; Wang, E.B. Synthesis and structure of inorganic organic hybrid molybdenum phosphorus, vanadium molybdenum phosphorus polyoxometalates compounds. *J. Northeast. Norm. Univ. (Nat. Sci.)* **2001**, *04*, 51–56. (In Chinese)
22. Guo, S.R.; Kong, Y.M.; Peng, J.; Wang, E.B. Research progress of polyoxometalate nanomaterials. *Chem. Bull.* **2007**, *10*, 748–758. (In Chinese)
23. Wei, Y.G. *Proceedings of the Collection of Abstracts of Papers of the Sixth Annual Conference of the Chinese Society of Crystallography and Its Members' Congress (Functional Molecular Crystals Branch)*, Guangzhou, China, 19 December 2016; Chinese Society of Crystallography: Beijing, China, 2016; p. 28.
24. Wei, Z.Y.; Li, Q.; Wei, Y.G. Synthesis and crystal structure of an isomer of the single-side pentaerythritol-modified alkoxo derivative of anderson-type chromohexamolybdate. *J. Mol. Sci.* **2017**, *33*, 391–395.
25. Song, Y.F.; Wei, Y.G. New development trend of polyoxometalates. *Chin. Sci. Bull.* **2018**, *63*, 3261–3262. (In Chinese)
26. Yu, F.L.; Liu, C.Y.; Xie, P.H.; Yuan, B.; Xie, C.X.; Yu, S.T. Oxidative-extractive deep desulfurization of gasoline by functionalized heteropoly acid catalyst. *RSC Adv.* **2015**, *5*, 85540–85546. [CrossRef]
27. Yao, L.; Zhang, L.Z.; Zhang, Y.; Wang, R.; Wongchitphimon, S.; Dong, Z.L. Self-assembly of rare-earth Anderson polyoxometalates on the surface of imide polymeric hollow fiber membranes potentially for organic pollutant degradation. *Sep. Purif. Technol.* **2015**, *151*, 155–164. [CrossRef]
28. Li, P.C. Study on green catalytic oxidation desulfurization of diesel oil. Master's Dissertation, Yantai University, Yantai, China, 2017. (In Chinese)
29. Yang, W.; Hou, Y.J.; An, H.Y. Synthesis and catalytic oxidation properties of Anderson type polyacid inorganic organic hybrid compounds. *J. Mol. Sci.* **2017**, *33*, 385–390. (In Chinese)
30. Sun, L.; Su, T.; Li, P.; Xu, J.; Chen, N.; Liao, W.; Deng, C.; Ren, W.; Lü, H. Extraction coupled with aerobic oxidative desulfurization of model diesel using a B-type Anderson polyoxometalate catalyst in ionic liquids. *Catal. Lett.* **2019**, *149*, 1888–1893. [CrossRef]
31. Ji, H.B.; She, Y.B. Development and Research on basic problems of green chemistry. *Prog. Chem. Eng.* **2007**, *26*, 605–614. (In Chinese)
32. Li, J.J.; Wu, F. *Textbook of Introduction to Green Chemistry*; Wuhan University Press: Wuhan, China, 2015; Volume 8.
33. Du, X.Y. Application and development of green chemical and environmental protection technology in industrial production. *Yunnan Chem. Technol.* **2019**, *46*, 123–124. (In Chinese)
34. Song, X.M. Application of green chemical technology in fine chemical industry. *Chem. Enterp. Manag.* **2020**, *03*, 116. (In Chinese)
35. SD, K.; Gokavi, G.S. Anderson type Hexamolybdochromate hydrazide by $KBrO_3$ in acidic medium. *Res. J. Chem.* **2016**, *6*, 17–23.
36. Yu, H.; Zhai, Y.Y.; Dai, G.Y.; Ru, S.; Han, S.; Wei, Y.G. Transition-Metal-Controlled Inorganic Ligand-Supported Non-Precious Metal Catalysts for the Aerobic Oxidation of Amines to Imines. *Chem.–A Eur. J.* **2017**, *23*, 13883–13887. [CrossRef]

37. Yu, H.; Ru, S.; Zhai, Y.Y.; Dai, G.Y.; Han, S.; Wei, Y.G. An Efficient Aerobic Oxidation Protocol of Aldehydes to Carboxylic Acids in Water Catalyzed by an Inorganic-Ligand-Supported Copper Catalyst. *ChemCatChem.* **2018**, *10*, 1253–1257. [CrossRef]
38. Zhai, Y.Y.; Zhang, M.Q.; Fang, H.; Ru, S.; Yu, H.; Zhao, W.; Wei, Y.G. An efficient protocol for the preparation of aldehydes/ketones and imines by an inorganic-ligand supported iron catalyst. *Org. Chem. Front.* **2018**, *5*, 3454–3459. [CrossRef]
39. Zhang, M.Q.; Zhai, Y.Y.; Ru, S.; Zang, D.; Han, S.; Yu, H.; Wei, Y.G. Highly practical and efficient preparation of aldehydes and ketones from aerobic oxidation of alcohols with an inorganic-ligand supported iodine catalyst. *Chem. Commun.* **2018**, *54*, 10164–10167. [CrossRef] [PubMed]
40. Wang, J.J.; Zhai, Y.Y.; Wang, Y.; Yu, H.; Zhao, W.; Wei, Y.G. Selective aerobic oxidation of halides and amines with an inorganic-ligand supported zinc catalyst. *Dalton Trans.* **2018**, *47*, 13323–13327. [CrossRef]
41. Sawant, J.D.; Patil, K.K.; Gokavi, G.S. Kinetics and mechanism of oxidation of paracetamol by an Anderson-type 6-molybdocobaltate (III) in acidic medium. *Transit. Met. Chem.* **2019**, *44*, 153–159. [CrossRef]
42. Wei, Z.Y.; Ru, S.; Zhao, Q.X.; Yu, H.; Zhang, G.; Wei, Y.G. Highly efficient and practical aerobic oxidation of alcohols by inorganic-ligand supported copper catalysis. *Green Chem.* **2019**, *21*, 4069–4075. [CrossRef]
43. Zhou, Z.; Dai, G.Y.; Ru, S.; Yu, H.; Wei, Y.G. Highly selective and efficient olefin epoxidation with pure inorganic-ligand supported iron catalysts. *Dalton Trans.* **2019**, *48*, 14201–14205. [CrossRef]
44. Yu, H.; Wang, J.J.; Wu, Z.K.; Zhao, Q.; Dan, D.; Han, S.; Tang, J.; Wei, Y.G. Aldehydes as potential acylating reagents for oxidative esterification by inorganic ligand-supported iron catalysis. *Green Chem.* **2019**, *21*, 4550–4554. [CrossRef]
45. Yu, H.; Wu, Z.K.; Wei, Z.Y.; Zhai, Y.; Ru, S.; Zhao, Q.; Wang, J.J.; Han, S.; Wei, Y.G. N-formylation of amines using methanol as a potential formyl carrier by a reusable chromium catalyst. *Commun. Chem.* **2019**, *2*, 15. [CrossRef]
46. Wu, Z.K.; Zhai, Y.Y.; Zhao, W.; Wei, Z.Y.; Yu, H.; Han, S.; Wei, Y.G. An efficient way for the N-formylation of amines by inorganic-ligand supported iron catalysis. *Green Chem.* **2020**, *22*, 737–741. [CrossRef]
47. Xu, J.; Zhu, Z.; Yuan, Z.; Su, T.; Zhao, Y.; Ren, W.; Zhang, Z.; Lü, H. Selective oxidation of 5-hydroxymethylfurfural to 5-formyl-2-furancar-boxylic acid over a Fe-Anderson type catalyst. *J. Taiwan Inst. Chem. Eng.* **2019**, *104*, 8–15. [CrossRef]
48. Wang, J.J.; Yu, H.; Wei, Z.Y.; Li, Q.; Xuan, W.M.; Wei, Y.G. Additive-Mediated Selective Oxidation of Alcohols to Esters via Synergistic Effect Using Single Cation Cobalt Catalyst Stabilized with Inorganic Ligand. *Research* **2020**, *2020*, 3875920. [CrossRef] [PubMed]
49. Luo, J.H.; Huang, Y.C.; Ding, B.; Wang, P.; Geng, X.; Zhang, J.; Wei, Y.G. Single-atom Mn active site in a triol-stabilized β-Anderson manganohexamolybdate for enhanced catalytic activity towards adipic acid production. *Catalysts* **2018**, *8*, 121. [CrossRef]
50. Yu, H.; Ru, S.; Dai, G.Y.; Zhai, Y.; Lin, H.; Han, S.; Wei, Y.G. An Efficient Iron (III)-Catalyzed Aerobic Oxidation of Aldehydes in Water for the Green Preparation of Carboxylic Acids. *Angew. Chem. Int. Ed.* **2017**, *56*, 3867–3871. [CrossRef] [PubMed]
51. She, S.; Mu, L.; Li, Q.; Huang, Z.H.; Wei, Y.G.; Yin, P.C. Unprecedented Halide-Ion Binding and Catalytic Activity of Nanoscale Anionic Metal Oxide Clusters. *ChemPlusChem* **2019**, *84*, 1645. [CrossRef]

MDPI
St. Alban-Anlage 66
4052 Basel
Switzerland
www.mdpi.com

Molecules Editorial Office
E-mail: molecules@mdpi.com
www.mdpi.com/journal/molecules

Disclaimer/Publisher's Note: The statements, opinions and data contained in all publications are solely those of the individual author(s) and contributor(s) and not of MDPI and/or the editor(s). MDPI and/or the editor(s) disclaim responsibility for any injury to people or property resulting from any ideas, methods, instructions or products referred to in the content.

www.ingramcontent.com/pod-product-compliance
Lightning Source LLC
LaVergne TN
LVHW070641100526
838202LV00013B/855